RESEARCH METHODS AND APPLICATIONS IN CHEMICAL AND BIOLOGICAL ENGINEERING

AAP Research Notes on Chemical Engineering

RESEARCH METHODS AND APPLICATIONS IN CHEMICAL AND BIOLOGICAL ENGINEERING

Edited by

Ali Pourhashemi

Sankar Chandra Deka

A. K. Haghi

AAP | APPLE ACADEMIC PRESS

Apple Academic Press Inc.
3333 Mistwell Crescent
Oakville, ON L6L 0A2
Canada

Apple Academic Press Inc.
1265 Goldenrod Circle NE
Palm Bay, Florida 32905
USA

Exclusive worldwide distribution by CRC Press, a member of Taylor & Francis Group

International Standard Book Number-13: 978-1-77188-768-7 (Hardcover)
International Standard Book Number-13: 978-0-42942-413-7 (eBook)

Library and Archives Canada Cataloguing in Publication

Title: Research methods and applications in chemical and biological engineering / edited by Ali Pourhashemi, Sankar Chandra Deka, A.K. Haghi.

Names: Pourhashemi, Ali, editor. | Deka, Sankar Chandra, editor. | Haghi, A. K., editor.

Series: AAP research notes on chemical engineering.

Description: Series statement: AAP research notes on chemical engineering | Includes bibliographical references and index.

Identifiers: Canadiana (print) 20190114924 | Canadiana (ebook) 20190114959 | ISBN 9781771887687 (hardcover) | ISBN 9780429424137 (ebook)

Subjects: LCSH: Chemical engineering—Research. | LCSH: Bioengineering—Research.

Classification: LCC TP165 .R43 2019 | DDC 660/.20721—dc23

Library of Congress Cataloging-in-Publication Data

Names: Pourhashemi, Ali, editor. | Deka, Sankar Chandra, 1965- editor. | Haghi, A. K., editor.

Title: Research methods and applications in chemical and biological engineering / editors, Ali Pourhashemi, Sankar Chandra Deka, A.K. Haghi.

Description: Toronto ; [Hackensack?] New Jersey : Apple Academic Press, 2020. | Includes bibliographical references and index. | Summary: "This research-oriented book consolidates and provides up-to-date experimental methods currently used in research for many branches of chemical and biological engineering. The book surveys essential ideas and research methodologies, concentrating on experiments as used in applications rather than on fine points of rigorous mathematics. Examples of important applications are reviewed in sufficient detail to provide the reader with a critical understanding of context and research methodology. Research Methods and Applications in Chemical and Biological Engineering presents a broad spectrum of chapters in the various branches of chemical and biological engineering, which demonstrate key developments in these rapidly changing fields. Chapters explore the design, development, operation, monitoring, control and optimization of chemical, physical and biological processes. Case studies are included in some chapters, building a real-world connection"-- Provided by publisher.

Identifiers: LCCN 2019020174 (print) | LCCN 2019980903 (ebook) | ISBN 9781771887687 (hardcover) | ISBN 9780429424137 (ebook)

Subjects: LCSH: Chemical engineering--Research. | Bioengineering--Research.

Classification: LCC TP165 .R47 2020 (print) | LCC TP165 (ebook) | DDC 660.6--dc23

LC record available at https://lccn.loc.gov/2019020174

LC ebook record available at https://lccn.loc.gov/2019980903

Apple Academic Press also publishes its books in a variety of electronic formats. Some content that appears in print may not be available in electronic format. For information about Apple Academic Press products, visit our website at **www.appleacademicpress.com** and the CRC Press website at **www.crcpress.com**

AAP RESEARCH NOTES ON CHEMICAL ENGINEERING

The AAP Research Notes on Chemical Engineering series will report on research development in different fields for academic institutes and industrial sectors interested in advanced research books. The main objective of the AAP Research Notes series is to report research progress in the rapidly growing field of chemical engineering.

Books in the AAP Research Notes on Chemical Engineering series:

Quantum-Chemical Calculations of Unique Molecular Systems
(2-volume set)
Editors: Vladimir A. Babkin, DSc, Gennady E. Zaikov, DSc, and A. K. Haghi, PhD

Chemical and Biochemical Engineering: New Materials and Developed Components
Editor: Ali Pourhashemi, PhD
Reviewers and editorial board members: Gennady E. Zaikov, DSc, and A. K. Haghi, PhD

Clearing of Industrial Gas Emissions: Theory, Calculation, and Practice
Usmanova Regina Ravilevna, PhD, and Gennady E. Zaikov, DSc

Research Methods and Applications in Chemical and Biological Engineering
Editors: Ali Pourhashemi, PhD, Sankar Chandra Deka, PhD, and A. K. Haghi, PhD

ABOUT THE EDITORS

Ali Pourhashemi, PhD

Ali Pourhashemi, PhD, is currently a professor of chemical and biochemical engineering at Christian Brothers University (CBU) in Memphis, Tennessee. He was formerly the department chair at CBU and also taught at Howard University in Washington, DC. He taught various courses in chemical engineering, and his main area has been teaching the capstone process design as well as supervising industrial internship projects. He is a member of several professional organizations, including the American Institute of Chemical Engineers. He is on the international editorial review board of the *International Journal of Chemoinformatics and Chemical Engineering* and is an editorial member of the *International Journal of Advanced Packaging Technology*. He has published many articles and presented them at many professional conferences.

Sankar Chandra Deka, PhD

Sankar Chandra Deka, PhD, is presently working as a Professor in the Department of Food Engineering and Technology, Tezpur University Assam, India. He has more than 29 years of teaching and research experience. He has guided the theses of more than 35 BTech/MSc/MTech students. He has already guided seven PhD students and another eight students are actively working under his guidance. He successfully handled more than 16 research projects funded by various funding agencies, namely, ICAR, DBT, DST, DRDO, MoFPI, UGC (New Delhi), and ASTEC (Guwahati). He has published more than 92 research papers in journals of national and international repute and about 20 book chapters. His areas of interests are food quality, food chemistry, fermented foods, and processing of locally important foods. He is one of the lead editors of the forthcoming book *Processing, Technologies and Functionality of Foods,* to be published by Apple Academic Press (distribution by CRC Press under Taylor and Francis Group).

A. K. Haghi, PhD

A. K. Haghi, PhD, is the author and editor of 165 books, as well as over 1000 published papers in various journals and conference proceedings. Dr. Haghi has received several grants, consulted for a number of major corporations, and is a frequent speaker to national and international audiences. Since 1983, he served as professor at several universities. He is currently Editor-in-Chief of the *International Journal of Chemoinformatics and Chemical Engineering* and *Polymers Research Journal* and on the editorial boards of many international journals. He is also a member of the Canadian Research and Development Center of Sciences and Cultures (CRDCSC), Montreal, Quebec, Canada. He holds a BSc in urban and environmental engineering from the University of North Carolina (USA), an MSc in mechanical engineering from North Carolina A&T State University (USA), a DEA in applied mechanics, acoustics and materials from the Université de Technologie de Compiègne (France), and a PhD in engineering sciences from Université de Franche-Comté (France).

CONTENTS

CONTRIBUTORS

Cristóbal Noé Aguilar
Food Research Department, Group of Bioprocesses and Bioproducts, School of Chemistry, Universidad Autonoma de Coahuila, 25280 Saltillo, Coahuila, México
Research Group of Science Food Department, Chemistry School, Autonomous University of Coahuila, Bvd. V. Carranza e Ing. J. Cáredenas V., Col. República, Saltillo 25280, Coahuila, México
Facultad de Ciencias Químicas, Universidad Autónoma de Coahuila, Unidad Saltillo, México

Miguel A. Aguilar-González
Food Research Department, Group of Bioprocesses and Bioproducts, School of Chemistry, Universidad Autonoma de Coahuila, 25280 Saltillo, Coahuila, México

Jorge A. Aguirre-Joya
Facultad de Ciencias Químicas, Universidad Autónoma de Coahuila, Unidad Saltillo, México

Olga B. Alvarez-Perez
Food Research Department, Group of Bioprocesses and Bioproducts, School of Chemistry, Universidad Autonoma de Coahuila, 25280 Saltillo, Coahuila, México

Aftab Ansari
Department of Physical Sciences, Nanoscience and Soft Matter Laboratory, Tezpur University, Napaam, Tezpur 784028, Assam, India

Juan Alberto Ascacio-Valdes
Research Group of Science Food Department, Chemistry School, Autonomous University of Coahuila, Bvd. V. Carranza e Ing. J. Cáredenas V., Col. República, Saltillo 25280, Coahuila, México

I. Azreen
Chemical Engineering Programme, Faculty of Engineering, University Malaysia Sabah, Jalan UMS, 88400 Kota Kinabalu, Sabah, Malaysia

Devrim Balköse
Department of Chemical Engineering, Izmir Institute of Technology, Gulbahce Urla, Izmir, Turkey

Y. A. Begum
Department of Food Engineering & Technology, Tezpur University, Assam, India

Sanjay Kumar Bharti
Institute of Pharmaceutical Sciences, Guru Ghasidas Vishwavidyalaya, Bilaspur 495009, Chhattisgarh, India

Ana M. T. D. P. V. Cabral
Faculty of Pharmacy, University of Coimbra, 3000-295 Coimbra, Portugal

Gloria Castellano
Departamento de Ciencias Experimentales y Matemáticas, Facultad de Veterinaria y Ciencias Experimentales, Universidad Católica de Valencia San Vicente Mártir, Guillem de Castro-94, E-46001 València, Spain

Elda Patricia Segura Ceniceros
Research Group of NanoBioscience, Chemistry School, Autonomous University of Coahuila, Saltillo 25280, Coahuila, México

Mónica L. Chávez-González
Research Group of NanoBioscience, Chemistry School, Autonomous University of Coahuila, Saltillo 25280, Coahuila, México

Alejandro Zugasti Cruz
Research Group of Science Food Department, Chemistry School, Autonomous University of Coahuila, Bvd. V. Carranza e Ing. J. Cáredenas V., Col. República, Saltillo 25280, Coahuila, México

Monoj Kumar Das
Department of Molecular Biology and Biotechnology, Tezpur University, Tezpur, Assam, India

Sankar Chandra Deka
Department of Food Engineering and Technology, Tezpur University, Napaam, Tezpur, Assam, India

V. B. Dementiev
Institute of Mechanics, Ural Division, Russian Academy of Sciences, Izhevsk, Russia

Hasan Demir
Department of Chemical Engineering, Osmaniye Korkut Ata University, Osmaniye, Turkey

María Lourdes Virginia Díaz-Jiménez
Center for Research and Advanced Studies of the National Polytechnic Institute, CINVESTAV, Saltillo 25900, Ramos Arizpe, Coahuila, Mexico

M. A. Esteso
U.D. Química Física, Universidad de Alcalá, 28871 Alcalá de Henares, Spain

A. S. Fazlin
Chemical Engineering Programme, Faculty of Engineering, University Malaysia Sabah, Jalan UMS, 88400 Kota Kinabalu, Sabah, Malaysia

Ariel García-Cruz
Research Group of NanoBioscience, Chemistry School, Autonomous University of Coahuila, Saltillo 25280, Coahuila, México

M. Ghosh
Department of Food Engineering and Technology, Tezpur University, Assam, India

José Luis Martínez Hernández
Research Group of NanoBioscience, Chemistry School, Autonomous University of Coahuila, Saltillo 25280, Coahuila, México

Anna Ilyina
Research Group of NanoBioscience, Chemistry School, Autonomous University of Coahuila, Saltillo 25280, Coahuila, México

Hema Joshi
Department of Botany, Hindu Girls College, Sonepat, Haryana, India

Rajpreet Kaur
Department of Chemistry, BBK DAV College for Women, Amritsar 143001, Punjab, India

Poonam Khullar
Department of Chemistry, BBK DAV College for Women, Amritsar 143001, Punjab, India

Miguel Ángel De León-Zapata
Food Research Department, Group of Bioprocesses and Bioproducts, School of Chemistry, Universidad Autonoma de Coahuila, 25280 Saltillo, Coahuila, México

Lluvia I López-López
Facultad de Ciencias Químicas, Universidad Autónoma de Coahuila, Unidad Saltillo, México

Dambarudhar Mahanta
Department of Physical Sciences, Nanoscience and Soft Matter Laboratory, Tezpur University, Napaam, Tezpur 784028, Assam, India

Debarshi Kar Mahapatra
Department of Pharmaceutical Chemistry, Dadasaheb Balpande College of Pharmacy, Nagpur 440037, Maharashtra, India

Divya Mandial
Department of Chemistry, BBK DAV College for Women, Amritsar 143001, Punjab, India

R. Mariani
Chemical Engineering Programme, Faculty of Engineering, University Malaysia Sabah, Jalan UMS, 88400 Kota Kinabalu, Sabah, Malaysia

José L. Martínez
Facultad de Ciencias Químicas, Universidad Autónoma de Coahuila, Unidad Saltillo, México

Arturo I. Martínez-Enríquez
Center for Research and Advanced Studies of the National Polytechnic Institute, CINVESTAV, Saltillo 25900, Ramos Arizpe, Coahuila, Mexico

Anurag Maurya
Department of Botany, Shivaji College, University of Delhi, New Delhi 110027, India

Romeo Rojas Molina
Food Research Department, Group of Bioprocesses and Bioproducts, School of Chemistry, Universidad Autonoma de Coahuila, 25280 Saltillo, Coahuila, México

S. Muchahary
Department of Food Engineering and Technology, Tezpur University, Assam, India

Ana Karina Pérez-Guzmán
Research Group of NanoBioscience, Chemistry School, Autonomous University of Coahuila, Saltillo 25280, Coahuila, México

Paulraj Rajamani
School of Environmental Sciences, Jawaharlal Nehru University, New Delhi 110067, India

Rodolfo Ramos-González
CONACYT, Autonomous University of Coahuila, Saltillo 25280, Coahuila, México

Anand Ramteke
Department of Molecular Biology and Biotechnology, Tezpur University, Tezpur, Assam, India

Ana C. F. Ribeiro
Department of Chemistry, Coimbra Chemistry Centre, University of Coimbra, 3004-535 Coimbra, Portugal

Daniela F. S. L. Rodrigues
Department of Chemistry, Coimbra Chemistry Centre, University of Coimbra, 3004-535 Coimbra, Portugal

Raúl Rodríguez-Herrera
Facultad de Ciencias Químicas, Universidad Autónoma de Coahuila, Unidad Saltillo, México

José Sandoval
Facultad de Ciencias Químicas, Universidad Autónoma de Coahuila, Unidad Saltillo, México

S. Sariah
Chemical Engineering Programme, Faculty of Engineering, University Malaysia Sabah, Jalan UMS, 88400 Kota Kinabalu, Sabah, Malaysia

Anamika Singh
Department of Botany, Maitreyi College, University of Delhi, Delhi, India

Neelu Singh
School of Environmental Sciences, Jawaharlal Nehru University, New Delhi 10067, India

Rajeev Singh
Department of Environmental Studies, Satyawati College, University of Delhi, Delhi, India

Lavanya Tandon
Department of Chemistry, BBK DAV College for Women, Amritsar 143001, Punjab, India

Francisco Torrens
Institut Universitari de Ciència Molecular, Universitat de València, Edifici d'Instituts de Paterna, P. O. Box 22085, E-46071 València, Spain

A. L. Urakov
Institute of Mechanics, Ural Division, Russian Academy of Sciences, Izhevsk, Russia
Institute of Thermology, Izhevsk, Russia
Izhevsk State Medical Academy, Izhevsk, Russia

N. A. Urakova
Institute of Thermology, Izhevsk, Russia
Izhevsk State Medical Academy, Izhevsk, Russia

Fatma Üstün
Department of Clinical Biochemistry, Ege University Hospital, Bornova, İzmir, Turkey

Janeth Ventura-Sobrevilla
Food Research Department, Group of Bioprocesses and Bioproducts, School of Chemistry, Universidad Autonoma de Coahuila, 25280 Saltillo, Coahuila, México

Luís M. P. Veríssimo
Department of Chemistry, Coimbra Chemistry Centre, University of Coimbra, 3004-535 Coimbra, Portugal

José Luis Villarreal-López
Facultad de Ciencias Químicas, Universidad Autónoma de Coahuila, Unidad Saltillo, México

A. Y. Zahrim
Chemical Engineering Programme, Faculty of Engineering, University Malaysia Sabah, Jalan UMS, 88400 Kota Kinabalu, Sabah, Malaysia

Y. Zulkiflee
Chemical Engineering Programme, Faculty of Engineering, University Malaysia Sabah, Jalan UMS, 88400 Kota Kinabalu, Sabah, Malaysia

ABBREVIATIONS

AgNPs	silver nanoparticles
AnPOME	anaerobically treated palm oil mill effluent
APP	ammonium polyphosphate
CF	cellulose fibers
CNTs	carbon nanotubes
COD	chemical oxygen demand
CPO	crude palm oil
DLS	dynamic light scattering
DVM	digital voltmeter
EF	electric field
EFB	empty fruit bunch
EO	essential oil
FFB	fresh fruit bunches
FLU	fluorescence
FRPP	flame-retardant polypropylene
FTIR	Fourier transforms infrared spectroscopy
GI	germination index
GMOs	genetically modified organisms
HFCS	high fructose corn syrup
HRR	heat release rate
IFR	intumescent flame retardant
LDH	layered double hydroxide
LOI	limiting oxygen index
MNP	magnetic nanoparticles
MS	mass spectrometry
NEMS	nanoelectromechanical system
NLCs	nanostructured lipid carriers
NMs	nanomaterials
NPs	nanoparticles
OVMT	organic vermiculite
POME	palm oil mill effluent
PVC	polyvinyl chloride
QD	quantum dot

QIT	quadrupole ion trap
RBC	red blood cell
ROS	reactive oxygen specie
RRG	relative root growth
RSG	relative seed germination
SEM	scanning electron microscopy
SLNs	solid–lipid nanoparticles
TCM	traditional Chinese medicine
TEM	transmission electron microscopy
TGA	thermogravimetric analysis
TOC	total organic carbon
TOF	time of flight
TSS	total suspended solids
UV–VIS	ultraviolet–visible

PREFACE

Chemical and biological engineers create and develop processes to change raw materials into the products that society depends on, such as food, chemicals, pharmaceuticals, paper, plastics, and personal care products. Chemical and process engineers help to manage natural resources, protect the environment, control health and safety procedures, and recycle materials while developing and managing the processes, which make the products we use. Knowledge in the fields of chemistry, physics, biology, mathematics, and engineering sciences is necessary in order to apply processes developed on a laboratory scale and to adapt them for production on an industrial scale.

Biological and chemical engineering is a multidisciplinary research area in the cross field between biology, chemistry, physics, and engineering. It encompasses major cross-disciplinary areas such as disease and health, materials, environmental technologies, and food technology. Biological engineering is an interdisciplinary application of engineering principles to analyze biological systems and to expand the knowledge of chemical engineering into the biochemical realm.

Chemical engineering is an optimal combination of the molecular sciences (chemistry and biology), the physical sciences (physical chemistry and physics), the analytical sciences (math and computer programming), and engineering. Moreover, chemical engineering provides students with real-world experience through laboratory classes, hands-on operation of pilot-scale equipment, and research projects that result in clean energy production and medical applications.

Chemical and biological engineering integrates chemistry and biology and uses this broad foundation along with engineering fundamentals to study the synthesis of new processes and products.

Chemical and biological engineering uses process engineering, transforms chemical and biological materials into high-value products and services, in a safe and cost-effective way. Chemical and biological engineers bring about the large-scale benefits of advances in chemistry, biotechnology, materials, and environmental sustainability to the real world.

Chemical or biological engineers could play a vital role in the creation and production of new medicines, nutritious foods, novel materials, better waste treatment methods, and a sustainable global future.

Research into bioprocess engineering is a major part of the wider chemical and biological engineering program and includes fermentation of functional foods, production of high-value pharmaceuticals, and bioconversion of waste into value-added chemicals.

This book is designed for exploring interdisciplinary research progress in life sciences, agriculture, food processing, and environment. It is also designed to accommodate students and researchers with broad interests within the field of chemical and biological engineering.

This research-oriented book is based on experimental methods and is aimed at the design, development, operation, monitoring, control, and optimization of chemical, physical, and biological processes.

This new research-oriented book provides innovative chapters covering the growth of educational, scientific, and industrial research activities among chemical and biological engineers and provides a medium for mutual communication between international academia and the industry. This book publishes significant research reporting new methodologies and important applications and latest coverage of physical chemistry and the development of new experimental methods. This collection presents to the reader a broad spectrum of chapters in the various branches of chemical and biological engineering, which demonstrate important developments in these rapidly changing fields. Case studies are included in some chapters, building a real-world connection. These case studies form a common thread throughout the book, motivating the reader and offering enhanced understanding.

PART I

Analytical and Experimental Methods

CHAPTER 1

CHEMICAL AND BIOLOGICAL SCREENING APPROACHES TO PHYTOPHARMACUTICALS

FRANCISCO TORRENS[1,*] and GLORIA CASTELLANO[2]

[1]*Institut Universitari de Ciència Molecular, Universitat de València, Edifici d'Instituts de Paterna, P. O. Box 22085, E-46071 València, Spain*

[2]*Departamento de Ciencias Experimentales y Matemáticas, Facultad de Veterinaria y Ciencias Experimentales, Universidad Católica de Valencia San Vicente Mártir, Guillem de Castro-94, E-46001 València, Spain*

[*]*Corresponding author. E-mail: torrens@uv.es*

ABSTRACT

Most drugs currently available in the market are natural products, or compounds derived from or inspired by plant-based natural products, which offer the structural diversity that is not rivaled by the creativity or synthetic ingenuity of medicinal chemists. The demand and amount of phytopharmaceuticals entering the global market is on a steady rise because of their growing acceptance. China, Korea, India, and Brazil remain the world leaders in research and production of medicinal plant products. In 1911–2000, Chinese were able to isolate different bioactive plant compounds via bioassay-guided approach. The large number of medicinal plants to be evaluated and documented in developing countries will benefit from the chemo/bioscreening approach. If the countries become major players with their vast biodiversity resources, they should begin the process of research and development of the resources via an aggressive program.

Possible bioactivity of a herbal product/extract can be speculated based on its phytochemical composition; for example, (1) plant polysaccharides are known to exhibit stimulating or suppressing effect on the immune system; (2) phenolic compounds showed strong antioxidant, anti-inflammatory, antiproliferative, and antiageing activities; (3) amino acids and proteins in herbs are usually regarded as natural nutritional supplements for patients recovering from diseases. 1,2,4-Trioxolane cycle spiro-conjugated with lactone is the key pharmacophore fragment in artemisinin. *Citrus* fruits contain organic acids, vitamins, minerals, citric and ascorbic acids, and flavonoids. Unripe *Citrus* fruit is more enriched than ripe fruit in dietary fiber, polyphenol, and flavonoids. Thymol is chemically related to anesthetic propofol. Propofol/sevoflurane protocols are used in anesthesia maintenance. However, propofol is recommended neither in a neonatal/pediatric population nor at home, because of narrow therapeutic index.

1.1 INTRODUCTION

Perdiguero analyzed complementary and alternative medicines (CAMs) on *social security pluralism* with a historical perspective.[1] Schwarcz reviewed modern inventions via commentaries on the fascinating chemistry of everyday life.[2] The critical approach to pseudosciences as didactic exercise, homeopathy, and successive dilutions were examined.[3] Jiménez-Liso group explored writing science for teaching and popularizing like the science on Procusto's bed.[4] They informed some difficulties and proposals to use broadsheets' science news at science class.[5] They revised socioscientific issues, their topics, and importance to science education.[6]

Liposomal incorporation of *Artemisia arborescens* essential oil (EO) and in vitro antiviral activity were informed.[7] Natural product (NP) isolation was introduced to organic chemistry students via steam distillation (SD) and liquid phase extraction (LPE) of camphor from sagebrush *A. tridentata*.[8] Ibrahim et al. provided a guide to chemical and biological screening approaches in the research and development of phytopharmaceuticals.[9] They discussed methods toward achieving quality NPs that meet basic regulatory requirements. Artesunate enhanced the cytotoxicity of 5-aminolevulinic acid (ALA)-based sonodynamic therapy (SDT) versus mouse mammary tumor cells in vitro.[10] The effects of artemisinin (ART) on the cytolytic activity of natural killer cells were informed.[11] Extracts obtained from *Pterocarpus angolensis* and *Ziziphus mucronata* exhibited antiplasmodial activity and inhibited heat shock protein 70 (Hsp70).[12] Some ART derivatives (ARTDs) target topoisomerase 1 and cause deoxyribonucleic acid (DNA) damage in silico and in vitro.[13] Is ART a panacea eligible for unrestrictive use?[14] Four new compounds were obtained from cultured cells of *A. annua*.[15] Efficacy of compounds from *A. lavandulaefolia* EO in control of the cigarette beetle was informed.[16] Quantitative structure–activity relationship (QSAR)-driven design and discovery of novel compounds were informed with antiplasmodial and transmission-blocking activities.[17]

Earlier publications in *Nereis*, and so on classified yams,[18] lactic acid bacteria,[19] fruits,[20] food spices,[21] and oil legumes[22] by principal component, cluster, and meta-analyses. The molecular classifications of 33 phenolic compounds derived from the cinnamic and benzoic acids from *Posidonia oceanica*,[23] 74 flavonoids,[24] 66 stilbenoids,[25] 71 triterpenoids and steroids from *Ganoderma*,[26] 17 isoflavonoids from *Dalbergia parviflora*,[27] 31 sesquiterpene lactones (STLs),[28,29] and ARTDs[30] were informed. A tool for interrogation of macromolecular structure was reported.[31] Mucoadhesive polymer hyaluronan favors transdermal penetration absorption of caffeine.[32,33] Polyphenolic phytochemicals in cancer prevention and therapy, bioavailability, and bioefficacy were reviewed.[34]

1.2 CHEMICAL AND BIOLOGICAL SCREENING APPROACHES TO PHYTOPHARMACEUTICALS

Well-documented practices, for example, the traditional Chinese medicine (TCM) yielded a number of drugs and drug candidates, for example, ART (cf. Fig. 1.1a), the active ingredient of sweet wormwood *Artemisia annua* used for treating malaria,[35] the alkaloids quinine (Fig. 1.1b), and cinchonine (Fig. 1.1c) from Peruvian tree quinine *Cinchona officinalis* (Rubiaceae) and other *Cinchona spp.* used for treating malaria,[36] nicotine (Fig. 1.1d) from cultivated tobacco *Nicotiana tabacum* L. (Solanaceae), gedunin (Fig. 1.1e) from neem *Azadirachta indica*, thymol (Fig. 1.1f) from common thyme *Thymus vulgaris*, eugenol from clove basil *Ocimum gratissimum* L., rotenone from Mexican yam bean or jicama *Pachyrhizus erosus*, paclitaxel (PTX, taxol, Fig. 1.1g) from Pacific yew *Taxus brevifolia*, canthin-6-one (Fig. 1.1h), berberine (Fig. 1.1i) from barberry *Berberis vulgaris* L. (Berberidaceae) and chelerythrine (Fig. 1.1j) from Senegal prickly-ash *Zanthoxylum zanthoxyloides* Watchman. Chemical compounds from plant sources make good drug candidates because they evolved as components of biosystems.

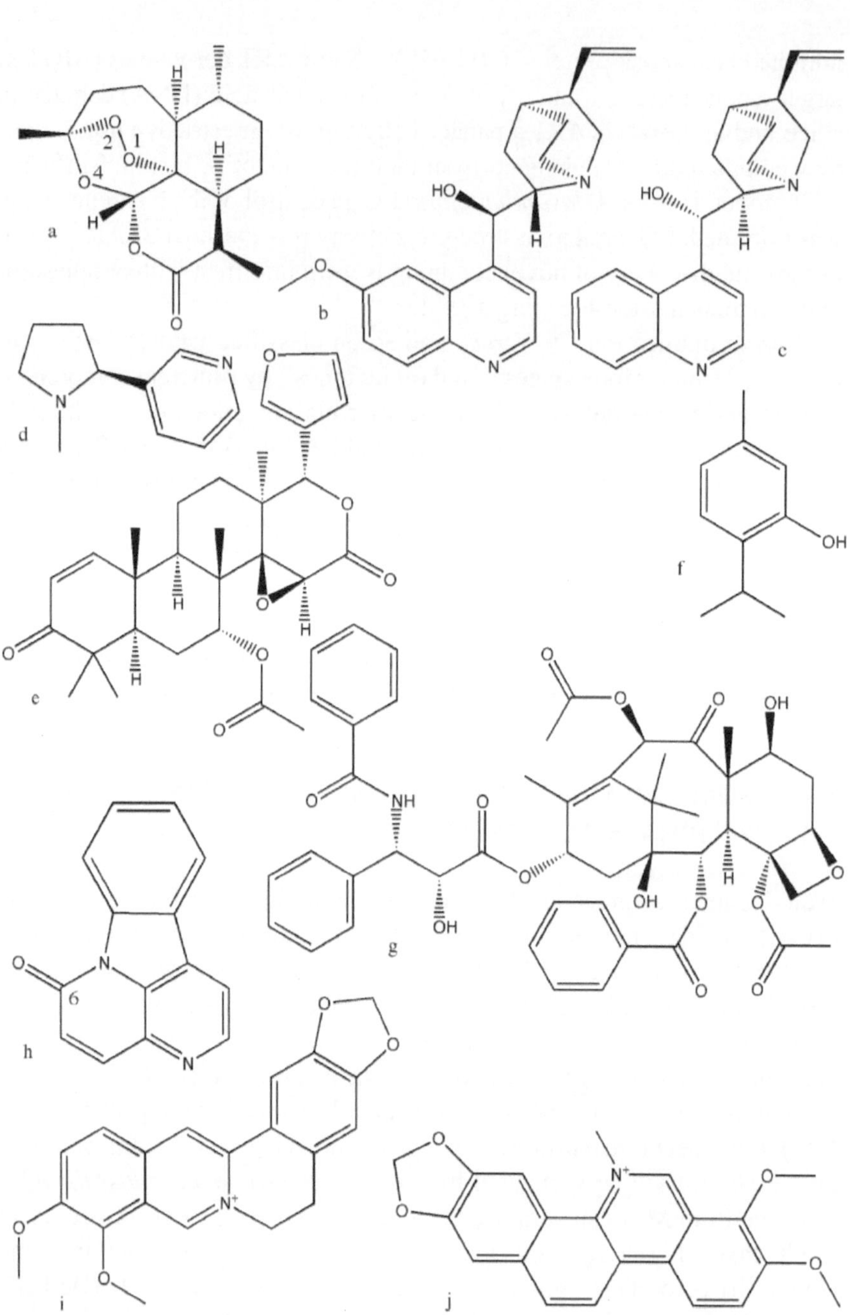

FIGURE 1.1 Molecular structures: (a) ART, (b) quinine, (c) cinchonine, (d) nicotine, (e) gedunin, (f) thymol, (g) PTX, (h) canthin-6-one, (i) berberine, and (j) chelerythrine.

The oral bioavailability of naturally occurring anticancer phytomolecules was reported.[37] *Langerstroemia fauriei* and *Lophira lanceolata* ethanolic extracts were recommended as substitute indicators in acid–base titrimetry.

1.3 CITRUS

Anti-photoaging effect of Jeju putgyul extracts was informed on human dermal fibroblasts and ultraviolet B-induced hairless mouse skin.[38] Chemotaxonomic classification was applied to the identification of *Citrus* TCMs.[39] Sulfur-mediated-alleviation of Al-toxicity in *C. grandis* seedlings was reported.[40] Composition and bioactivity of EO from *C. grandis* leaf were published.[41] Preparation, characterization, and bioactivities of topical antiageing ingredients were informed in a *C. junos* callus extract.[42] The benefits of the *Citrus* flavonoid diosmin was reported on human retinal pigment epithelial cells under high-glucose conditions.[43] Dissection of the mechanism for compatible and incompatible graft combinations of *C. grandis* (*Hongmian miyou*) was published.[44] Antibacterial activity of emulsified pomelo (*C. grandis* Osbeck) peel EO and water-soluble chitosan was informed on *Staphylococcus aureus* and *Escherichia coli*.[45]

1.4 ARTEMISIN PHARMACOPHORE

1,2,4-Trioxolane cycle spiro-conjugated with sesquiterpene δ-lactone (activated δ-lactone) is the key pharmacophore fragment in ART.

1.5 DISCUSSION

Most drugs currently available in the market are NPs, or compounds derived from or inspired by plant-based NPs, which offer the structural diversity that is not rivalled by the creativity or synthetic ingenuity of medicinal chemists. The demand and amount of phytopharmaceuticals entrying the global market is on a steady rise because of their growing acceptance. China, Korea, India, and Brazil remain the world leaders in research and production of medicinal plant NPs. Between 1911 and 2000, Chinese were able to isolate different bioactive plant compounds, for example, alkaloid, steroid, triterpene, and limonoid, via bioassay-guided approach. The large number of medicinal plants to be evaluated and documented in developing countries will benefit from the chemical and biological screening approach. If the countries become major players with their vast biodiversity resources, they should begin the process of research and development of the resources via an aggressive program that ensures quality NPs. Possible bioactivity of a herbal NP/extract can be speculated based on its phytochemical composition; for example: (1) plant polysaccharides are known to exhibit stimulating or suppressing effect on the immune system, for example, *Ganoderma lucidum* (Leyssex Fr.) Karst and *Cordycep sinensis* (Berk.) Sacc.; (2) phenolic compounds, for example, flavonoids (flavones, isoflavones, flavonols, flavonones, and xanthones) and non-flavonoids (lignins and stilbenes), showed strong antioxidant, anti-inflammatory, antiproliferative and antiageing activities; and (3) amino acids (AAs) and proteins in herbs are usually regarded as NP nutritional supplements for patients recovering from diseases.

Thymol (2-isopropyl-5-methylphenol, IPMP, cf. Fig. 1.2a) is chemically related to the anesthetic propofol (2,6-diisopropylphenol, Diprivan, Fig. 1.2b).

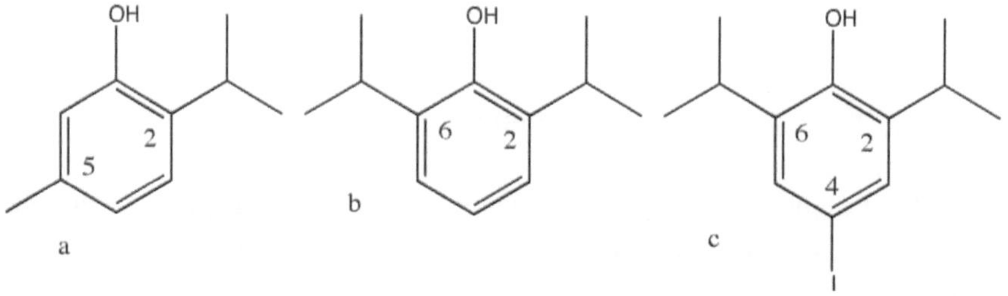

FIGURE 1.2 Molecular structures: (a) thymol, (b) propofol, and (c) 4-iodopropofol.

Citrus fruits, for example, oranges, contain organic acids, vitamins (Vits), minerals, citric and ascorbic acids, and flavonoids. Unripe *Citrus* fruit is more enriched than ripe fruit in dietary fiber, polyphenol, and flavonoids, for example, hesperedin and naringin.

1.6 FINAL REMARKS

From the preceding results and discussion, the following final remarks can be drawn:

1. 1,2,4-Trioxolane cycle spiro-conjugated with sesquiterpene δ-lactone (activated δ-lactone) is the key pharmacophore fragment in artemisinin.
2. Thymol is chemically related to the anesthetic propofol. Propofol/sevoflurane protocols are used in anaesthesia maintenance. However, propofol is recommended neither in a neonatal/pediatric population nor at home, because of narrow therapeutic index: overdose causes cardiac/respiratory arrests (Michael Jackson's death), requiring machines only available in hospitals.
3. Further work will deal with anesthetic alcohols/propofol family, for example, 4-iodopropofol (2,6-diisopropyl-4-iodophenol, Fig. 1.2c), which presents similar effects to propofol on isolated receptors, acting as γ-aminobutyric-acid type-A positive modulator/Na^+-channel blocker but, in animals, it shows anxiolytic/anticonvulsant effects, lacking propofol sedative/hypnotic profile.

ACKNOWLEDGMENTS

The authors thank support from Generalitat Valenciana (Project No. PROMETEO/2016/094) and Universidad Católica de Valencia *San Vicente Mártir* (Project No. UCV.PRO.17-18.AIV.03).

KEYWORDS

- **herbal product**
- **medicinal plant**
- **research and development**
- **quality parameter**
- **organic peroxide**
- **cycle**
- **complementary medicine**

REFERENCES

1. Perdiguero Gil, E. A Propósito de las Medicinas Alternativas y Complementarias: Sobre el Pluralismo Asistencial. In *Antropología y Enfermería. Campos de Encuentro: Un Homenaje a Dina Garcés, II*; Martorell, M. A., Comelles, J. M., Bernal, M., Eds.; Universitat Rovira i Virgili: Tarragona, Spain, 2010; pp 278–300.
2. Schwarcz, J. *Radars, Hula Hoops, and Playful Pigs: 67 Digestible Commentaries on the Fascinating Chemistry of Everyday Life*; ECW: Toronto, ON, 1993.
3. Abellán, G.; Rosaleny, L. E.; Carnicer, J.; Baldoví, J. J.; Gaita-Ariño, A. La aproximación crítica a las pseudociencias como ejercicio didáctico: Homeopatía y diluciones sucesivas. *An. Quím.* **2014,** *110*, 211–217.
4. González García, F.; Jiménez-Liso, M. R. Escribir ciencia para enseñar y divulgar o la ciencia en el lecho de Procusto. *Alambique* **2005,** *2005* (43), 8–20.
5. Jiménez-Liso, M. R.; Hernández-Villalobos, L.; Lapetina, J. Dificultades y propuestas para utilizar las noticias científicas de la prensa en el aula de ciencias. *Rev. Eureka Enseñ. Divul. Cien.* **2010,** *7*, 107–126.
6. Díaz Moreno, N.; Jiménez-Liso, M. R. Las controversias sociocientíficas: Temáticas e importancia para la educación científica. *Rev. Eureka Enseñ. Divul. Cien.* **2012,** *9*, 54–70.
7. Sinico, C.; De Logu, A.; Lai, F.; Valenti, D.; Manconi, M.; Loy, G.; Bonsignore, L.; Fadda, A. M. Liposomal Incorporation of *Artemisia arborescens* L. Essential Oil and In vitro Antiviral Activity. *Eur. J. Pharm. Biopharm.* **2005,** *1*, 161–168.
8. McLain, K. A.; Miller, K. A.; Collins, W. R. Introducing Organic Chemistry Students to Natural Product Isolation Using Steam Distillation and Liquid Phase Extraction of Thymol, Camphor, and Citral, Monoterpenes Sharing a Unified Biosynthetic Precursor. *J. Chem. Educ.* **2015,** *92*, 1226–1228.
9. Ibrahim, J. A.; Egharevba, H. O.; Gamaniel, K. S. Chemical and Biological Screening Approaches to Phytopharmaceuticals. *Int. J. Sci.* **2017,** *6* (10), 22–31.
10. Osaki, T.; Uto, Y.; Ishizuka, M.; Tanaka, T.; Yamanaka, N.; Kurahashi, T.; Azuma, K.; Murahata, Y.; Tsuka, T.; Itoh, N.; Imagawa, T.; Okamoto, Y. Artesunate Enhances the Cytotoxicity of 5-Aminolevulinic Acid-Based Sonodynamic Therapy Against Mouse Mammary Tumor Cells In Vitro. *Molecules* **2017,** *22*, 533.
11. Houh, Y. K.; Kim, K. E.; Park, S.; Hur, D. Y.; Kim, S.; Kim, D.; Bang, S. I.; Yang, Y.; Park, H. J.; Cho, D. The Effects of Artemisinin on the Cytolytic Activity of Natural Killer (NK) Cells. *Int. J. Mol. Sci.* **2017,** *18*, 1600.
12. Zininga, T.; Anokwuru, C. P.; Sigidi, M. T.; Tshisikhawe, M. P.; Ramaite, I. I. D.; Traoré, A. N.; Hoppe, H.; Shonhai, A.; Potgieter, N. Extracts Obtained from *Pterocarpus angolensis* DC and *Ziziphus mucronata* Exhibit Antiplasmodial Activity and Inhibit Heat Shock Protein 70 (Hsp70) Function. *Molecules* **2017,** *22*, 1224.
13. Kadioglu, O.; Chan, A.; Qiu, A. C. L.; Wong, V. K. W.; Colligs, V.; Wecklein, S.; Rached, H. F. H.; Efferth, T.; Hsiao, W. L. W. Artemisinin Derivatives Target Topoisomerase 1 and Cause DNA Damage *In silico* and *In vitro*. *Front. Pharmacol.* **2017,** *8*, 711.
14. Yuan, D. S.; Chen, Y. P.; Tan, L. L.; Huang, S. Q.; Li, C. Q.; Wang, Q.; Zeng, Q. P. Artemisinin: A Panacea Eligible for Unrestrictive Use? *Front. Pharmacol.* **2017,** *8*, 737.
15. Zhu, J.; Xiao, P.; Qian, M.; Chen, C.; Liang, C.; Zi, J.; Yu, R. Four New Compounds Obtained from Cultured Cells of *Artemisia annua*. *Molecules* **2017,** *22*, 2264.

16. Zhou, J.; Zou, K.; Zhang, W.; Guo, S.; Liu, H.; Sun, J.; Li, J.; Huang, D.; Wu, Y.; Du, S.; Borjigidai, A. Efficacy of Compounds Isolated from the Essential Oil of *Artemisia lavandulaefolia* in Control of the Cigarette Beetle, *Lasioderma serricorne*. *Molecules* **2018,** *23*, 343.
17. Lima, M. N. N.; Melo-Filho, C. C.; Cassiano, G. C.; Neves, B. J.; Alves, V. M.; Braga, R. C.; Cravo, P. V. L.; Muratov, E. N.; Calit, J.; Bargieri, D. Y.; Costa, F. T. M.; Andrade, C. H. QSAR-driven Design and Discovery of Novel Compounds with Antiplasmodial and Transmission Blocking Activities. *Front. Pharmacol.* **2018,** *9*, 146.
18. Torrens-Zaragozá, F. Molecular Categorization of Yams by Principal Component and Cluster Analyses. *Nereis* **2013,** *2013* (5), 41–51.
19. Torrens-Zaragozá, F. Classification of Lactic Acid Bacteria Against Cytokine Immune Modulation. *Nereis* **2014,** *2014* (6), 27–37.
20. Torrens-Zaragozá, F. Classification of Fruits Proximate and Mineral Content: Principal Component, Cluster, Meta-Analyses. *Nereis* **2015,** *2015* (7), 39–50.
21. Torrens-Zaragozá, F. Classification of Food Spices by Proximate Content: Principal Component, Cluster, Meta-Analyses, *Nereis* **2016,** *2016* (8), 23–33.
22. Torrens, F.; Castellano, G. From Asia to Mediterranean: Soya Bean, Spanish Legumes and Commercial *Soya Bean* Principal Component, Cluster and Meta-Analyses. *J. Nutr. Food Sci.* **2014,** *4* (5), 98.
23. Castellano, G.; Tena, J.; Torrens, F. Classification of Polyphenolic Compounds by Chemical Structural Indicators and its Relation to Antioxidant Properties of *Posidonia oceanica* (L.) Delile. *MATCH Commun. Math. Comput. Chem.* **2012,** *67*, 231–250.
24. Castellano, G.; González-Santander, J. L.; Lara, A.; Torrens, F. Classification of Flavonoid Compounds by Using Entropy of Information Theory. *Phytochemistry* **2013,** *93*, 182–191.
25. Castellano, G.; Lara, A.; Torrens, F. Classification of Stilbenoid Compounds by Entropy of Artificial Intelligence. *Phytochemistry* **2014,** *97*, 62–69.
26. Castellano, G.; Torrens, F. Information Entropy-Based Classification of Triterpenoids and Steroids from *Ganoderma*. *Phytochemistry* **2015,** *116*, 305–313.
27. Castellano, G.; Torrens, F. Quantitative Structure–Antioxidant Activity Models of Isoflavonoids: A Theoretical Study. *Int. J. Mol. Sci.* **2015,** *16*, 12891–12906.
28. Castellano, G.; Redondo, L.; Torrens, F. QSAR of Natural Sesquiterpene Lactones as Inhibitors of Myb-Dependent Gene Expression. *Curr. Top. Med. Chem.* **2017,** *17*, 3256–3268.
29. Torrens, F.; Redondo, L.; León, A.; Castellano, G. Structure–Activity Relationships of Cytotoxic Lactones as Inhibitors and Mechanisms of Action. *Curr. Drug Discov. Technol.*; Submitted for Publication.
30. Torrens, F.; Redondo, L.; Castellano, G. Artemisinin: Tentative Mechanism of Action and Resistance. *Pharmaceuticals* **2017,** *10*, 20.
31. Torrens, F.; Castellano, G. A Tool for Interrogation of Macromolecular Structure. *J. Mater. Sci. Eng. B* **2014,** *4* (2), 55–63.
32. Torrens, F.; Castellano, G. Mucoadhesive Polymer Hyaluronan as Biodegradable Cationic/Zwitterionic-Drug Delivery Vehicle. *ADMET DMPK* **2014,** *2*, 235–247.
33. Torrens, F.; Castellano, G. Computational Study of Nanosized Drug Delivery from *Cyclo*dextrins, Crown Ethers and Hyaluronan in Pharmaceutical Formulations. *Curr. Top. Med. Chem.* **2015,** *15*, 1901–1913.

34. Estrela, J. M.; Mena, S.; Obrador, E.; Benlloch, M.; Castellano, G.; Salvador, R.; Dellinger, R. W. Polyphenolic Phytochemicals in Cancer Prevention and Therapy: Bioavailability *Versus* Bioefficacy. *J. Med. Chem.* **2017,** *60*, 9413–9436.
35. Tu, Y. Y.; Ni, M. Y..; Zhong, Y. R.; Li, L. N.; Cui, S. L.; Zhang, M. Q.; Wang, X. Z.; Liang, X. T. Studies on the constituents of *Artemisia annua* L. (Authors' Translation). Article in Chinese. *Acta Pharm. Sinica* **1981,** *16*, 366–370.
36. López Piñero, J. M.; Calero, F. *De Pulvere Febrifugo Occidentalis Indiae (1993) de Gaspar Caldera de Heredia y la Introducción de la Quina en Europa*; Cuadernos Valencianos de Historia de la Medicina y de la Ciencia, Ser. A, No. 39, Universitat de València–CSIC: València, Spain, 1992.
37. Sharma, S.; Gupta, M.; Sharma, A.; Agarwal, S. M. Oral Bioavailability of Naturally Occurring Anticancer Phytomolecules. *Lett. Drug Des. Discov.* **2018,** *15*, 1–1.
38. Choi, S. H.; Choi, S. I.; Jung, T. D.; Cho, B. Y.; Lee, J. H.; Kim, S. H.; Yoon, S. A.; Ham, Y. M.; Yoon, W. J.; Cho, J. H.; Lee, O. H. Anti-photoaging Effect of Jeju putgyul (Unripe Citrus) Extracts on Human Dermal Fibroblasts and Ultraviolet B-Induced Hairless Mouse Skin. *Int. J. Mol. Sci.* **2017,** *18*, 2052.
39. Zhao, S. Y.; Liu, Z. L.; Shu, Y. S.; Wang, M. L.; He, D.; Song, Z. Q.; Zeng, H. L.; Ning, Z. C.; Lu, C.; Lu, A. P.; Liu, Y. Y. Chemotaxonomic Classification Applied to the Identification of two Closely-Related *Citrus* Tcms Using UPLC-Q-TOF-MS-based Metabolomics. *Molecules* **2017,** *22*, 1721.
40. Guo, P.; Li, Q.; Qi, Y. P.; Yang, L. T.; Ye, X.; Chen, H. H.; Chen, L. S. Sulfur-mediated-Alleviation of Aluminum-Toxicity in *Citrus grandis* Seedlings. *Int. J. Mol. Sci.* **2017,** *18*, 2570.
41. Tsai, M. L.; Lin, C. D.; Khoo, K. A.; Wang, M. Y.; Kuan, T. K.; Lin, W. C.; Zhang, Y. N.; Wang, Y. Y. Composition and Bioactivity of Essential Oil from *Citrus grandis* (L.) Osbeck *Mato peiyu* Leaf. *Molecules* **2017,** *22*, 2154.
42. Adhikari, D.; Panthi, V. K.; Pangeni, R.; Kim, H. J.; Park, J. W. Preparation, Characterization, and Biological Activities of Topical Anti-Aging Ingredients in a *Citrus junos* callus Extract. *Molecules* **2017,** *22*, 2198.
43. Liu, W. Y.; Liou, S. S.; Hong, T. Y.; Liu, I. M. The Benefits of the Citrus Flavonoid Diosmin on Human Retinal Pigment Epithelial Cells Under High-Glucose Conditions. *Molecules* **2017,** *22*, 2251.
44. He, W.; Wang, Y.; Chen, Q.; Sun, B.; Tang, H. R.; Pan, D. M.; Wang, X. R. Dissection of the Mechanism for Compatible and Incompatible Graft Combinations of *Citrus grandis* (L.) Osbeck (*Hongmian miyou*). *Int. J. Mol. Sci.* **2018,** *19*, 505.
45. Chen, G. W.; Lin, Y. H.; Lin, C. H.; Jen, H. C. Antibacterial Activity of Emulsified Pomelo (*Citrus grandis* Osbeck) Peel Oil and Water-soluble Chitosan on *Staphylococcus aureus* and *Escherichia coli*. *Molecules* **2018,** *23*, 840.

CHAPTER 2

MOLECULAR DEVICES AND MACHINES: HYBRID ORGANIC–INORGANIC STRUCTURES

FRANCISCO TORRENS[1,*] and GLORIA CASTELLANO[2]

[1]*Institut Universitari de Ciència Molecular, Universitat de València, Edifici d'Instituts de Paterna, P. O. Box 22085, E-46071 València, Spain*

[2]*Departamento de Ciencias Experimentales y Matemáticas, Facultad de Veterinaria y Ciencias Experimentales, Universidad Católica de Valencia San Vicente Mártir, Guillem de Castro-94, E-46001 València, Spain*

[*]*Corresponding author. E-mail: torrens@uv.es*

ABSTRACT

Sustainable or *green nanotechnology* is commonly used in the development of clean technologies. This work discusses different aspects of molecular devices and machines, and hybrid organic–inorganic structures in dissimilar fields. There is a general interest in photocatalysis, in the field of sustainable chemistry, aiming to transform solar into chemical energy, in order to obtain products or processes that are thermally forbidden, or take place with low yield. *Qubits* are information. It is reasonable to assume that religious root born of material conditions of every society guided certain people by the way of placing growth avidity in front of harmony with environment, which caused fast development of important philosophical structures that rested on technical advance, but also placed the part of humanity that took the chief role in them in a limit situation in which its own survival capacity is threatened. Concept *sustainable development* connects directly with West anthropocentric cultural tradition.

2.1 INTRODUCTION

Venturi group reviewed molecular devices, machines, concepts, and perspectives for the nanoworld.[1] They revised the electrochemistry of functional supramolecular systems.[2] They developed sustainability via scientific and ethical issues.[3] Hazard screening methods for nanomaterials (NMs) were comparatively studied.[4]

In earlier publications, fractal hybrid-orbital analysis,[5,6] resonance,[7] molecular diversity,[8] periodic table of the elements,[9,10] law, property, information entropy, molecular classification, simulators,[11–19] labor risk prevention, preventive healthcare at work with NMs,[20–22] science and ethics of developing sustainability via nanosystems and devices,[23] and *green nanotechnology* as an approach toward environment safety[24] were reviewed. Many researches were performed in different areas in the past many years related to nanotechnology. The present work discusses dissimilar aspects of molecular devices and machines, and hybrid organic–inorganic structures in diverse fields. There is a general interest in photocatalysis, in the field of sustainable chemistry, aiming to transform solar into chemical energy, in order to obtain products or processes that are thermally forbidden, or take place with low yield. *Qubits* are information. The aim of this work is to initiate a debate by suggesting a number of questions (Q), which can arise when addressing subjects of molecular devices, machines, and so on in different fields, and providing, when possible, answers (A) and hypotheses (H).

2.2 MOLECULAR DEVICES AND MACHINES: HYBRID ORGANIC–INORGANIC STRUCTURES

Venturi group reviewed molecular devices and machines and hybrid structures, proposing (H/Q).[25]

H1. (Feynman, 1959). There is plenty of room at the bottom.[26]
H2. (Feynman, 1959). Scientists will be able to manipulate/control individual atoms/molecules.
Q1. (Whatmore, 2006). Nanotechnology, what is it?
Q2. (Whatmore, 2006). Nanotechnology, should we be worried?
H3. (Taniguchi). He invented the term *nanotechnology*.
Q3. What is nanotechnology?

H4. Modern nanotechnology began with scanning tunneling (STM) and atomic force (AFM) microscopes (1981/1982).
H5. (Fischer, 1894). *Lock-and-key* principle.
H6. (Ehrlich). *Receptor* idea.
H7. (Werner). Coordination chemistry.
H8. (Pedersen, 1967). He synthesized crown ethers.
H9. (Cram; Lehn; Vögtle). They synthesized shape- and ion-selective receptors.
H10. (Lehn, 1995). *Supramolecular chemistry*: chemistry beyond molecule, bearing on organized entities of higher complexity resulting from association of two/more chemical species held together by intermolecular forces.
H11. (Stoddart, 2007). He created molecular machines and complex self-assembled structures.
H12. (Willner, 2000). He developed sensors and biological and electronic interfacing methods.
H13. If interaction between units is lower than other energies, system is a supramolecular species.
H14. (Venturi, 2008). If supramolecular system is molecular-components assembly with weak interaction, properties of each component are learned from isolated component's study/choosing suitable molecule model.
H15. Such types of supramolecular structure are fragile because of disassembling by external factors.
H16. (Michi, 2005; Zerbetto, 2007). They defined *mechanical device, machine,* and *motor*.
H17. (Venturi, 2008). Molecular-machines features: energy-input kind; operation monitoring; components-motion type; cyclic operation repeating possibility; cycle period; function.
H18. (Venturi, 2008, 2010, 2014). Main interactions in rotaxanes/catenanes are: charge transfer, H-bonding, hydrophobic/philic, π–π stacking, electrostatic, metal–ligand bonding.
H19. (Venturi, 2007, 2014). Advantages from rotaxanes/catenanes use to design/construct molecular machines: mechanical-bond presence enables variety of component mutual arrangements assuring system stability; intercomponent-motion amplitude is limited in three directions by interlocked architecture; intercomponent-interactions strength determines specific-arrangement (*co-conformation*) stability; intercomponent interactions are modulated by external stimulation.

H20. Rotaxanes main large-amplitude motions: translation (*molecular shuttle*; Venturi, 2007); rotation.

H21. (Feynman, 1959). Physics principles do not speak versus possibility of manoeuvring things atom-by-atom.

H22. (Drexler, 1986, 1992). To construct device (*nanorobot*) to build atom-by-atom any structure.

H23. (Venturi, 2008). Aspects regarding unrealistic *nanorobot* character: fingers must be made themselves by atoms; atoms will stick on fingers; continuous movement of each nanostructure will determine nanoengineering-accuracy obstruction.

H24. (Venturi, 2007). Molecule-by-molecule approach is sustained by: molecules are stable; in nature nanodevices and machines are made by molecules (Goodsell, 2004); chemical processes driven in laboratory are based on molecules; molecules have different sizes/shapes/properties modified by photo/electrochemical inputs; molecules self-assemble (*supramolecular chemistry*).

Q4. (Venturi, 2001). Which energy to make artificial molecular-level machines work?

H25. (Venturi, 2001, 2007). Main sources of energy: chemical/electro/photochemical energies.

H26. (Venturi, 2001). pH/redox reactions advantages: they represent most simple/common way to supply energy to system and they allow an appropriate energy storage/transport/delivery.

H27. (Ciamician, 1908). He investigated the photochemical reactions.

H28. (Ciamician, 1908). *Photosynthesis* is based on plants ability to use the sunlight producing energy.

H29. (Ciamician, 1908). *The Chemical Action of Light*: to obtain fuel by artificial photochemical reactions from solar energy.

H30. (Balzani, 2006; Venturi, 2008, 2009). Photoreactions advantages: energy-amount control by exciting-light wavelength/intensity related to species absorption spectrum, transmit energy without physical connection with source, fast switch on/off light sources, light polarization, reduce working spaces/time, induce excitation with nanoresolution by near field, address high number of individual nanodevices, and renewable energy sources.

H31. (Balzani, 2008, 2014). A–L–B supramolecular system [A: light-absorbing molecular unit; B: another molecular unit involved in light-powered processes; L: connecting unit (*bridge*)].

H32. (Venturi, 2007). Performed functions from different domains: signal transfer, energy conversion, mechanical movements, and informational processes.
H33. General aspects: simple-systems design/construction; understanding of principles/processes underlying performance; overcoming difficulties aroused when connected to macroworld.
H34. Logic-gates advantages: fluorescence (FLU) operates without wire, light bridges molecules/macroworld gap, and FLU is easily detected.
H35. Pseudorotaxanes dethreading/rethreading is based on: pseudorotaxane incorporates *photosensitizer*, dethreading occurs as result of visible (VIS) photosensitizer excitation in presence of sacrificial electron donor, and rethreading occurs when O_2 enters into solution.
H36. In distinct *stations,* photoinduced shuttling occurs in steps: destabilization of stable translational isomer, ring displacement, electronic reset, and nuclear reset.
H37. (Balzani, 2006; Venturi, 2006). Optimized rotaxane acts like autonomous *four-stroke* linear motor where intramolecular processes correspond to: fuel injection/combustion, piston displacement, exhaust removal, and piston replacement.
H38. Azobenzene pseudorotaxane characteristics: light-controlled threading/dethreading, no waste-products formation, repeat dethreading/rethreading cycle, and changes in macrocycle-based FLU.
H39. Nonsymmetric molecular axle contains units: photosensitive azobenzene, central ammonium (*recognition site*), and nonphotoactive methyl*cyclo*pentyl.
H40. Photoinduced shuttling movement of ring between stations triggered by VIS excitation of unit P is obtained in solution at room temperature in distinct ways by: mechanism involving energy contribution from low-energy fuels, mechanism exploiting kinetic assistance of external electron relay, and intramolecular photochemical mechanism.
H41. Intramolecular mechanism is based on operations: light excitation of photoactive unit P (Step 1) is followed by electron transfer from P excited state to station A_1 being encircled by ring R (Step 2) with station-A_1 *deactivation* and such photoinduced electron transfer must compete with P intrinsic excited-state decay (Step 3); ring moves from reduced station A_1 to A_2 (Step 4) a step that must compete with back electron transfer from A_1 to oxidized photoactive P^+ (Step 5); back electron transfer from *free* reduced station A_1 to P^+ (Step 6); ring back movement from A_2 to A_1 (Step 7).

H42. Low efficiency is compensated by advantages: system operation relies on intramolecular processes, artificial molecular motor does not need external species, fuel (*sunlight*) is free, and it can work at single-molecule level.
H43. Dendrimers points: cooperation among photoactive components allowing dendrimer to perform useful functions (*light harvesting*); photophysical-properties changes being exploited for sensing with signal amplification; luminescence signals offering handle to understand dendritic structures and superstructures.
H44. (Balzani, 2004, 2007). Photoactive units are directly incorporated/appended with covalent/coordination bonds in different regions of dendritic structure in: core, branches, surface, core and branches, core and surface, branches and surface, and core, branches, and surface.
H45. Dansyl-functionalized dendrimers titration with Co^{2+} shows that their absorption spectra are unaffected but quenching of FLU of peripheral dansyl units occurs, FLU quenching occurs by static mechanism involving metal ions coordination in dendrimers inside, metal ion coordination by dendrimers is fully reversible, and FLU-quenching amplification occurs with rising dendrimer generation.
H46. Nonconventional applications of molecular devices/machines: behavior is exploited for processing information at molecular level/chemical-computers construction; mechanical features are used to transport nanoobjects/gate molecular-level channels/nanorobotics.

2.3 SUSTAINABLE CHEMISTRY AND PHOTOCATALYSIS: METAL NANOPARTICLES

González-Béjar discussed the general interest of photocatalysis, in the field of sustainable chemistry, aiming to transform solar into chemical energy, in order to obtain products or processes that are thermally forbidden, or take place with low yield.[27] She included an introduction of general concepts in photocatalysis, and a brief description of the main characteristics of metal NPs (MNPs) and their capabilities as photocatalysts upon irradiation. She exemplified syntheses of organic compounds to illustrate the capabilities, with special emphasis on supported MNPs for heterogeneous photocatalysis.

2.4 NANOTECHNOLOGY PROMISES: FUTURE AND IDEOLOGY IN ITS RHETORICS

Ureña López proposed the following questions, answers, and hypotheses on nanotechnology promises[28]:

Q1. What is nanotechnology?
A1. People go toward a technological realism.
Q2. What is that thing called *nanotechnology*?
A2. It is a series of technologies being performed at nanometric scale, especially materials science.
H1. The problems of identity of nanotechnology.
H2. Nanotechnology made promises in literature, journalism, public institutions, and scientific activity.
He revised the following promises of nanotechnology in literature:
H3. (Drexler, 1986). *Engines of Creation*.[29]
H4. (Crichton, 2002). *Prey*.[30]
He revised the following positive and negative promises of nanotechnology in journalism.
H5. (Kurzweil, 2005). *The Singularity Is Near* (a positive promise).[31]
H6. (Sandberg, 2014). *Five biggest threats* to human existence are: nuclear war, bioengineered pandemic, superintelligence, nanotechnology, and unknown unknowns.
He proposed the following additional questions, answers, and hypotheses:
Q3. Why do people invest in nanotechnology?
A3. Because of: strong optimism; no dystopia; it does not explain the adverse effects.
Q4. *Bottom-up* (demanded by *citizens, it is asked for*)?
A4. Two companies (IBM, HP), National Nanotechnology Initiative (NNI), and so on proposed it.
H7. (Drexler). It was by him that the US Government went into nanotechnology.
H8. (Drexler, 1986). He founded in Silicon Valley the Foresight Institute.
H9. (Bueno, 2004). Smalley versus Drexler: The Drexler–Smalley debate on nanotechnology.[32]
Q5. (Bueno, 2004). Incommensurability at work?
H10. Investments passed from information/communications technology to nanotechnology to synthetic biology.

Q6. The scientific/technological (SR/TR) realism, is it rhetorical?
A6. The objective of SR (TR) is to mitigate in science (technology) what is untrue.

2.5 ASSEMBLIES OF QUBITS: ORGANIZING THE NANOSCALE

Quantum binary digits (*bits*) (*qubits*) are information: Qubit is the basic unit of quantum information. Because of being quantum, it is special and marvelous but, for that very reason, it is information sensitive to noise. It is fragile information, which easily gets lost or distortion. Gaita Ariño discussed the difficulty that presents, as it also occurs in a multitudinarious assembly, that a great number of qubits efficiently *communicate* with each other, without succumbing to noise and other distraction sources.[33]

2.6 DISCUSSION

It is important the general concepts of photocatalysis, and a brief description of the main characteristics of MNPs and their capabilities as photocatalysts upon irradiation. It is significant the exemplified syntheses of organic compounds to illustrate the capabilities, with special emphasis on supported MNPs for heterogeneous photocatalysis.

Qubits are information: Qubit is the basic unit of quantum information. Because of being quantum, it is special and marvellous but, for that very reason, it is information sensitive to noise. It is fragile information, which easily gets lost or distortion.

It is reasonable to assume that religious root born of material conditions of every society guided certain people by the way of placing growth avidity in front of harmony with environment, which caused fast development of important philosophical structures that rested on technical advance, but also placed the part of humanity that took the chief role in them in a limit situation in which its own survival capacity is threatened. Concept *sustainable development* connects directly with West anthropocentric cultural tradition.

2.7 FINAL REMARKS

From the present discussion, the following final remarks can be drawn:

1. It is important the general concepts of photocatalysis, and a brief description of the main characteristics of metal nanoparticles and their capabilities as photocatalysts upon irradiation. It is significant the exemplified syntheses of organic compounds to illustrate the capabilities, with special emphasis on supported metal nanoparticles for heterogeneous photocatalysis.
2. It is difficult, as it also occurs in a multitudinarious assembly, that a great number of *qubits* efficiently *communicate* with each other, without succumbing to noise and other distraction sources.
3. It is reasonable to assume that religious root born of material conditions of every society guided certain people by the way of placing growth avidity in front of harmony with environment, which caused fast development of important philosophical structures that rested on technical advance, but also placed the part of humanity that took the chief role in them in a limit situation in which its own survival capacity is threatened.
4. Concept *sustainable development* connects directly with West anthropocentric cultural tradition.

ACKNOWLEDGMENTS

The authors thank support from Generalitat Valenciana (Project No. PROMETEO/2016/094) and Universidad Católica de Valencia *San Vicente Mártir* (Project No. UCV.PRO.17-18.AIV.03).

KEYWORDS

- **supramolecular chemistry**
- **molecular electronics and photonics**
- **molecular wire and switch**
- **logic gate**
- **rotaxane**
- **calixarene wheel**
- **molecular extension-cable system**

REFERENCES

1. Balzani, V.; Credi, A.; Venturi, M. *Molecular Devices and Machines: Concepts and Perspectives for the Nanoworld*; Wiley-VCH: Weinheim, Germany, 2008.
2. Ceroni, P.; Credi, A.; Venturi, M., Eds. *Electrochemistry of Functional Supramolecular Systems*; Wiley: New York, NY, 2010.
3. Venturi, M. *Developing Sustainability: Some Scientific and Ethical Issues*. In *Sustainable Nanosystems Development, Properties, and Applications*; Putz, M. V., Mirica M. C., Eds.; IGI Global: Hershey, PA, 2017; pp 657–680.
4. Sheehan, B.; Murphy, F.; Mullins, M.; Furxhi, I.; Costa, A. L.; Simeone, F. C.; Mantecca, P. Hazard Screening Methods for Nanomaterials: A Comparative Study. *Int. J. Mol. Sci.* **2018,** *19*, 649–1-22.
5. Torrens, F. Fractals for Hybrid Orbitals in Protein Models. *Complexity Int.* **2001,** *8*, Torren01–1-13.
6. Torrens, F. Fractal Hybrid-Orbital Analysis of the Protein Tertiary Structure. *Complexity Int.*; Submitted for Publication.
7. Torrens, F.; Castellano, G. Resonance in Interacting Induced-Dipole Polarizing Force Fields: Application to Force-Field Derivatives. *Algorithms* **2009,** *2*, 437–447.
8. Torrens, F.; Castellano, G. Molecular Diversity Classification *via* Information Theory: A Review. *ICST Trans. Complex Syst.* **2012,** *12* (10–12), e4–1-8.
9. Torrens, F.; Castellano, G. Reflections on the Nature of the Periodic Table of the Elements: Implications in Chemical Education. In *Synthetic Organic Chemistry*; Seijas, J. A., Vázquez Tato, M. P., Lin, S. K., Eds.; MDPI: Basel, Switzerland, 2015; Vol. 18; pp 8–1-15.
10. Putz, M. V., Ed. *The Explicative Handbook of Nanochemistry*; Apple Academic–CRC: Waretown, NJ, in press.
11. Torrens, F.; Castellano, G. Reflections on the Cultural History of Nanominiaturization and Quantum Simulators (Computers). In *Sensors and Molecular Recognition*; Laguarda Miró, N., Masot Peris, R., Brun Sánchez, E., Eds.; Universidad Politécnica de Valencia: València, Spain, 2015; Vol. 9; pp 1–7.
12. Torrens, F.; Castellano, G. Ideas in the History of Nano/Miniaturization and (Quantum) Simulators: Feynman, Education and Research Reorientation in Translational Science. In *Synthetic Organic Chemistry*; Seijas, J. A., Vázquez Tato, M. P., Lin, S. K., Eds.; MDPI: Basel, Switzerland, 2016; Vol. 19; pp 1–16.
13. Torrens, F.; Castellano, G. Nanominiaturization and Quantum Computing. In Sensors and Molecular Recognition; Costero Nieto, A. M., Parra Álvarez, M., Gaviña Costero, P., Gil Grau, S., Eds.; Universitat de València: València, Spain, 2016; Vol. 10; pp 31–1-5.
14. Torrens, F.; Castellano, G. Nanominiaturization, Classical/Quantum Computers/ Simulators, Superconductivity and Universe. In *Methodologies and Applications for Analytical and Physical Chemistry*; Haghi, A. K., Thomas, S., Palit, S., Main, P., Eds.; Apple Academic–CRC: Waretown, NJ, in press.
15. Torrens, F.; Castellano, G. Superconductors, Superconductivity, BCS Theory and Entangled Photons for Quantum Computing. In *Physical Chemistry for Engineering and Applied Sciences: Theoretical and Methodological Implication*; Haghi, A. K., Aguilar, C. N., Thomas, S., Praveen, K. M., Eds.; Apple Academic–CRC: Waretown, NJ, in press.

16. Torrens, F.; Castellano, G. EPR Paradox, Quantum Decoherence, Qubits, Goals and Opportunities in Quantum Simulation. In *Theoretical Models and Experimental Approaches in Physical Chemistry: Research Methodology and Practical Methods*; Haghi, A.K., Ed.; Apple Academic–CRC: Waretown, NJ; Vol. 5, in press.
17. Torrens, F.; Castellano, G. Nanomaterials, Molecular Ion Magnets, Ultrastrong and Spin–Orbit Couplings in Quantum Materials. In *Physical Chemistry for Chemists and Chemical Engineers: Multidisciplinary Research Perspectives*; Vakhrushev, A. V., Haghi, R., de Julián-Ortiz, J. V., Allahyari, E., Eds.; Apple Academic–CRC: Waretown, NJ, in press.
18. Torrens, F.; Castellano, G. Nanodevices and Organization of Single Ion Magnets and Spin Qubits. In *Chemical Science and Engineering Technology: Perspectives on Interdisciplinary Research*; Balköse, D., Ribeiro, A. C. F., Haghi, A. K., Ameta, S. C., Chakraborty, T., Eds.; Apple Academic–CRC: Waretown, NJ, in press.
19. Torrens, F.; Castellano, G. Superconductivity and Quantum Computing via Magnetic Molecules. In *New Insights in Chemical Engineering and Computational Chemistry*; Haghi, A. K., Ed.; Apple Academic–CRC: Waretown, NJ, in press.
20. Torrens, F.; Castellano, G. *Book of Abstracts, Certamen Integral de la Prevención y el Bienestar Laboral, València, Spain, September 28–29, 2016*; Generalitat Valenciana–INVASSAT: València, Spain, 2016; p 3.
21. Torrens, F.; Castellano, G. Nanoscience: From a Two-Dimensional to a Three-Dimensional Periodic Table of the Elements. In *Methodologies and Applications for Analytical and Physical Chemistry*; Haghi, A. K., Thomas, S., Palit, S., Main, P., Eds.; Apple Academic–CRC: Waretown, NJ, in press.
22. Torrens, F.; Castellano, G. El Trabajo con Nanomateriales: Historia Cultural, Filosofía Reduccionista/Positivista y Ética. In *Tecnología, Ciencia y Sociedad*; Gherab-Martín, K. J., Ed.; Global Knowledge Academics: València, Spain, in press.
23. Torrens, F.; Castellano, G. *Developing Sustainability* via *Nanosystems and Devices: Science–Ethics*. In *Chemical Science and Engineering Technology: Perspectives on Interdisciplinary Research*; Balköse, D., Ribeiro, A. C. F., Haghi, A. K., Ameta, S. C., Chakraborty, T., Eds.; Apple Academic–CRC: Waretown, NJ, in press.
24. Torrens, F.; Castellano, G. Green Nanotechnology: An Approach towards Environment Safety. In *Nanoscience/Nanotechnology/Nanomaterials*; Haghi, A. K., Ed.; Apple Academic–CRC: Waretown, NJ, in press.
25. Venturi, M.; Iorga, M. I.; Putz, M. V. Molecular Devices and Machines: Hybrid Organic–Inorganic Structures. *Curr. Org. Chem.* **2017,** *21*, 2731–3759.
26. Feynman, R.P. There is Plenty of Room at the Bottom. *Caltech Eng. Sci.* **1960,** *23*, 22–36.
27. González-Béjar, M. Química sostenible y fotocatálisis: Nanopartículas metálicas como fotocatalizadores para la síntesis de compuestos orgánicos. *An. Quím.* **2018,** *114*, 31–38.
28. Ureña López, S. *Book of Abstracts, III Congreso de Pensamiento Crítico y Divulgación Científica: Ciencia e Ideología, March 6–7, 2018, València, Spain*; Universitat de València: València, Spain, 2018; CFP10.
29. Drexler, K. E. *Engines of Creation: The Coming Era of Nanotechnology*; Doubleday: New York, NY, 1986.
30. Crichton, M. *Prey: A Novel*; HarperCollins: New York, NY, 2002.

31. Kurzweil, R. *The Singularity Is Near: When Humans Transcend Biology*; Viking: London, UK, 2005.
32. Bueno, O. The Drexler–Smalley Debate on Nanotechnology: Incommensurability at Work? *HYLE Int. J. Philos. Chem.* **2004,** *10* (2), 83–98.
33. Gaita Ariño, A., *Book of Abstracts, 10 a la Menos 9: Festival de Nanociencia y Nanotecnología, València, Spain, April 20, 2018*; Universitat de València: València, Spain, 2018; O–2.

CHAPTER 3

CULTURAL INTERBREEDING IN INDIGENOUS AND SCIENTIFIC ETHNOPHARMACOLOGY

FRANCISCO TORRENS[1,*] and GLORIA CASTELLANO[2]

[1]*Institut Universitari de Ciència Molecular, Universitat de València, Edifici d'Instituts de Paterna, P. O. Box 22085, E-46071 València, Spain*

[2]*Departamento de Ciencias Experimentales y Matemáticas, Facultad de Veterinaria y Ciencias Experimentales, Universidad Católica de Valencia San Vicente Mártir, Guillem de Castro-94, E-46001 València, Spain*

[*]*Corresponding author. E-mail: torrens@uv.es*

ABSTRACT

Ethnopharmacology is a field of research meandering between medicine and food science, and displays a medical–food science crossover. The history of medicine shows an European–American ethnopharmacology crossbreeding. More attention should be directed toward the cultural aspects of ethnopharmacology, which or, more generally, the information that forms the basis of research and development activities, is not well recognized as an important element of the industrial development pipeline, which contradicts the common perception that ethnopharmacology is seen as an important source for medicines, which reflects the classical separation of the natural sciences into defined disciplines, with ethnopharmacology being more strongly linked to pharmacology. Within 50 years, ethnopharmacology gains a profile and while conventionally linked usually to *traditional knowledge*, drug discovery, and some areas of pharmacology, the analysis highlights its emerging importance in the context of disease prevention, but also the development of research driven by the needs and interests of the fast developing economies, most notably of Asia. Research strategies of lead compounds are presented. Idea *impact factor* makes the fallacy to consider the best who presents the greatest diffusion. The pharmaceutical industry prefers investing in medicines that people should take the rest of their lives than in the cure of diseases. The great hope against hunger is precision agriculture, based on genetically modified organisms and without the problems of traditional cultivation.

3.1 INTRODUCTION

The use of different local herbs, vegetables, and fruits by humans is believed to contribute notably to human health in preventing and/or curing many diseases. Plants were a natural source of therapeutic agents for several diseases. Ethnopharmacology is a field of research meandering between medicine and disease prevention (food science). Heinrich and Jäger reviewed ethnopharmacology.[1]

Conceiving science as a unique social and cultural product of Western civilization is unfortunate. Many kinds of knowledge and practices related to health and disease enjoyed currency across space and time in all human societies. Exchanges between the diverse visions of nature advanced the wellbeing of humankind; medical exchanges can and should become advantageous and profitable for all. The history of medicine shows a crossbreeding between European and American ethnopharmacologies.[2] The *Revista de Fitoterapia* published a special issue on the cultural interbreeding in ethnopharmacology.[3] *Dendrobium officinale* orchid extract prevents ovariectomy-induced osteoporosis in vivo and inhibits receptor activator of nuclear factor κ-B ligand (RANKL)-induced osteoclast differentiation in vitro.[4] Underutilized wild edible plants were reviewed as a source of food and medicine. A bibliometric analysis of ethnopharmacology was informed.[5]

Earlier publications in *Nereis*, and others classified yams,[6] lactic acid bacteria,[7] fruits,[8] food spices,[9] and oil legumes[10] by principal component, cluster, and meta-analyses. The molecular classifications of 33 phenolic compounds derived from the cinnamic and benzoic acids from *Posidonia oceanica*,[11] 74 flavonoids,[12] 66 stilbenoids,[13] 71 triterpenoids and steroids from *Ganoderma*,[14] 17 isoflavonoids from *Dalbergia parviflora*,[15] 31 sesquiterpene lactones (STLs),[16,17] and ARTDs[18] were informed. A tool for interrogation of macromolecular structure was reported.[19] Mucoadhesive polymer hyaluronan favors transdermal penetration absorption of caffeine.[20,21] Polyphenolic phytochemicals in cancer prevention and therapy, bioavailability, and bioefficacy were reviewed.[22] From Asia to Mediterranean, soya bean, Spanish legumes, and commercial *soya bean* principal component, cluster and meta-analyses were informed.[23] Natural antioxidants from herbs and spices improved the oxidative stability and frying performance of vegetable oils.[24] The relationship between vegetable oil composition and oxidative stability was revealed via a multifactorial approach.[25] Chemical and biological screening approaches to phytopharmaceuticals were informed.[26] The aim of this work is to initiate a debate by suggesting a number of questions (Q), which can arise when addressing subjects of cultural interbreeding in ethnopharmacology from indigenous to scientific knowledge, in different fields, and providing, when possible, answers (A), and hypotheses (H).

3.2 MEDICINAL PLANTS, MIDWIVES, AND INDIGENOUS ETHNOVETERINARY MEDICINE

Jardins du Monde–Médicus Mundi–Rxiin Tnamet raised the following Qs on medicinal plants and midwives[27,28]:

Q1. Why is it good to know the use of medicinal plants?

They raised additional questions on how medicinal plants are handled.

Q2. How are medicinal plants handled?
Q3. How do plants act?
Q4. How are they cultivated?
Q5. How are they harvested?
Q6. How are plants dried?
Q7. How are dried plants stored?
Q8. How to prepare cooks, poultices, handkerchiefs, and gargles?

Jardins du Monde–Vétérinaires sans Frontières raised the following Qs on indigenous ethnoveterinary medicine in Guatemala High Plateau[29]:

Q9. What do birds give people?
Q10. What does pig give people?
Q11. What do ruminants give people?
Q12. What do equines give people?

They raised additional questions on the kitchen garden of medicinal plants.

Q13. How to harvest medicinal plants?
Q14. How to dry medicinal plants?
Q15. How to store medicinal plants?

3.3 TRADITIONAL AND MODERN MEDICINE IN LATIN AMERICA: HEALTH CULTURE

Greifeld raised the following Qs on traditional/modern medicine in Latin America and health cultural aspects[30]:

Q1. How do different peoples look for an appropriate solution to enhance wellbeing with biomedicine and traditional-medicine specialist?
Q2. People choose between a biomedical doctor or another healer, who is right/adequate for them?
Q3. Does it correspond to specific and different life realities did nobody bother was no interest for Qs?
Q4. What is happening during a therapy in traditional medicine?
Q5. (Greifeld and Rossbach, 1989). Did the patient either use prescribed remedies or not?
Q6. How to restore wellbeing with siblings, in-laws, parents, and grandparents become part of healing?
Q7. What does it make medical anthropology so incredibly exciting?

3.4 POPULAR AND ORTHODOX MEDICINES SURPRISING COINCIDENCE: COSPEITO POOL

Anllo Naveiras and Ortiz Núñez proposed the following H/Qs on popular/orthodox medicines coincidence[31]:

H1. 39% aged over 60 years, which is of interest as information on traditional botanical remedies.
Q1. Is the Terra Chá Region of ethnobotanical interest?
Q2. Do the traditional uses coincide with established medicinal utilization of the plants?
Q3. Did they know any medicinal plants?
Q4. What properties did they attribute to everyone?

Q5. What part of the plant is used?
Q6. How is the plant prepared for use?
Q7. What is its mode of application or administration?
Q8. What is its perceived efficacy?

3.5 KEYS ON THE SPANISH ORDER TO REGULATE HOMEOPATHY

García proposed the following Qs, As, and Hs on keys on the Spanish order to regulate homeopathy[32]:

Q1. What does the Spanish order tell people on the regularization of homeopathic medicines?
A1. It gives 3 months to laboratories to communicate intention to initiate regularization proceedings.
Q2. What documentation should homeopathic laboratories bring as a dowry?
A2. In 6 months, it will be enough that they bring information that proves quality and safety.
H1. Showing that they are safe products, homeopathy will be allowed to be sold as medicine.
Q3. To buy a medicine not curing is like buying a pen not writing; why, being there pens that do?
H2. Homeopathic medicines must pay taxes but cheaper because they pay by dilutions families.

She proposed the following three measures (Ms) to put a stop to homeopathy in Spain:

M1. Not put them a red carpet with regard to taxes.
M2. Ban publicizing homeopathic medicines.
M3. Force that they be dispensed under only medical prescription.
Q4. How much homeopathy is sold in Spain?
A4. Main manufacturer says 18%, polls say 5%, and most chemists say 0.2% invoicing.

She proposed the following two conclusions (Cs):

C1. State Spanish disagreement with Europe.
C2. Be coherent with institutional ineffectiveness statements and establish inconvenient normative.

3.6 DISCUSSION

Perhaps in the future, more attention should be directed toward the cultural aspects of ethnopharmacology, which or, more generally, the information that forms the basis of research and development activities, is not well recognized as an important element of the industrial development pipeline, which contradicts the common perception that ethnopharmacology is seen as an important source for medicines, which reflects the classical separation of the natural sciences into defined disciplines, for example, chemistry and pharmacology, with ethnopharmacology being more strongly linked to pharmacology (and food science). Within 50 years, ethnopharmacology gained a profile and while conventionally linked usually to *traditional knowledge*, drug discovery, and some areas of pharmacology, the analysis highlights not only its emerging importance in the context of disease prevention (food science) but also the development of research driven by the needs and interests of the fast developing economies, most notably of Asia. Figure 3.1 shows research strategies of lead compounds.

Measures exist to put a stop to homeopathy: not put them a red carpet with regard to taxes; ban publicizing homeopathic medicines; force that they be dispensed under only medical prescription.

Idea *impact factor* (number of quotations of an article) makes the fallacy to consider the best who presents the greatest diffusion. Figures say nothing of scientific work, so bureaucrats opinion, neither. The pharmaceutical industry prefers investing in medicines that people should take the rest of their lives than in the cure of diseases. The great hope against hunger is precision agriculture, based on genetically modified organisms (GMOs) and without the problems of traditional cultivation.

3.7 FINAL REMARKS

From the previous results and discussion, the following final remarks can be drawn:

1. Some surprising outcomes include the important link to food sciences, and the relevance of some biological assays, which, in reality, are not considered to be of pharmacological relevance.

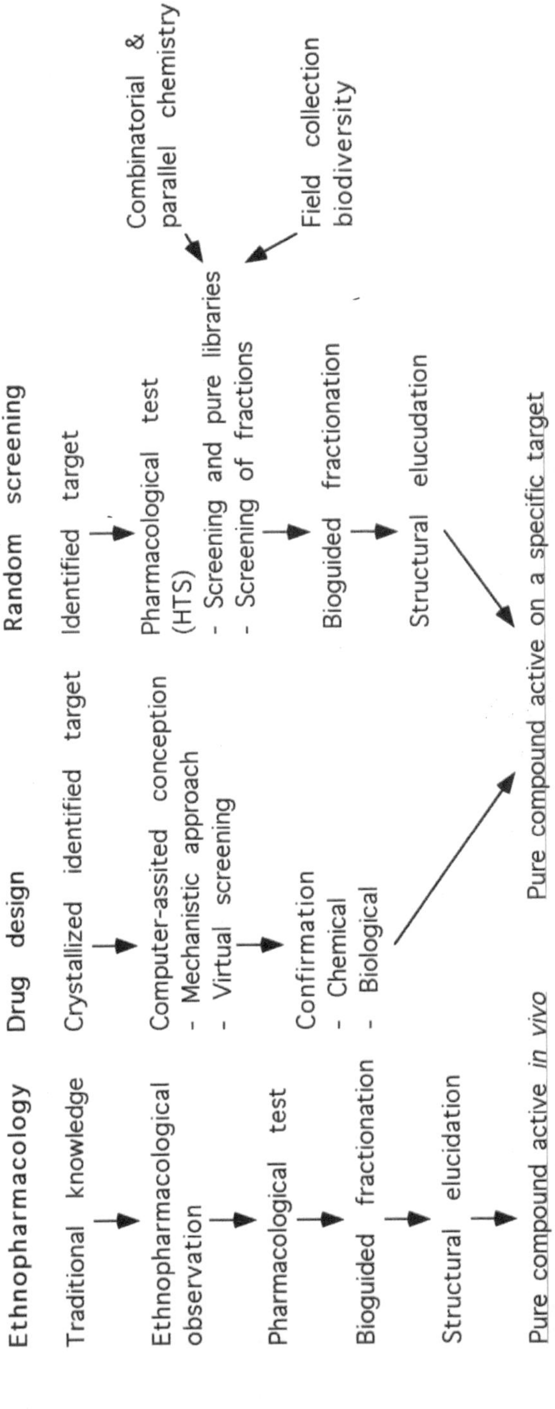

FIGURE 3.1 Research strategies of new lead compounds. HTS, high-throughput screening.

2 Ethnopharmacology is a field of research at the crossroads of several disciplines (most notably pharmacology and food science).
3. The diversity of ethnopharmacology clearly is one of its key strengths but also a challenge of it, which sees itself to be inter- or trans-disciplinary.
4. State Spanish disagreement with Europe on homeopathy.
5. Be coherent with Spanish institutional statements on homeopathy ineffectiveness and establish an inconvenient normative framework.
6. Idea *impact factor* makes the fallacy to consider the best who presents the greatest diffusion. Figures say nothing of scientific work, so bureaucrats opinion, neither.
7. The pharmaceutical industry prefers investing in medicines that people should take the rest of their lives than in the cure of diseases.
8. The great hope against hunger is precision agriculture, based on GMOs and without the problems of traditional cultivation. It is important to pay attention to Green Parties and non-governmental organizations, for example, Green Peace, and their sources of income, for example, The Rockefeller Foundation.

ACKNOWLEDGMENTS

The authors thank support from Generalitat Valenciana (Project No. PROMETEO/2016/094) and Universidad Católica de Valencia *San Vicente Mártir* (Project No. UCV.PRO.17-18.AIV.03).

KEYWORDS

- **ethnobotany**
- **ethnomedicine**
- **ethnoveterinary**
- **medical anthropology**
- **medicinal plant**
- **traditional pharmacopoeia**
- **botany**

REFERENCES

1. Heinrich, M.; Jäger, A. K. *Ethnopharmacology*; Wiley: Chichester, UK, 2015.
2. López Piñero, J. M.; Pardo Tomás, J. *Nuevos Materiales y Noticias sobre la* Historia de las Plantas de Nueva España *de Francisco Hernández*; Cuadernos Valencianos de Historia de la Medicina y de la Ciencia, Ser. A No. XLIV; Universitat de València–CSIC: València, Spain, 1994.
3. Aguirre Marco, C. P.; Fresquet Febrer, J. L. Foreword. *Rev. Fitoter*. **2005,** *5* (S1), 8–8.
4. Wang, Q.; Zi, C. T.; Wang, J.; Wang, Y. N.; Huang, Y. W.; Fu, X. Q.; Wang, X. J.; Sheng, J. *Dendrobium officinale* Orchid Extract Prevents Ovariectomy-Induced Osteoporosis In vivo and Inhibits RANKL-Induced Osteoclast Differentiation In vitro. *Front. Pharmacol.* **2018,** *8*, 966–1-13.
5. Yeung, A. W. K.; Heinrich, M.; Atanasov, A. G. Ethnopharmacology—A Bibliometric Analysis of a Field of Research Meandering Between Medicine and Food Science? *Front. Pharmacol.* **2018,** *9*, 215–1-15.
6. Torrens-Zaragozá, F. Molecular Categorization of Yams by Principal Component and Cluster Analyses. *Nereis* **2013,** *2013* (5), 41–51.
7. Torrens-Zaragozá, F. Classification of Lactic Acid Bacteria Against Cytokine Immune Modulation. *Nereis* **2014,** *2014* (6), 27–37.
8. Torrens-Zaragozá, F. Classification of Fruits Proximate and Mineral Content: Principal Component, Cluster, Meta-Analyses. *Nereis* **2015,** *2015* (7), 39–50.
9. Torrens-Zaragozá, F. Classification of Food Spices by Proximate Content: Principal Component, Cluster, Meta-Analyses. *Nereis* **2016,** *2016* (8), 23–33.
10. Torrens, F.; Castellano, G. From Asia to Mediterranean: Soya bean, Spanish Legumes and Commercial *Soya Bean* Principal Component, Cluster and Meta-Analyses. *J. Nutr. Food Sci.* **2014,** *4* (5), 98–98.

11. Castellano, G.; Tena, J.; Torrens, F. Classification of Polyphenolic Compounds by Chemical Structural Indicators and its Relation to Antioxidant Properties of *Posidonia oceanica* (L.) Delile. *MATCH Commun. Math. Comput. Chem.* **2012,** *67*, 231–250.
12. Castellano, G.; González-Santander, J. L.; Lara, A.; Torrens, F. Classification of Flavonoid Compounds by Using Entropy Of Information Theory. *Phytochemistry* **2013,** *93*, 182–191.
13. Castellano, G.; Lara, A.; Torrens, F. Classification of Stilbenoid Compounds by Entropy of Artificial Intelligence. *Phytochemistry* **2014,** *97*, 62–69.
14. Castellano, G.; Torrens, F. Information Entropy-Based Classification of Triterpenoids and Steroids from *Ganoderma. Phytochemistry* **2015,** *116*, 305–313.
15. Castellano, G.; Torrens, F. Quantitative Structure–Antioxidant Activity Models of Isoflavonoids: A Theoretical Study. *Int. J. Mol. Sci.* **2015,** *16*, 12891–12906.
16. Castellano, G.; Redondo, L.; Torrens, F. QSAR of Natural Sesquiterpene Lactones as Inhibitors of Myb-Dependent Gene Expression. *Curr. Top. Med. Chem.* **2017,** *17*, 3256–3268.
17. Torrens, F.; Redondo, L.; León, A.; Castellano, G. Structure–activity Relationships of Cytotoxic Lactones as Inhibitors and Mechanisms of Action. *Curr. Drug Discov. Technol.* Submitted for publication.
18. Torrens, F.; Redondo, L.; Castellano, G. Artemisinin: Tentative Mechanism of Action and Resistance. *Pharmaceuticals* **2017,** *10*, 20–4-4.
19. Torrens, F.; Castellano, G. A Tool for Interrogation of Macromolecular Structure. *J. Mater. Sci. Eng. B* **2014,** *4* (2), 55–63.
20. Torrens, F.; Castellano, G. Mucoadhesive Polymer Hyaluronan as Biodegradable Cationic/Zwitterionic-Drug Delivery Vehicle. *ADMET DMPK* **2014,** *2*, 235–247.
21. Torrens, F.; Castellano, G. Computational Study of Nanosized Drug Delivery from *Cyclo*dextrins, Crown Ethers and Hyaluronan in Pharmaceutical Formulations. *Curr. Top. Med. Chem.* **2015,** *15*,1901–1913.
22. Estrela, J. M.; Mena, S.; Obrador, E.; Benlloch, M.; Castellano, G.; Salvador, R.; Dellinger, R. W. Polyphenolic Phytochemicals in Cancer Prevention and Therapy: Bioavailability Versus Bioefficacy. *J. Med. Chem.* **2017,** *60*, 9413–9436.
23. Torrens, F.; Castellano, G. From Asia to Mediterranean: Soya Bean, Spanish Legumes and Commercial *Soya Bean* Principal Component, Cluster and Meta-Analyses. *J. Nutr. Food Sci.* **2014,** *4* (5), 98–98.
24. Redondo-Cuevas, L.; Castellano, G.; Raikos, V. Natural Antioxidants from Herbs and Spices Improve the Oxidative Stability and Frying Performance of Vegetable Oils. *Int. J. Food Sci. Technol.* **2017,** *52*, 2422–2428.
25. Redondo-Cuevas, L.; Castellano, G.; Torrens, F.; Raikos, V. Revealing the Relationship Between Vegetable Oil Composition and Oxidative Stability: A Multifactorial Approach. *J. Food Compos. Anal.* **2018,** *66*, 221–229.
26. Torrens, F.; Castellano, G. Chemical and Biological Screening Approaches to Phytopharmaceuticals. In *Research Methods and Applications in Chemical and Biological Engineering*; Pourhashemi, A., Deka, S.C., Haghi, A.K., Eds.; Apple Academic-CRC: Waretown, NJ, in press.

27. Jardins du Monde, Médicos del Mundo and Rxiin Tnamet. *Plantas Medicinales y Comadronas: Manual para el Personal de Salud.* Médicos del Mundo España: Guatemala (Guatemala), 2002.
28. Nocolas, J. P. Démarche de l'ethnopharmacologie appliquée: Exemple de la valorisation de plantes médicinales dans le cadre d'O.N.G. sur l'altiplano du Guatemala. *Rev. Fitoter.* **2005,** 5 (S1), 128–133.
29. Jardins du Monde and Vétérinaires sans Frontières. *Etnoveterinaria Indígena en el Altiplano de Guatemala: Alternativas de Producción Animal Sostenible.* Vétérinaires sans Frontières-Espagne: Guatemala (Guatemala), in press.
30. Greifeld, K. Traditional and Modern Medicine in Latin America: Cultural Aspects of Health. *Rev. Fitoter.* **2005,** *5* (S1), 134–136.
31. Anllo Naveiras, J.; Ortiz Núñez, S. Sorprendente coincidencia entre medicina popular y ortodoxa Laguna de Cospeito (Galicia /España). *Rev. Fitoter.* **2005,** *5* (S1), 144–145.
32. García, M. Claves sobre la orden para regular la homeopatía. *20 Minutos (València)* **2018,** *19* (4169), 9.

CHAPTER 4

ULTRAVIOLET–VISIBLE ABSORPTION SPECTROSCOPY: LINK WITH FLUORESCENCE

FRANCISCO TORRENS[1,*] and GLORIA CASTELLANO[2]

[1]*Institut Universitari de Ciència Molecular, Universitat de València, Edifici d'Instituts de Paterna, P. O. Box 22085, E-46071 València, Spain*

[2]*Departamento de Ciencias Experimentales y Matemáticas, Facultad de Veterinaria y Ciencias Experimentales, Universidad Católica de Valencia San Vicente Mártir, Guillem de Castro-94, E-46001 València, Spain*

[*]*Corresponding author. E-mail: torrens@uv.es*

ABSTRACT

In the 20th century, analytical techniques that were based on the interaction of light with matter began to be used. The comparison between incident and transmitted light provides information on the characteristics of the analyzed sample. This chapter discusses: (1) the relationship between electromagnetic radiations and their effects, (2) the relationship between spectroscopic techniques and their obtained information, (3) the fact that functional groups –OH and $–NH_2$ present an auxochromic effect on chromophore benzene, particularly compared to peak B, and (4) a comparison of ultraviolet–visible with fluorescence spectroscopies.

4.1 INTRODUCTION

In the 19th century, before present analytical techniques existed, organic chemists could determine the molecular formula of a pure substance from percentile composition.[1,2] In the 20th century, analytical techniques that were based on light–matter interaction began to be used.[3] Incident–transmitted-light comparison provides information on analyzed-sample characteristics.[4] Radiation–effect relationship is listed in Table 4.1.

TABLE 4.1 Relationship Between Electromagnetic Radiations and Their Effects.

Radiation	Effect
X and cosmic rays	Ionizations of molecules
UV–VIS	Electronic transitions between atomic and molecular orbitals
Infrared	Deformation of chemical bonds
Microwaves	Rotations of chemical bonds
Radio frequencies	Electronic or nuclear spin transitions in the atoms of the molecule

Spectroscopic technique-obtained information relationship is cataloged in Table 4.2.

TABLE 4.2 Relationship Between Spectroscopic Techniques and Their Obtained Information.

Spectroscopic technique	Obtained information
X-rays	Molecule total structure including stereochemistry from atoms relative positions
UV–VIS	Existence of molecule chromophores and/or conjugation from observed absorptions
Infrared	Functional groups from the observed absorptions
Mass spectrometry	Molecular formula and substructures from the observed ions
Nuclear magnetic resonance	Functional groups, substructures, connectivities, stereochemistry, and so on from data of observed chemical shifts, peaks areas, and coupling constants

4.2 INSTRUMENTS FOR ULTRAVIOLET–VISIBLE ABSORPTION SPECTROSCOPY

4.2.1 PHOTOMETERS

Photometers are simple instruments used to measure absorbance A, which utilize absorption or interferential filters to select *wavelength* λ. They are used practically in visible (VIS) region.

Advantages: They are simple, rather economic, robust, and of easy maintenance. They can be transported, which converts them in a useful apparatus to perform field spectroscopic analyses.

Disadvantages: They cannot be used to obtain absorption spectra.

4.2.2 SPECTROMETERS

Spectrometers are instruments used to measure absorbance that utilize a monochromatic selector to choose wavelength. They can be used in ultraviolet (UV), VIS, and infrared (IR) regions. They can be of single or double beam.

4.3 ULTRAVIOLET–VISIBLE ABSORPTION SPECTROSCOPY

4.3.1 FOUNDATION OF ULTRAVIOLET–VISIBLE ABSORPTION SPECTROSCOPY

Peaks of UV–VIS absorption are closely related with bond type. Spectrometry of UV–VIS is restricted to a limited number of functional groups (*chromophores*). It is a useful tool to identify functional groups in a molecule. It allows quantifying compounds that contain absorber groups.

Radiation UV–VIS absorption by a molecule can be considered in two stages:

(1) Electronic excitation:

$$M + h\nu \rightarrow M^{*} \text{ and} \tag{4.1}$$

(2) Relaxation:

$$M^{*} \rightarrow M + \text{heat}\cdot \tag{4.2}$$

Developed heat is undetectable. Half-life of excited species M^{*} is 10^{-9}–10^{-8} s.

Relaxation can also occur by: (1) decomposition of M^{*} in two new species (*photochemical reaction*) and (2) re-emission of fluorescence (FLU, F) or phosphorescence (P, F lasts lesser time than P).

Radiation UV–VIS absorption results from bond–electrons excitation; absorption peaks can be related to bond types (identification of functional groups in a molecule).

However, the most important use is quantifying compounds that contain absorber groups.

Three types of electronic transitions exist: (1) those that include π, σ, and n electrons, (2) those that include d and f electrons, and (3) those of charge transfer electrons.

Four types of electrons exist capable of absorbing UV–VIS radiation: (1) organic ions and molecules, and some inorganic anions, (2) *all* organic compounds are capable of absorbing radiation because all contain valence electrons (π, σ, n), which can be excited to above energy levels, (3) those that participate directly in formation of bond between atoms, and (4) nonbonding electrons or those that participate in no bond, which are mainly localized around atoms (e.g., O, halogens, S, N).

4.3.2 POSSIBLE TRANSITIONS IN ULTRAVIOLET–VISIBLE ABSORPTION SPECTROSCOPY

Transitions $\sigma \rightarrow \sigma^*$. Saturated compounds with simple bonds. $\lambda < 150$ nm [vacuum UV (VUV), not much accessible]. Methane $\lambda_{max} = 125$ nm; single C–H bonds. Ethane $\lambda_{max} = 135$ nm; single C–C and C–H bonds, the latter of greater energy than C–C, so that λ_{max} is lower. Transitions $\sigma \rightarrow \sigma^*$ require greater energy than any other permitted transition.

Transitions $n \rightarrow \sigma^*$. Saturated compounds with shared electron pairs. λ: in 150–250 nm. Molar absorptivity ε: low to medium, 100–3000 $L \cdot mol^{-1} \cdot cm^{-1}$. In presence of polar solvents, λ_{max} shifts to shorter λ (*hypsochromic* or *blue shift*). In not many functional groups, $n \rightarrow \sigma^*$ transitions are easily detected (cf. Table 4.3).

Transitions $n \rightarrow \pi^*$ and $\pi \rightarrow \pi^*$. Both require presence of the unsaturated functional groups that contribute π-electrons. They produce absorption peaks in an experimentally accessible spectral region. These unsaturated absorber centers are called *chromophores*.

Transitions $n \rightarrow \pi^*$. Unsaturated compounds with unshared electron pairs. λ: in 200–700 nm. ε: low, 10–100 $L \cdot mol^{-1} \cdot cm^{-1}$. On rising solvent polarity, λ_{max} shifts to shorter λ (*blue shift*). In many functional groups, $n \rightarrow \pi^*$ transitions are easily detected.

Transitions $\pi \rightarrow \pi^*$. Unsaturated compounds. λ: in 200–700 nm. ε: medium to high, 1000–10,000 L·mol^{-1}·cm^{-1}. On rising solvent polarity, λ_{max} shifts to longer λ (*bathochromic* or *red shift*). In many functional groups, $\pi \rightarrow \pi^*$ transitions are easily detected (cf. Table 4.4).

TABLE 4.3 Some Examples[a] of Absorption Because of $n \rightarrow \sigma^*$ transitions.

Compound	λ_{max} (nm)	ε_{max} (L·mol^{-1}·cm^{-1})
H_2O	167	1480
CH_3OH	184	150
CH_3Cl	173	200
CH_3I	258	365
$(CH_3)_2S$[b]	229	140
$(CH_3)_2O$	184	2520
CH_3NH_2	215	600
$(CH_3)_3N$	227	900

[a]Samples in vapor state.

[b]In ethanol as solvent.

TABLE 4.4 Characteristics of the Absorption of Some Common Chromophores.

Chromophore	Example	Solvent	λ_{max} (nm)	ε_{max} (L·mol^{-1}·cm^{-1})	Transition type
Alkene	$C_6H_{13}CH{=}CH_2$	*n*-heptane	177	13,000	$\pi \rightarrow \pi^*$
Alkine	$C_5H_{11}C{\equiv}C{-}CH_3$	*n*-heptane	178	10,000	$\pi \rightarrow \pi^*$
			196	2000	–
			225	160	–
Carbonyl	$CH_3C({=}O)CH_3$	*n*-hexane	186	1000	$n \rightarrow \sigma^*$
			280	16	$n \rightarrow \pi^*$
	$CH_3C({=}O)H$	*n*-hexane	180	high	$n \rightarrow \sigma^*$
			293	12	$n \rightarrow \pi^*$
Carboxyl	$CH_3C({=}O)OH$	ethanol	204	41	$n \rightarrow \pi^*$
Amide	$CH_3C({=}O)NH_2$	water	214	60	$n \rightarrow \pi^*$
Azo	$CH_3N{=}NCH_3$	ethanol	339	5	$n \rightarrow \pi^*$
Nitro	CH_3NO_2	*iso*octane	280	22	$n \rightarrow \pi^*$
Nitroso	C_4H_9NO	ethyl ether	300	100	–
			665	20	$n \rightarrow \pi^*$
Nitrate	$C_2H_5ONO_2$	dioxane	270	12	$n \rightarrow \pi^*$

4.3.3 ABSORPTION BY DIFFERENT MOLECULAR SYSTEMS

4.3.3.1 ABSORPTION BY AROMATIC SYSTEMS

Aromatics UV spectra are characterized by three groups of bands because of transitions; for example, benzene, λ_{max} = 184 nm, ε = 60,000 L·mol^{-1}·cm^{-1}; band E_2 λ_{weak} = 204 nm, ε = 7000 L·mol^{-1}·cm^{-1}; band B $\lambda_{weakest}$ = 256 nm, ε = 200 L·mol^{-1}·cm^{-1}. Bands result affected by ring substitution and solvents.

Effect of solvents. They tend to reduce or remove fine structure of sharp aromatics peaks.

Effect of substituents. Auxochrome: functional group that does not absorb in UV region but presents effect of shifting chromophore peaks to longer λ (*red shift*) and increasing their intensities (*hyperchromic*). It presents at least a pair of *n* electrons capable of interacting with ring π electrons. Table 4.5 shows that –OH and $-NH_2$ present an auxochromic effect on chromophore benzene, particularly compared to peak B.

TABLE 4.5 Characteristics of the Absorption of Some Aromatic Compounds.

Compound		**Peak E_2**		**Peak B**	
		λ_{max} (nm)	ε_{max} (L·mol^{-1}·cm^{-1})	λ_{max} (nm)	ε_{max} (L·mol^{-1}·cm^{-1})
Benzene	C_6H_6	204	7900	256	200
Toluene	$C_6H_5CH_3$	207	7000	261	300
m-Xylene	$C_6H_4(CH_3)_2$	–	–	263	300
Chlorobenzene	C_6H_5Cl	210	7600	265	240
Phenol	C_6H_5OH	211	6200	270	1450
Phenolate ion	$C_6H_5O^-$	235	9400	287	2600
Aniline	$C_6H_5NH_2$	230	8600	280	1430
Anilinium ion	$C_6H_5NH_3^+$	203	7500	254	160
Thiophenol	C_6H_5SH	236	10,000	269	700
Nafthalene	$C_{10}H_8$	286	9300	312	289
Styrene	$C_6H_5CH{=}CH_2$	244	12,000	282	450

4.3.3.2 ABSORPTION BY INORGANIC ANIONS

Inorganic anions present UV absorption peaks that are consequence of $n \rightarrow \pi^*$ transitions [e.g., nitrate (313 nm), carbonate (217 nm), nitrite (360 and 280 nm), azide (230 nm), and trithiocarbonate (500 nm)].

4.3.3.3 *ABSORPTION BY THE ELEMENTS OF THE FIRST AND SECOND SERIES OF THE TRANSITION METALS*

Ions and complexes of the 18 elements of first series of transition metals tend to absorb VIS radiation in one or all oxidation states. Electrons transitions between different energy levels of *d* orbitals (3*d* in first and 4*d* in second series).

4.3.3.4 *ABSORPTION BY LANTHANOID AND ACTINOID IONS*

Lanthanoid and actinoid ions absorb in UV–VIS region. Electrons transitions between different energy levels of *f* orbitals (4*f* in lanthanoids and 5*f* in actinoids). Spectra formed by absorption-peaks characteristic, well defined and narrow, not much affected by milieu chemical factors or type of ligand associated with the metal ion. They differ from most spectra of inorganic and organic absorbers.

4.4 APPLICATIONS OF UV–VIS ABSORPTION SPECTROSCOPY IN CHEMICAL ANALYSIS

Qualitative. Identification of chromophores from spectrum: A versus λ.

Quantitative. To measure chromophore concentration:

$$A_{\lambda} = \varepsilon_{\lambda} bc \quad (4.3)$$

4.5 DISCUSSION

On comparing UV–VIS with FLU spectroscopies, UV–VIS presents the advantage of its generality as analytical technique: (1) UV–VIS can be applied to a greater number of molecules than FLU, and (2) in UV–VIS, linear calibration extends to greater concentrations than in FLU. On the other hand, FLU presents other advantages: (1) it is more sensitive and can be applied to lower sample concentrations (10^{-10} M in FLU vs. 10^{-5} M in UV–VIS), (2) greater selectivity (in FLU, two parameters are controlled vs. only one in UV–VIS), and (3) lower bands superposition in mixtures because of the lower number of FLU bands.

4.6 FINAL REMARKS

From the present results and discussion, the following final remarks can be drawn:

1. The comparison between incident and transmitted light provided information on the characteristics of the analyzed sample.
2. A relationship between electromagnetic radiations and their effects was found.
3. A relationship between spectroscopic techniques and their obtained information was established.
4. Functional groups –OH and $–NH_2$ presented an auxochromic effect on chromophore benzene, particularly compared to peak B.
5. A comparison of ultraviolet–visible with fluorescence spectroscopies showed that both analytical techniques are complementary with regard to generality, selectivity, sensitivity, and bands superposition in mixtures.

ACKNOWLEDGMENTS

The authors thank support from Generalitat Valenciana (Project No. PROMETEO/2016/094) and Universidad Católica de Valencia *San Vicente Mártir* (Project No. UCV.PRO.17-18.AIV.03).

KEYWORDS

- **chemical analysis**
- **electromagnetic radiation**
- **electromagnetic spectrum**
- **spectroscopic technique**
- **electronic transition**
- **photometer**
- **spectrometer**

REFERENCES

1. Skoog, D. A.; Holler, F. J.; Crouch, S. R. *Principles of Instrumental Analysis;* Cengage Learning: Boston, MA, 2018.
2. Valls Planells, O.; del Castillo García, B., Eds. *Técnicas Instrumentales en Farmacia y Ciencias de la Salud;* Universidad Norbert Wiener: Lima, Peru, 2009.
3. Robinson, J. W.; Frame, E. M. S.; Frame, G. M., II. *Undergraduate Instrumental Analysis;* CRC: Boca Raton, FL, 2014.
4. Miñones Trillo, J. *Manual de Técnicas Instrumentales;* Círculo Editor Universo: Barcelona, Spain, 1978.

CHAPTER 5

MASS SPECTROMETRY: THE ONLY METHOD WITH MANY IONIZATION TECHNIQUES

FRANCISCO TORRENS[1,*] and GLORIA CASTELLANO[2]

[1]*Institut Universitari de Ciència Molecular, Universitat de València, Edifici d'Instituts de Paterna, P. O. Box 22085, E-46071 València, Spain*

[2]*Departamento de Ciencias Experimentales y Matemáticas, Facultad de Veterinaria y Ciencias Experimentales, Universidad Católica de Valencia San Vicente Mártir, Guillem de Castro-94, E-46001 València, Spain*

Corresponding author. E-mail: torrens@uv.es

ABSTRACT

This chapter addresses some elementary aspects of the *mass spectrum*, discusses tools used by experienced practitioners, and provides some glimpses into current advances in the science and art of deriving unambiguous answers from unknown spectra. Two techniques improve the *signal-to-noise ratio* of a mass spectrum: (1) *filtering* ions *via quadrupole*, double quadrupole, and triple quadrupole tandem mass spectrometry (MS) and (2) use of accurate *isotopic masses*. The *quantitative variable* is the intensity of the peaks, which is proportional to the concentration of the analyte in the sample. The *qualitative variable* is the ratio between the two most intense peaks. MS presents high *sensitivity*: 0.04 for small molecules and two atomic units of mass for large proteins. It shows extremely low limits of detection and quantification, which can be used in concentrations of traces and ultra-traces. Modern techniques and scientific rigor are applied to research. Tu and co-workers determined the proteins bonded to artemisinin via MS. Thanks to the technique, it was determined and analyzed artemisinin proteins and created the effective medicine versus malaria.

5.1 INTRODUCTION

The phenomenal growth of *mass spectrometry* (MS) as a diverse analytical tool underscored the need for useful references: courses, books, and different learning tools. In earlier publications, it was informed ultraviolet–visible (UV–VIS) *absorption* spectroscopy, its link with *fluorescence* (FLU, F),[1] FLU *excitation*, *emission*, and its comparison with UV–VIS absorption.[2] MS is an analytical technique that allows determining the distribution of the molecules of a substance versus their mass.[3–9] The *mass spectrometer* is a device that allows analyzing, with great accuracy, the composition of different chemical elements and atomic isotopes, separating the atomic nuclei versus their *mass/charge* (*m*/*z*) *ratio*.[10] It was used to determine, for example, the structure of artemisinin (ART, cf. Fig. 5.1), effective versus malaria, which resulted the sesquiterpenic lactone (STL) with an unusual peroxide group $–O_1–O_2–$.

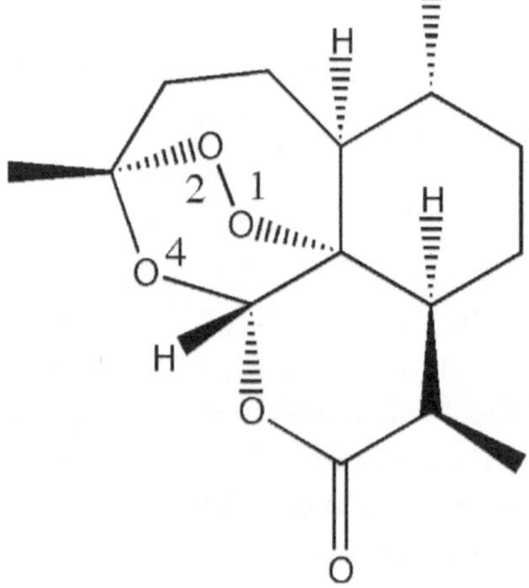

FIGURE 5.1 The endoperoxide bridge $–O_1–O_2–$ at the heart of the artemisinin antiparasitic activity.

The endoperoxide bond $–O_1–O_2–$ is cleaved when it comes into contact with Fe^{II}, realizing reactive radicals that ultimately destroy the *Plasmodium* malaria parasite (cf. Fig. 5.2).[11]

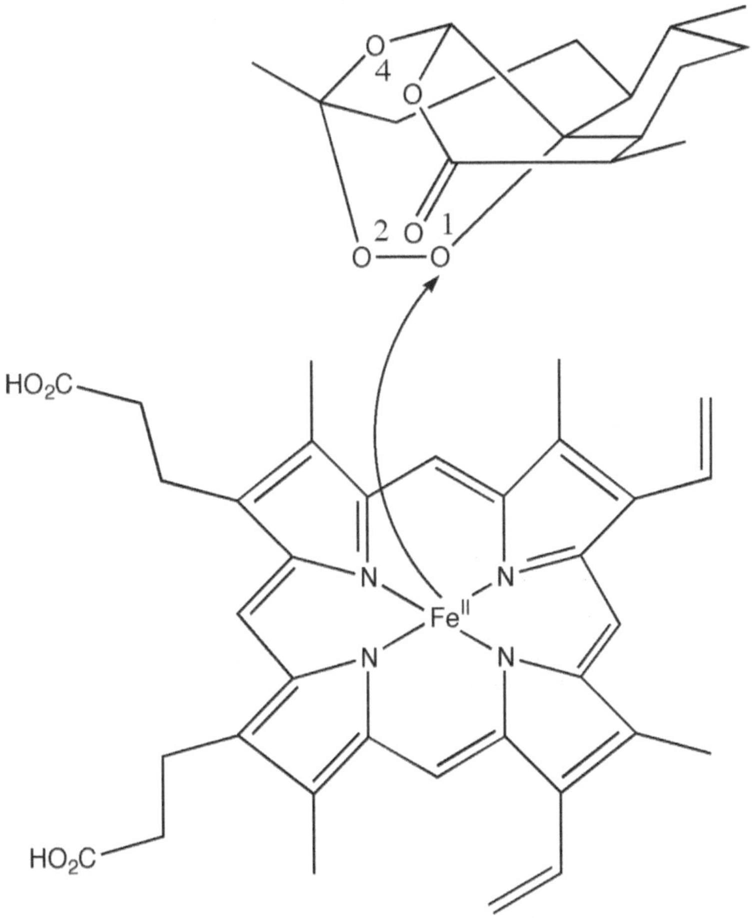

FIGURE 5.2 Cleavage of the endoperoxide bond $–O_1–O_2–$ in artemisinin by Fe^{II} in a heme group.

The mass spectrometer measures *m/z* ratios of ions, heating a beam of material of the compound to be analyzed to vaporize and ionize the different atoms. The beam of ions produces an unusual specific pattern in the detector that allows analyzing the compound.

5.2 THE MASS SPECTROMETER

An interest application of the movement of charged particles in a magnetostatic field is constituted by the *mass spectrometers*, which utility lies in the precise determination of the mass of cations, that is, atoms that lost a number of electrons small enough in order that the resultant variation with respect to the mass of the atom can be neglected. Via a simple device of this type, Thomson (1911) discovered the existence of the *isotopes*, atoms with the same chemical properties but with different masses.

Among the different models of mass spectrometers based on the action of the magnetic field (MF) B on the charged particles, that of Dempster will be described. Among other interesting models are those of Aston and Bainbridge.

The ions formed in the source A, which pass through a fine hole P, are accelerated through the F_1 aperture via a voltage V so that they acquire a velocity v that will result:

$$v=\sqrt{\frac{2zV}{m}}. \tag{5.1}$$

A uniform MF B perpendicular to the plane of the paper makes that the ions, on penetrating by F_1, describe a semicircular trajectory to arrive in the F_2 aperture where they are taken by an electrometer. In order that the ions of different masses describe all the same trajectory from F_1 to F_2 in MF B, which value is maintained constant, the voltage V is regulated so that the velocity v be adequate. Taking into account that the distance $F_1 - F_2 = D$ is the diameter of the trajectory and that its radius results $R = mv/zB$, expressing v versus V one obtains:

$$D= 2\mathrm{R}=\frac{2mv}{zB}=2\sqrt{\frac{m}{z}\frac{2V}{B^2}}, \tag{5.2}$$

where the specific charge z/m and mass m of such ions will be provided, respectively, by:

$$\frac{z}{m}=\left(\frac{8}{D^2B^2}\right)V \tag{5.3}$$

$$m=z\left(\frac{D^2B^2}{8}\right)\frac{1}{V}. \tag{5.4}$$

5.3 OPERATION

The operation of the *mass spectrometer* is based on the fact that different molecules present dissimilar masses, information that in a mass spectrometer is used to determine which molecules are present in a sample; for example, table salt NaCl is vaporized and the ions are analyzed in the first part of the mass spectrometer, which produces Na^+ and Cl^+ ions, which present specific ion weights and a charge, which means that, because of it, they will show movement under the influence of a certain electric field (EF).

The ions above are sent to an acceleration compartment and passed through a metal sheet. An MF is applied at a side of the compartment that attracts every ion with the same force (assuming equal charge), and they are deflected into a detector. Naturally, the lightest ions will be deflected more than the heaviest ions, because the force applied to every ion is the same but lighter ions present lesser mass! The detector measures exactly how far every ion is deflected and, from the datum, *m/z ratio* is calculated. With the information, it is possible to determine with a great level of certainty which is the chemical composition of the original sample.

There are many types of mass spectrometers that not only analyze the ions but also produce their different types. However, all use EF and MF to change the ionic trajectory in a certain way.

5.4 COMPONENTS

A mass spectrometer presents three basic components: the ion source, mass analyzer, and detector.

5.4.1 THE SOURCE OF IONS

The *ion source* is the element of the mass spectrometer that ionizes the material to be analyzed. Later, MF or EF transport the ions to the total analyzer. The ionization techniques were fundamental to determine which type of samples can be analyzed by MS. The electron and molecular ionization are used for gases and vapors. Two techniques, frequently used with liquid and solid biosamples, include ionization by electrospray (after

Fenn) and desorption/ionization by matrix-assisted light amplification by stimulated emission of radiation (laser) desorption ionization (MALDI, after Karas, and Hillenkamp). The sources of inductively coupled plasma (ICP) are used, above all, for the analysis of metals in an extensive range of samples. Other techniques include the ionization via fast atom bomb (FAB), thermospray, atmospheric-pressure chemical ionization (APCI), secondary-ion MS (SIMS), and so on.

5.4.2 THE ANALYZER OF MASS

The *mass analyzer* is the most flexible piece of the *mass spectrometer*. It uses an EF or MF to affect the trajectory or velocity of the particles charged in a certain way. All total analyzers use Lorentz forces in one or another way in the determination of *m/z*, either static or dynamically. Besides the original types of the magnetic sector, other types of analyzers are nowadays in a common use, for example, time of flight (TOF), *quadrupole* ion trap (QIT), quadrupole and Fourier transform ion cyclotron resonance (FT-ICR) mass analyzers. In addition, there are many more total experimental analyzers and their exotic combinations. The instruments of the sector change the direction of the ions that are flying through the total analyzer. The ions, because of MF or EF, result deflected in their trajectories versus their mass and charge, in such a way that the more they result deflected, the lighter they will be. The analyzer directs the particles to the detector varying an EF or MF that is based on *m/z ratio*.

Nowadays, different methods exist to *filter* the ions with regard to their *m/z* ratio. The most commonly used is *quadrupole*. It consists of four long rods arranged in a square formation, electrically connected to each other in opposed pairs (*poles*), to which a variable radio frequency voltage that is in tune with a certain ion is applied. When there is tuning between the ion that passes through them and the applied frequency, such ion continues its way deflecting all not tuning others outside the quadrupole and, so, they do not impact on the detector. The configuration of the present mass spectrometers consists of making use of one, two or till three quadrupoles in series. Nonquadrupolar configurations also exist, for example, hexapoles, although their operation is analogous to the quadrupole.

The analyzer can be used to select a narrow range of *m/z* or explore via a range of *m/z*, in order to catalog the present ions in two ways.

1. Perhaps the easiest to understand is TOF analyzer that joins typically with MALDI ion sources. It accelerates the ions to the same kinetic energy via an EF and measures the times that take to reach the detector. Although the kinetic energy is the same, the velocity is different, so that the most highly charged ion of lighter will arrive the first to the detector.
2. The total quadrupole analyzers and oscillating EFs of QIT selectively stabilize or destabilize the ions in a narrow window of *m/z* values. The FT MS is formed detecting the current of the image produced by the ions, cyclotroning them in presence of an MF. To select the most suitable analyzer for an experiment depends on the type of information that is obtained from such experiment.

Two techniques improve the *signal-to-noise* (S/N) *ratio* of a *mass spectrum*: (1) *filtering* ions via quadrupole (QMS), double quadrupole (QQMS), and triple quadrupole (TQMS) tandem MS and (2) use of accurate *isotopic masses*, for example, $^{1}H = 1.0078$ atomic units of mass (a.u.m.), $^{12}C = 12.0000$ a.u.m. and $^{16}O = 15.9949$ a.u.m.

5.4.3 DETECTOR

The final element of the total spectrometer is the *detector*, which registers the induced charge or current produced when an ion passes near or hits a surface. In an exploration instrument, the signal is produced in the detector during the trajectory of the ion (in which *m/z*) and will produce a mass spectrum: a record of the *m/z* in which the ions are present. Typically, a certain type of multiplier of electrons (*electron multiplier*) is used, although other detectors (e.g., *Faraday caps*) were utilized.

The operation of the electron multiplier is based on the *cascade effect* produced on impacting a certain ion (or ions) on it. Applying a difference of potential between its ends, the amplification factor is risen, which will be determined by the number of amplifier sub-stages that compose the detector. Usually, it is a component subjected to exhaustion that must be replaced with time on losing amplification efficiency. As the number of ions that leave the total analyzer in a particular moment is really small, a significant amplification is generally needed to get a minimally processable signal. *Micro-channel plate* (MCP) detectors are commonly used in modern commercial instruments. In FT MS, the *detector* is a pair of metal plates in the total region of the analyzer that the ions pass only near. No direct current (DC) is produced, only a weak current of the image of the alternating current (AC) occurs in a circuit between the plates.

5.4.4 ADVANTAGES OF A SPECTROMETER OF MASSES

Some of the main advantages of a *mass spectrometer* are as follows:

1. It allows getting satisfactory results in no much time.
2. To obtain qualitative and quantitative results of the same sample.
3. It works with all types of molecules either large or small.

5.5 DISCUSSION

Two techniques improve S/N *ratio* of a *mass spectrum*: (1) *filtering* ions via QMS, QQMS, and TQMS, and (2) use of accurate *isotopic masses*. The *quantitative variable* in MS is the intensity of the peaks, which is proportional to the concentration of the analyte in the sample. The *qualitative variable* in MS is the ratio r between the two most intense peaks; for example, consider C-atom, with isotopic abundances ^{12}C (98.93%) and ^{13}C (1.07%). A molecule with one C-atom will show two peaks: M and $M+1$, for example, CH_4 with $M = 12 + 4 = 16$ a.u.m. and $M + 1 = 13 + 4 = 17$ a.u.m., with relative intensities 98.93:1.07 and a ratio $r = (M + 1) / M = 1.07{:}98.93 \approx 1{:}100$. A molecule with two C-atoms will show three peaks M, $M+1$, and $M+2$, for example, CH_3–CH_3 with $M = 12 + 12 + 6 = 30$ a.u.m., $M + 1 = 13 + 12 + 6 = 31$ a.u.m. and $M + 2 = 13 + 13 + 6 = 32$ a.u.m., with relative intensities $98.93^2{:}2 \times 98.93 \times 1.07{:}1.07^2 \approx 9787{:}212{:}1$ and a ratio $r = (M + 1) / M \approx 212{:}9787 \approx 1{:}50$. A molecule with three C-atoms will show four peaks M, $M+1$, $M+2$, and $M+3$, for example, CH_3–CH_2–CH_3 with $M = 12 + 12 + 12 + 8 = 44$ a.u.m., $M + 1 = 13 + 12 + 12 + 8 = 45$ a.u.m., $M+2 = 13 + 13 + 12 + 8 = 46$ a.u.m., and $M+3 = 13 + 13 + 13 + 8 = 47$ a.u.m., with relative intensities $98.93^3{:}3 \times 98.93^2 \times 1.07{:}3 \times 98.93 \times 1.07^2{:}1.07^3 \approx 968242{:}31417{:}340{:}1$ and a ratio $r = (M + 1) / M \approx 31417{:}968242 \approx 1{:}30$. In general, the number of C-atoms in the molecule can be obtained from the equation:

$$n = r\frac{100-\rho}{\rho}, \qquad (5)$$

where r is the observed ratio $(M + 1) / M$ and ρ is the percent ^{13}C isotope abundance.

MS presents high *sensitivity* S: It is a technique for small molecules with $S = 0.04$ a.u.m. It was extended to large proteins with $S = 2$ a.u.m. It shows extremely low limits of detection and quantification, which can be used in concentrations of traces and ultra-traces.

Modern techniques and scientific rigor are applied to research. Tu and co-workers determined the proteins bonded to ART via MS.[12–15] Thanks to the technique, it was determined and analyzed ART proteins, and created the effective medicine versus malaria.[16–18]

5.6 FINAL REMARKS

From the present results and discussion, the following final remarks can be drawn:

1. Two techniques improve the *signal-to-noise ratio* of a *mass spectrum*: (a) *filtering* ions *via quadrupole*, double quadrupole and triple quadrupole tandem MS, and (b) use of accurate *isotopic masses*.
2. The *quantitative variable* is the intensity of the peaks, which is proportional to the concentration of the analyte in the sample.
3. The *qualitative variable* is the ratio between the two most intense peaks.
4. MS presents high *sensitivity*: 0.04 for small molecules and two atomic units of mass for large proteins.
5. MS shows extremely low limits of detection and quantification, which can be used in concentrations of traces and ultra-traces.
6. Modern techniques and scientific rigor are applied to research.
7. Tu and co-workers determined the proteins bonded to artemisinin via MS. Thanks to the technique, it was determined and analyzed artemisinin proteins and created the effective medicine versus malaria.

ACKNOWLEDGMENTS

The authors thank support from Generalitat Valenciana (Project No. PROMETEO/2016/094) and Universidad Católica de Valencia *San Vicente Mártir* (Project No. UCV.PRO.17-18.AIV.03).

KEYWORDS

- **ionization technique**
- **mass analyzer**
- **magnetic field**
- **electric field**
- **quadrupole**
- **matrix-assisted laser desorption ionization**
- **isotopic mass**

REFERENCES

1. F. Torrens and G. Castellano, Ultraviolet-Visible Absorption Spectroscopy: Link with Fluorescence. In *Research Methods and Applications in Chemical and Biological Engineering*; Pourhashemi, A., Deka, S.C., Haghi, A.K., Eds.; Apple Academic-CRC: Waretown, NJ, in press.
2. Torrens, F.; Castellano, G. Fluorescence Excitation and Emission: Comparison with Absorption. In *Analytical Methods in Chemistry Research with Applications*; Haghi, A. K., Ed.; Apple Academic–CRC: Waretown, NJ, in press.
3. Skoog, D. A.; Holler, F. J.; Crouch, S. R. *Principles of Instrumental Analysis*; Cengage Learning: Boston, MA, 2018.
4. Valls Planells, O.; del Castillo García, B., Eds. *Técnicas Instrumentales en Farmacia y Ciencias de la Salud*; Universidad Norbert Wiener: Lima, Peru, 2009.
5. Robinson, J. W.; Frame, E. M. S.; Frame, G. M., II. *Undergraduate Instrumental Analysis*; CRC: Boca Raton, FL, 2014.
6. Miñones Trillo, J. *Manual de Técnicas Instrumentales*; Círculo Editor Universo: Barcelona, Spain, 1978.
7. Levine, I. N. *Molecular Spectroscopy*; Wiley: Chichester, UK, 1975.
8. Turro, N. J.; Ramamurthy, V.; Scaiano, J. C. *Modern Molecular Photochemistry of Organic Molecules*; Viva Books: Delhi, India, 2017.
9. Valeur, B.; Barberan-Santos, M. N. *Molecular Fluorescence: Principles and Applications*; Wiley–VCH: Weinheim, Germany, 2012.
10. Balogh, M. P. The Nature and Utility of Mass Spectra. *LC•GC Eur.* **2010,** *23* (2), 82–93
11. Brown, G. Artemisinin and a New Generation of Antimalarial Drugs. *Educ. Chem.* **2006,** *2006* (7), 1–7.
12. Collaboration Research Group for Qinghaosu. A Novel Kind of Sequiterpene Lactone—Artemisinin. *Chin. Sci. Bull.* **1977,** *22*, 142.
13. Anonymous. Antimalaria Studies on Qinghaosu. *Chin. Med. J.* **1979,** *92*, 811–816.

14. Tu, Y. Y.; Ni, M. Y.; Zhong, Y. R.; Li, L. N.; Cui, S. L.; Zhang, M. Q.; Wang, X. Z.; Liang, X. T. Studies on the Constituents of *Artemisia annua* L. (authors' translation). Article in Chinese. *Acta Pharm. Sin.* **1981,** *16*, 366–370.
15. Tu, Y. Y.; Ni, M. Y.; Zhong, Y. R.; Li, L. N.; Cui, S. L.; Zhang, M. Q.; Wang, X. Z.; Ji, Z.; Liang, X. T. Studies on the Constituents of *Artemisia annua* Part II. *Planta Med.* **1982,** *44*, 143–145.
16. Zhang, J., Ed. *A Detailed Chronological Record of Project 523 and the Discovery and Development of* Qinghaosu *(Artemisinin)*; Strategic Book: Houston, TX, 2013.
17. Klayman, D. L. *Qinghaosu* (artemisinin): An Antimalarial Drug from China. *Science* **1985,** *228*, 1049–1055.
18. Krishna, S.; Bustamante, L.; Haynes, R. K.; Staines, H. M. Artemisinins: Their Growing Importance in Medicine. *Trends Pharmacol. Sci.* **2008,** *29*, 520–527.

CHAPTER 6

FLAME RETARDANT AND SMOKE SUPPRESSANT ADDITIVES FOR POLYPROPYLENE: VERMICULITE AND ZINC PHOSPHATE

FATMA ÜSTÜN[1], HASAN DEMIR[2], and DEVRIM BALKÖSE[3,*]

[1]*Department of Clinical Biochemistry Ege University Hospital, Bornova, İzmir, Turkey*

[2]*Department of Chemical Engineering, Osmaniye Korkut Ata University, Osmaniye, Turkey*

[3]*Department of Chemical Engineering, Izmir Institute of Technology, Gulbahce Urla, Izmir, Turkey*

Corresponding author. E-mail: devrimbalkose@gmail.com

ABSTRACT

A flame retardant composite similar to a commercial flame retardant masterbatch was aimed to be developed in this study. The commercial masterbatch was characterized by its functional groups, the elemental composition, crystal structure, morphology, and thermal stability, and limiting oxygen index (LOI). It was a polypropylene matrix composite filled with a clay mineral most probably montmorrilonite and a phosphate compound. Thus vermiculite which is a clay mineral and a phosphate compound and zinc phosphate were added to polypropylene matrix in order to improve flame retardancy. Fillers were coated with stearic acid to make them compatible with polypropylene. It was observed that the composites having 20% zinc phosphate had the highest degradation onset temperature (261°C). The remaining mass at 600°C was 21% for commercial masterbatch, 22% for 20% zinc phosphate, 33% for 10% zinc phosphate, and 10% vermiculite and 43% for 20% vermiculite composite. The additives in the composites had no significant effect on the activation energy of polypropylene degradation, but they led the formation of more solid carbonized compounds. It was understood that 20% vermiculite composite had the highest smoke suppression capacity and formed less gaseous and more solid degradation products. The LOI was 22% for flame retardant masterbatch, 19% for polypropylene and composites having vermiculite and 18% for zinc phosphate. In conclusion the composites prepared were not as succesful as commercial masterbatch in flame retardancy, however, the composite with 20% vermiculite had higher smoke suppression capability and the composite with 20% zinc phosphate had highest degradation onset temperature (261°C).

6.1 INTRODUCTION

Polymers have an important place in the economy of modern industry because they are cheap and durable materials. Today, the polymers with improved flame retardant properties by additive materials had a wide application area such as household materials, transporting, packaging, electrical engineering. Magnesium hydroxide,[1] magnesium oxide and calcium oxide pair,[2] lantanium oxide,[3] boron compounds,[4] ammonium polyphosphate (APP) montmorrillonite pair,[5] and organo bentonite[6] are used flame retardant fillers for polymers. Halogen-free flame retardants which are not environmentally harmful are industrially important.[7] Flame retardant compositions added to polymers include nanoclays,[8] modified clays,[9] surface modified silica with phosphorous compounds,[10] sepiolite,[11] natural zeolite,[12] pristine and modified vermiculite,[13–16] montmorrilonite,[17] hydrotalcite and nanozinc phosphate,[18] zinc borate and zinc phosphate.[19]

Some of the additives are added to intumescent sytems such as APP as acid source and blowing agent and carbonific agent pentaeritritol[12] improving the residual protective layer,[11] increasing the char residue[17]. Some of the additives are dispersed in exfoliated form and increased the diffusion path for oxygen resulting in flame retardancy.[14] If clays are added, the active sites on clay layers can catalyze the initial decomposition and the ignition of the composites. On the other hand, the active sites can also catalyze the formation of a protective coating char on the samples and the dehydrogenation and crosslinking of polymer chains.[9]

Minerals with low concentrations of selected flame retardants introduced into polypropylene improve the flammability and thermal degradation behavior.[8] The presence of metal oxides, such as silicon dioxide, aluminum, iron and magnesium oxides, in the added minerals improves the flame retardancy.[13,17]

The combustion behavior and thermal-oxidative degradation of polypropylene and clay nanocomposite were investigated by Qin et al.[9] The influence of compatibilizer, alkylammonium, organoclay, protonic clay, and pristine clay was considered, respectively. The decrease of heat release rate (HRR) was mainly due to the delay of thermal-oxidative decomposition of the composites.[9]

Surface modification of various silicas by phosphorous agents such as diethylphosphatoethyltriethoxysilane was carried out with the aim to use these particles as flame retardant additive in a polypropylene (PP) matrice at low content (10 wt%).[10] Quite surprisingly the untreated fillers induce the most significant reduction of peak of HRR (50% decrease) while the surface modification by phosphorous agents does not lead to the expected effect on the fire behavior of PP composite. This phenomenon was related to the morphology and rheological behavior of the various PP composites. Indeed, the higher the storage modulus at low frequencies is, the better the fire behavior is, because of the induced barrier effect. Moreover, the addition of nonmodified silicas leads to a decrease of 10 s of the time to ignition. This phenomenon was related to the formation of bubbles after the PP melting during cone calorimeter test.[10]

The pyrolysis, flammability, and fire behavior of PP containing an intumescent flame retardant (IFR) and surface modified (with HCl treatment and cetyltrimethylammonium bromide) sepiolite nanoparticles were investigated by Pappalardo et al.[11] The synergistic effect between the nanofiller and the flame retardant resulted from the improved properties of the residual protective layer. The limiting oxygen index (LOI) value of PP is increased from 20.8 to 25.7 when 0.5% sepiolite or organosepiolite was added to 12% IFR in polypropylene.[11]

An intumescent system consisting of APP as an acid source and blowing agent, pentaerythritol (PER) as a carbonific agent and natural zeolite (clinoptilolite, Gordes II) in pristine and surface modified form as a synergistic agent was used to enhance flame retardancy of polypropylene (FR-PP).[12] Zeolite (1–10%) was incorporated into flame retardant formulation. Filler content was fixed at 30 wt% of total amount of flame retardant PP composites. The LOI values reached a maximum value of 41% for mercapto silane treated APP:PER (2:1) PP composite containing 5 wt% zeolite.[11]

Morphological, thermal, and mechanical properties of blends prepared from polypropylene (PP) and 1, 3, and 5 wt% of vermiculite (VMT) were studied by Gomes et al.[13] The PP/VMT blends showed an increase in the temperature of maximum rate of mass loss (T_{max}). T_{max} increased from 411°C to 424°C when 5% VMT is added. Vermiculite clay (VMT) which was organically modified with a quaternary organic salt increased the T_{max} to 448°C when it was added 5% to PP matrix.[14] The organo VMT was exfoliated and increased the path length of the gas molecules through the polymer matrix. Thus T_{max} was increased.

Polypropylene-vermiculite nanocomposites with an intercalated or exfoliated structure can also be achieved by simple melt mixing of maleic anhydride-modified vermiculite with polypropylene.[15] TGA results showed that vermiculite improved the thermal stability of PP considerably. Dynamic mechanical analysis profiles demonstrate that the nanocomposites exhibit the formation of a new microphase consisting of ternary PP/MA/vermiculite.[15]

Organic vermiculite (OVMT) prepared from vermiculite (VMT), with high aspect ratio and orderly arranged platelets intercalated by octadecyl trimethyl ammonium bromide (OTAB) was used as a synergistic agent on the flame retardancy of a polypropylene/intumescent flame retardant (PP/IFR) system by Chen et al.[16] The results of LOI and UL-94 testing showed that low loading of OVMT improved the flame retardancy and retarded dripping for PP/IFR composites. OVMT, with 1% loading increased the char residue of PP/IFR composites and could act as an effective additive for improvement in flame retardancy. It was concluded that OVMT with 1% loading showed a synergistic effect with IFR in the combustion of the PP/IFR composites.[16]

Synergistic effects of pristine vermiculite (VMT), exfoliated vermiculate (EVMT), and montmorillonite (MMT) on the intumescent flame retardance of polypropylene were investigated systematically with the usual fire testing methods.[17] The LOI of flame-retardant polypropylene (FRPP) filled with 30 wt% IFRs composed of APP and PER were increased from 30 to 33 vol% for VMT and MMT and to 36 vol% for EVMT when 1 wt% IFR was substituted for clay. The thermogravimetric analysis results show that EVMT had the best performance for increasing the char residue of

FRPP higher than 650°C compared with VMT and MMT. The high content of iron and the small particle size of EVMT may have been responsible for its high synergistic effect at a low filling level.[17]

The synergistic effect of layered double hydroxide (LDH) and prepared nano-zinc phosphate was studied by melt compounding of PP on a Brabender Plastograph by Navare et al.[18] Hydrotalcite as LDH (1–3%) and nano-zinc phosphate (1–3%) acted as synergistic fillers in PP nanocomposites improving the flammability as compared to pristine PP. The rate of burning (flammability) decreased with increasing amount of LDH nano-particles in the PP nanocomposites. The char formation ability of LDH and nano-zinc phosphate might have resulted in the delayed flame propagation and given more residues of burning as compared to virgin PP.[18]

The synergistic effects of zinc borate (ZB)–zinc phosphate (ZP) on the thermal stability of polyvinyl chloride (PVC) were investigated using thermal techniques by Erdogdu et al.[19] The induction and stability time values of PVC plastigels were determined at 140°C and 160°C. The PVC plastigels having only ZP and ZB retarded dehydrochlorination of PVC compared with the unstabilized sample. However, the plastigels with both ZB and ZP had a superior synergistic effect on char formation of PVC.[19]

The flame retardent zinc phosphate can be synthesized from different water-soluble compounds by precipitation reactions. It was synthesized from zinc nitrate, sodium hydroxide, and phosphoric acid or casein by Ramos et al.[20] Macroporous particles were obtained after hydrothermal synthesis and thermal treatment. Nano zinc phosphate was precipitated by mixing zinc nitrate and sodium phosphate solutions having etoxylated surfactants by Navare et al.[18] Various phosphate compounds such as phosphoric acid, sodium di-hydrogen phosphate, disodium hydrogen phosphate, sodium pyrophosphate, or sodium triphosphate were used in zinc phosphate synthesis.[21,22]

In the present study, a flame retardant masterbatch supplied by an industrial end-user company was characterized in detail in order to prepare a similar composition from locally available materials. Since the analysis of masterbatch indicated that it was polypropylene containing a layered compound and a phosphorous compound, composite materials from locally available vermiculate and zinc phosphate were prepared and tested for flame retardancy.

6.2 EXPERIMENTAL

6.2.1 MATERIALS

A halogen-free flame retardant polypropylene masterbatch with 4.5% active component supplied by Ges Electric Company Izmir was characterized to prepare a similar masterbatch from local materials. Polypropylene MH-418 from Petkim Izmir was used in preparation of composites. Vermiculate provided by Food Engineering Department of Ege University and zinc phosphate supplied by Pigment AŞ İzmir were used as flame retardant additives. Stearic acid (90%) from Sigma Aldrich and Ethanol (99%) from Pancreas were used for surface modification of zinc phosphate and vermiculate.

6.2.2 METHODS

6.2.2.1 ASH OF MASTERBATCH

The ash of mastebatch was obtained by heating at 10°C/min rate up to 600°C in Setaram Labsys TG analyzer under air flow.

6.2.2.2 SURFACE MODIFICATION OF VERMICULATE AND ZINC PHOSPHATE

The vermiculate and zinc phosphate were ground separately in Laborotary Vibrating Cup Mill (Fristch) and the fraction below 63 μm was used in preparation of polypropylene composites. A total of 30 g powder was added to a solution of 600 mg stearic acid in 100 cm^3 ethanol and mixed for 45 min. Then the mixture was dried in a rotary evaporator (Buchi) at 175 mbar at 50°C. Thus the surface of the particles was made hydrophobic to be compatible with polypropylene.

6.2.2.3 CHARACTERIZATION

The masterbatch, the masterbatch ash obtained by heating it up to 600°C with 10°C/min heating rate in air, composite films prepared from polypropylene, vermiculate, and zinc phosphate were characterized in this study. X-ray diffraction diagrams of polypropylene, masterbatch film, vermiculate and zinc phosphate was obtained with Philips Xpert XRA-480 X-ray diffractometer using CuK_α radiation. TG analyses were made by heating the samples at 10°C/min rate up to 600°C with SETERAM LABSYS TGA under continous air flow. FTIR spectra of the ash of masterbatch obtained after TG analysis, vermiculate and zinc phosphate were obtained using KBr disc technique with Shimadzu FTIR 8210. FTIR spectra of the masterbatch and PP films prepared by pressing their pellets placed between two glass plates heated to 180°C were also taken by Shimadzu FTIR 8210 using transmission method. SEM micrographs of vermiculate, zinc phosphate and masterbatch film fracture surfaces were obtained by Philiphs XL-305 FEG–SEM. EDX analysis of samples was made by using the same instrument. LOI of the masterbatch and the composites was determined according to ASTM D2863.

6.2.2.4 PREPARATION OF COMPOSITES

Stearic acid coated vermiculate and zinc phosphate was used as flame retardant fillers. Composites sheets were prepared from polypropylene pellets with 2% stearic acid. The composition of the sheets was selected by considering analysis of the masterbatch. Thus composites with 20% vermiculate, 20% zinc phosphate and 10% vermiculate and 10% zinc phosphate were prepared by using Thermo Haake (557-8310) thorque rheometer. Polypropylene was melted at 190°C at 50 rpm and solid ingredients were added to molten polymer and mixed for 10 min at 50 rpm. The cooled mixtures were taken to the Carver Press and 150 mm × 150 mm × 3 mm sheets were prepared by pressing at 13.7 MPa at 190°C for 10 min and then cooling by circulating water at 25°C.

6.3 RESULTS

6.3.1 CHARACTERIZATION OF MASTERBATCH

The masterbatch was characterized by FTIR analysis, X-ray diffraction, SEM, EDX, and TG analysis.

6.3.1.1 FTIR ANALYSIS

The FTIR spectrum of the masterbatch film is seen in Figure 6.1. At 973 cm^{-1}—CH_3 rocking and C–C chain stretching; at 998 cm^{-1} —CH_3 rocking, CH_2 wagging, and CH bending, at 1168 cm^{-1}—C–C chain stretching, CH_3 rocking, and CH bending; at 1220 cm^{-1}—CH bending and C–C chain stretching; at 1256 cm^{-1}—CH bending, CH_2 twisting, and CH_3 rocking; at 1377 cm^{-1}—CH_3 symmetric bending and CH_2 wagging; at 1450 cm^{-1}—CH_2 deformation bands[23] that were observed in the FTIR spectrum of masterbatch indicated that it had PP.

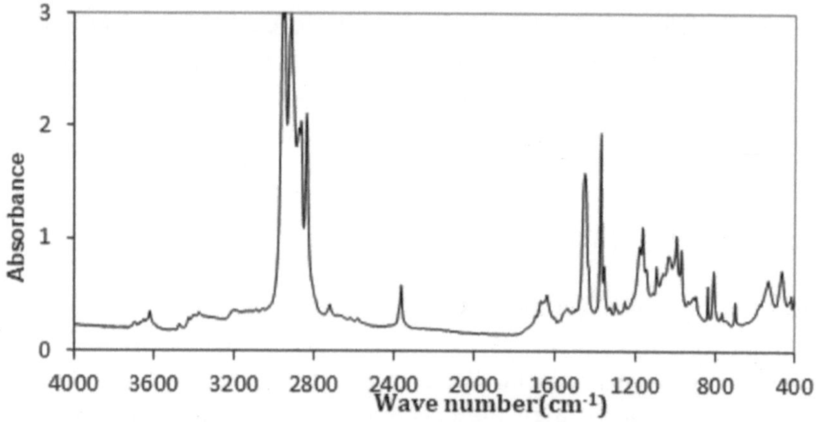

FIGURE 6.1 FTIR spectrum of masterbatch film.

The polymer phase and organic additives were eliminated from the masterbatch by oxidative thermal degradation and the remaining ash was analyzed by FTIR spectroscopy. The FTIR spectrum of the masterbatch ash in Figure 6.2 had vibrations of isolated OH group at 3700 cm^{-1}, hydrogen bonded OH groups at 3400 cm^{-1}, H_2O bending vibrations at 1640 cm^{-1} and Si–O or Al–O stretching vibrations at 1045 cm^{-1}.

6.3.1.2 X-RAY DIFFRACTION DIAGRAM OF POLYPROPYLENE AND MASTERBATCH

The X-ray diffraction diagrams of polypropylene and the masterbatch in Figure 6.3, confirms that the matrix phase is polypropylene in α monoclinic structure, since there are the same polypropylene peaks at 2θ values of 14.2°, 17°,18.8°, 21.2° and 21.9° for planes (110), (040), (130), (111), and (131) respectively.[24] The sharp diffraction peaks at 2θ 9.5°, 12.5°, 17.5°, 27.5°, and 46° in the X-ray diffraction diagram of the masterbatch indicated there are inorganic additives in masterbatch. The X-per pro software indicated the peaks at 2θ 9.1°, 12.2°, 17.6°, 24.3° and 27.03° belonged to montmorrilonite. Thus it can be concluded that montmorrilonite was dispersed in PP in the masterbatch.

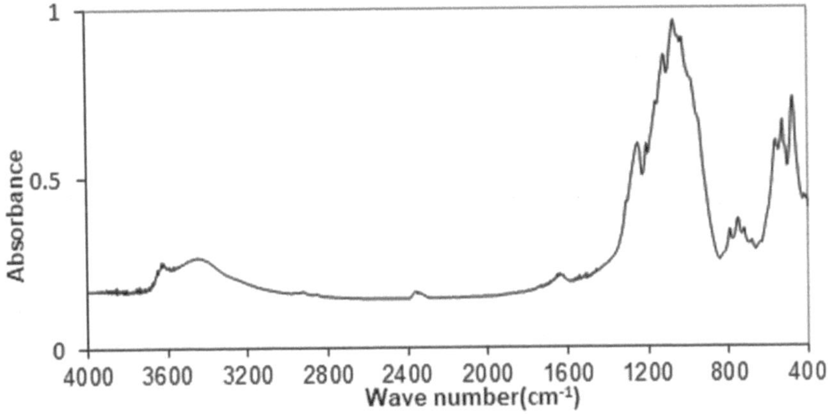

FIGURE 6.2 FTIR spectrum of ash of masterbatch.

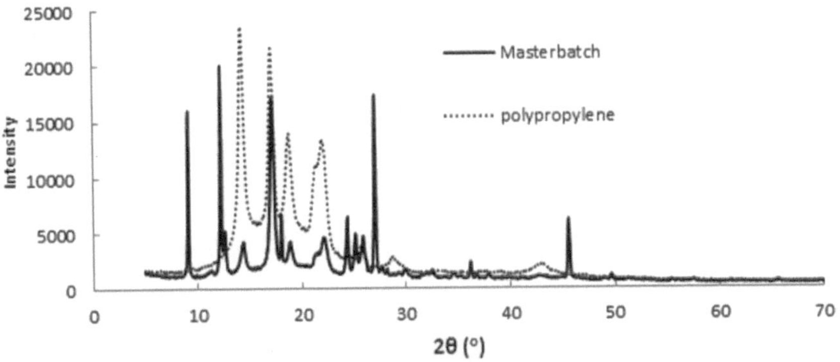

FIGURE 6.3 X-ray diffraction diagrams of masterbatch and polypropylene.

6.3.1.3 MORPHOLOGIES OF MASTERBATCH AND MASTERBATCH ASH

In Figures 6.4 and 6.5, the SEM micrographs of the master batch film fracture surface and masterbatch ash are seen respectively. As seen in Figure 6.4, there are solid sharp-edged particles embedded in continuous polymer matrix. The masterbatch ash had both layered and spherical particles as seen in Figure 6.5.

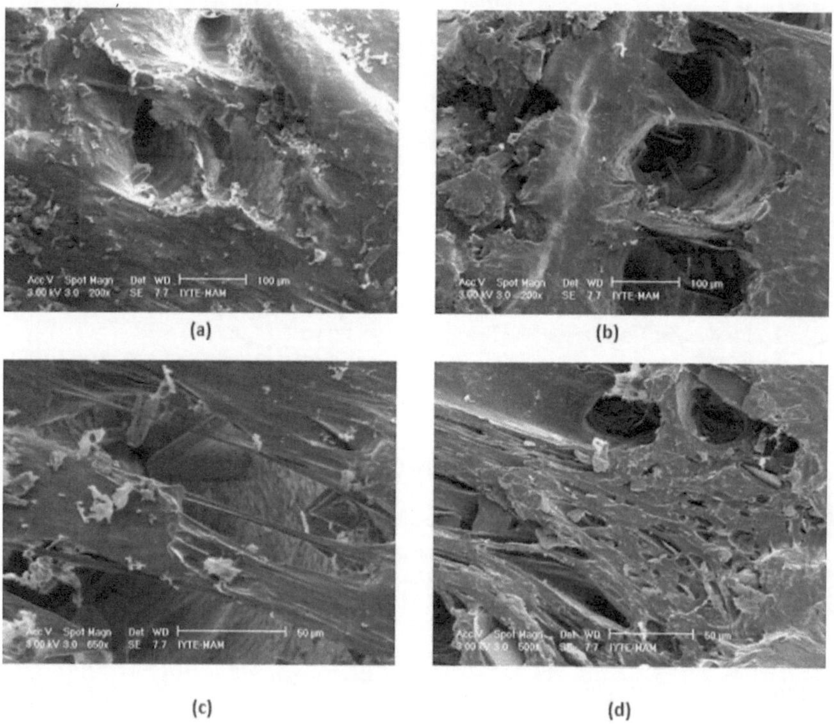

FIGURE 6.4 SEM micrographs of the fracture surface of the masterbatch film at (a) and (b) 200×, (c) 650× (d) 500× magnification at different points.

6.3.1.4 EDX ANALYSIS OF MASTERBATCH AND MASTERBATCH ASH

The elemental composition excluding H element of masterbatch, layered fraction of masterbatch, and spherical fraction of masterbatch and overall average of masterbatch are shown in Table 6.1. The values reported are the averages of at least five measurements. The error in measurements is around 5%.

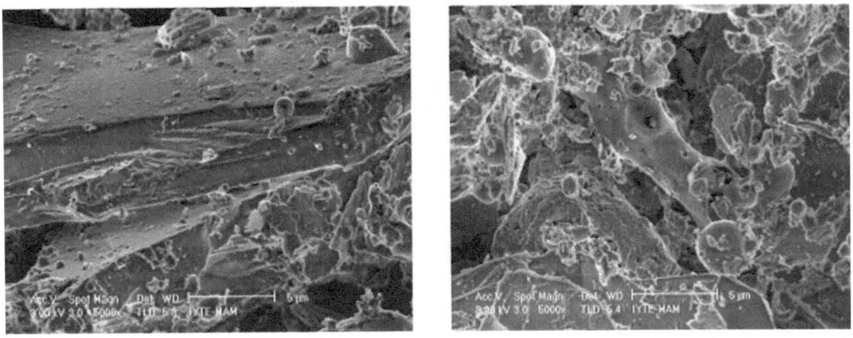

FIGURE 6.5 SEM micrographs of masterbatch ash at 5000× magnification at two different regions.

TABLE 6.1 Elemental Composition of Masterbatch, Layered, and Spherical Fractions and Overall Average of Masterbatch Ash, Vermiculate, and Zinc Phosphate, Mass%.

Element	Masterbatch (%)	Layered fraction of ash (%)	Spherical fraction of ash (%)	Overall average ash (%)	Vermiculite (%)	Zinc phosphate (%)
C	88.09	3.84	10.63	6.42	3.10	3.55
O	7.08	37.5	41.34	36.37	35.77	13.93
F	0	0	0	0	4.48	1.45
Ni	0	0	0	0	5.99	1.76
Al	1.44	19.52	6.36	10.88	7.13	0.30
Si	1.07	22.78	3.31	13.99	17.34	0.14
P	1.10	4.58	19.35	12.62	2.65	7.18
Mg	0.20	0	0	0	15.96	2.17
K	–	6.55	1.70	4.22	1.14	0.07
Ca	0.60	1.96	9.97	9.04	0.64	0.95
Zn	0	3.24	7.65	6.43	2.32	48.87
Na	0	0	0	0	0.99	18.19
Fe	0	0	0	0	2.51	1.50

The EDX analysis of masterbatch indicated it contained mainly C (88.9%) and O (7.08%) and minor amounts of Al (1.44%), Si (1.07%), P (1.10%), and Ca (0.60%) elements. The particles with layered and spherical geometry in masterbatch ash had different compositions. While the layered particles had mainly Al (19.52%) and Si (22.78%) elements, spherical particles were rich in C (10.63%), Ca (9.97%), Zn (7.65%), and P (19.35%) elements. The overall average of the masterbatch ash had mainly C (6.52%), Al (10.88%), Si (13.99%), P (12.62%), zinc (6.43%), and O (36.37%).

6.3.1.5 TG ANALYSIS OF POLYPROPYLENE AND THE MASTERBATCH

The TG curves of polypropylene and masterbatch in air are shown in Figure 6.6. Important points of the TG curves are reported in Table 6.2. The onset of mass loss is at 235°C for polypropylene (Curve 1 in Fig. 6.6) and 20% mass loss occurs at 336°C. Above 500°C, 27% mass remains as carbonized residue. A similar TG curve for the same polypropylene with the same TG analyzer was reported by Demir et al.[12] previously. The onset of the mass loss is 250°C for the masterbatch (Curve 2 in Fig. 6.6) and at 317°C 80%, at 415°C 25%, at 600°C 21% mass remains. Since EDX analysis indicated a very low amount of C in the masterbatch ash the inorganic filler content of the masterbatch is 21%.

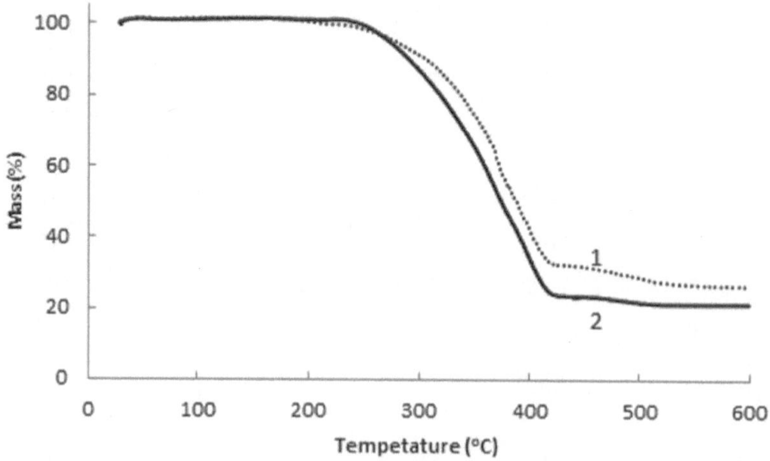

FIGURE 6.6 TG curve of (1) polypropylene and (2) masterbatch.

TABLE 6.2 Thermal Decomposition and Activation Energy of Polypropylene Degradation in Polypropylene, Masterbatch and Composites with 20% Vermiculate, 10% Vermiculate+, 10% Zinc Phosphate, and 20% Zinc Phosphate for the 265–350°C range and r^2 Values of Broido's Plot.

Sample	1% Mass loss temperature (°C)	20% Mass loss temperature (°C)	Remaining mass at 600°C (%)	Activation energy (kJ/mol)	r^2 of Broido's plot
Polypropylene	235	330	26	72.2	0.99
Masterbatch	250	317	21	73.2	0.99
20% vermiculate/PP	235	330	43	72.6	0.99
10% vermiculate + 10% zinc phosphate/PP	260	352	33	64.8	0.98
20% zinc phosphate/PP	261	325	22	84.8	0.98

The masterbatch was a PP composite filled with a layered inorganic material at 21% level. There were two types of inorganic material rich in Al, Si, Zn, and P, respectively. Based on this analysis polypropylene composites were prepared from vermiculite which is a layered clay rich in Al and Si and zinc phosphate which is rich in Zn and phosphate.

6.3.2 CHARACTERIZATION OF VERMICULATE AND ZINC PHOSPHATE

The FTIR spectra of vermiculate and zinc phosphate are seen in Figures 6.8 and 6.6. The vermiculate had H bonded vibrations at 3419 cm^{-1}, H_2O bending vibrations at 1637 cm^{-1}.[25] 989 cm^{-1}, 663 cm^{-1}, and 455 cm^{-1} are the fingerprint vibrations of vermiculite. However, the vermiculate was an organo modified vermiculate since it had antisymmetric and symmetric stretching vibrations at 2940 and 2920 cm^{-1} and bending vibrations at 1383 cm^{-1} and 1462 cm^{-1} of CH_2 groups as seen in Figure 6.7.

The isolated OH group vibration at 3541 cm^{-1}, hydrogen bonded OH vibration at 3327 cm^{-1}, H_2O bending vibration at 1630 cm^{-1}, at 1109 cm^{-1}, 1070 cm^{-1}, and 1004 cm^{-1} antisymmetric stretching vibrations of the PO_4 group, at 943 cm^{-1} PO_4 group symmetric stretching vibration are observed in the FTIR spectrum of zinc phosphate in Figure 6.8.[26]

X-ray diffraction diagram of vermiculite is seen in Figure 6.9. There are sharp peaks at 2θ values of 6.07°, 8°, 12°, 18.5°, 25°, 31°, 37°, 44°, and 65°. These peaks were not identical to that reported in literature for vermiculate.[27] The peak at 2θ value of 6.07° corresponding to 1.46 nm belongs to [001] planes of vermiculite under consideration. Pristine vermiculite had [001] diffractions at 2θ value of 6.46° corresponding to 1.37 nm in Gomes et al's study.[13] Hilleer et al.[27] reported the interlayer spacing as 1.47 nm and Calle et al.[28] determined it as 1.49 nm. Thus the interlayer spacing of the vermiculate in the present study was close to that reported for pristine vermiculate reported by previous investigators.

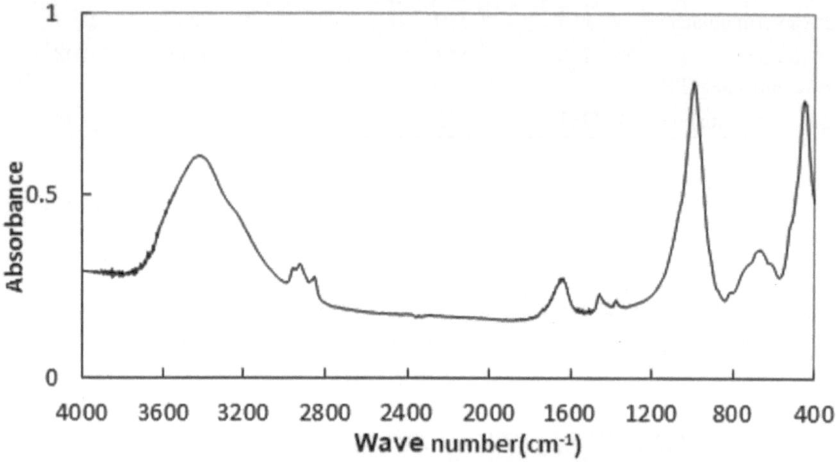

FIGURE 6.7 FTIR spectrum of vermiculite.

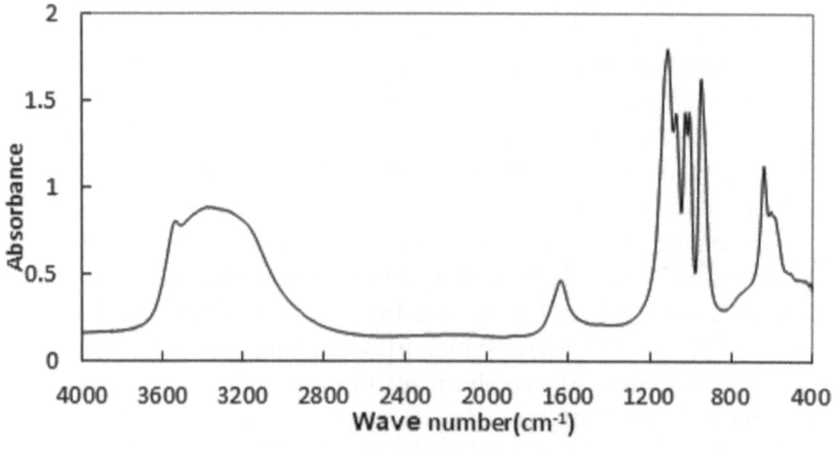

FIGURE 6.8 FTIR spectrum of zinc phosphate.

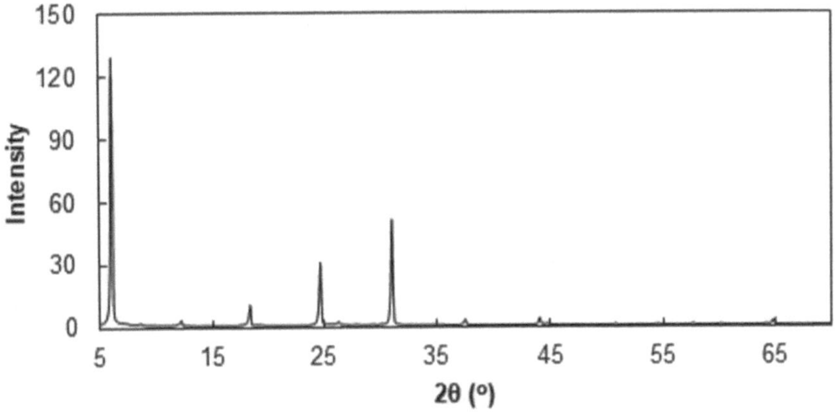

FIGURE 6.9 X-ray diffraction diagram of vermiculite.

The X-ray diffraction diagram of zinc phosphate seen in Figure 6.10 is very similar to that was reported for α-hopeite and β-hopeite ($Zn_3(PO_4)_2 \cdot 4H_2O$).[29] They both have orthorhombic crystals with a = 1.05 nm, b = 1.83 nm, and c = 0.50 nm, but the orientation of 1 mol of water are different in two forms.[29,30] The most intense diffraction peaks observed in Figure 6.10 are at 2θ values of 9.67°, 19.41°, and 31.37° identical by that reported by Herske et al.[30] for β hopeite. The peaks at 2θ values of 31.6°, 34.2°, and 36.10° indicated the presence of ZnO as impurity in the zinc phosphate according to JCPDS card number 79-0206.

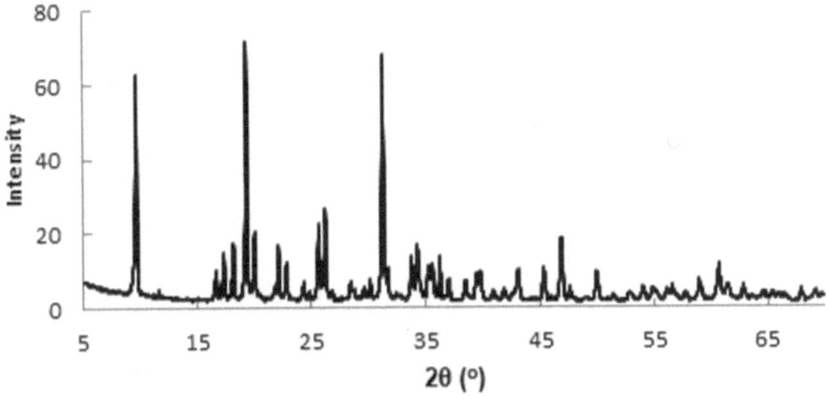

FIGURE 6.10 X-ray diffraction diagram of zinc phosphate.

SEM micrographs of vermiculate and zinc phosphate are shown in Figure 6.11. The layered structure of vermiculite can be seen in Figure 6.12a. Zinc phosphate consisted of aggregated small crystals.

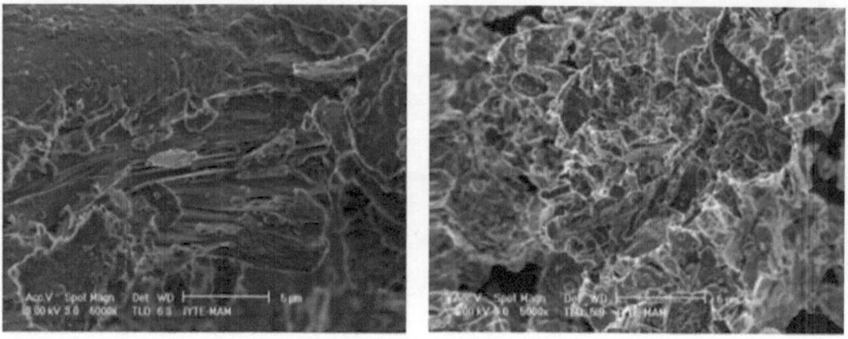

(a) (b)

FIGURE 6.11 SEM micrographs of (a) vermiculite and (b) zinc phosphate at 5000× magnification.

EDX analysis of vermiculate and zinc phosphate are reported in Table 6.1. Considering main elements present in vermiculate its empirical formula was predicted as $5MgO.Al_2O_3.5SiO_2$ using the data in Table 6.1. A representative EDX spectrum for zinc phosphate is shown in Figure 6.12. The empirical formula is found as $Zn_3(PO_4)_2$ from average elemental composition. There were impurities present both in vermiculate and zinc phosphate. Vermiculate has 4.24% Ca and 3.28% Ni and zinc phosphate had considerable amount of Na (18.19%), minor amount of iron (1.5%). This high value of sodium in zinc phosphate could be the misinterpretation of EDX spectrum. The K_α of Na and L_α of Zn are very close to each other at 1.041 keV and 1.012 keV, respectively.[31] This may create incorrect determination of Zn and Na content.

TG curves of vermiculite and zinc phosphate are seen in Figure 6.13. Vermiculate lost 4% of its mass at 121°C and it has 5.6% mass loss at 600°C. Zinc phosphate lost 5% of its mass at 164°C, 6.3% at 320°C, and 6.8% of its mass at 600°C. The loss in mass attributed to the elimination of water from their structure.

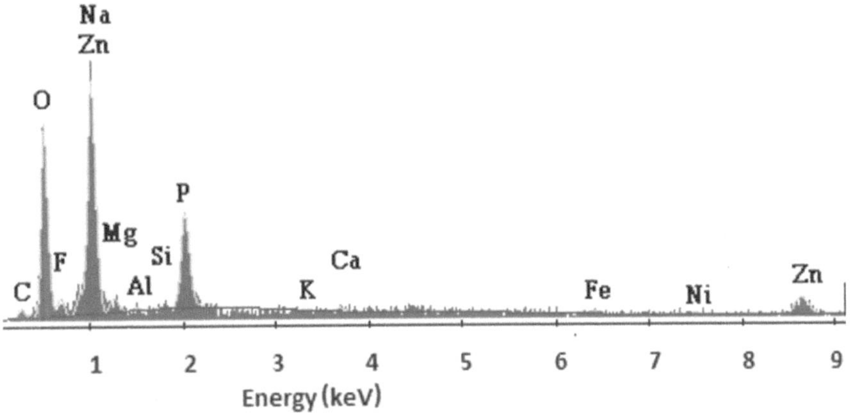

FIGURE 6.12 **(See color insert.)** EDX spectrum of zinc phosphate.

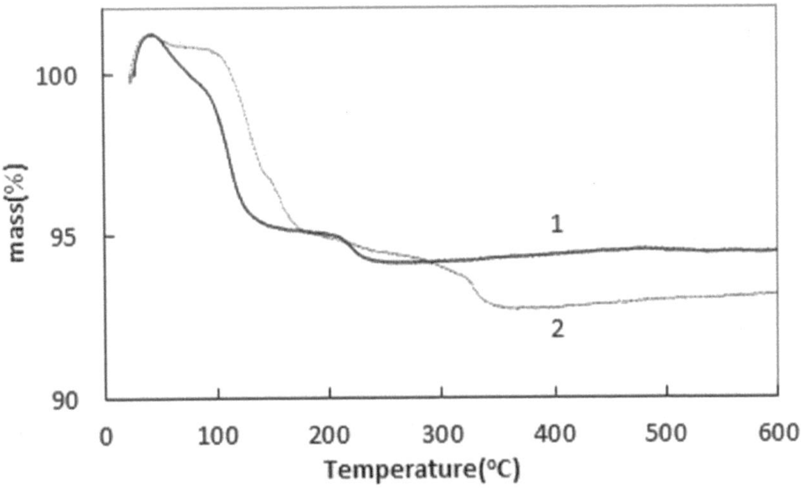

FIGURE 6.13 TG curves of (1) vermiculite and (2) zinc phosphate.

6.3.3 CHARACTERIZATION OF COMPOSITES

6.3.3.1 TG ANALYSIS

Vermiculate and zinc phosphate were added to polypropylene to obtain a similar composite with the commercial masterbatch. They were coated with stearic acid to make them compatible with polypropropylene. Their thermal stability and flame retardancy were determined.

The TG curves of composites having 20% vermiculate, 20% zinc phosphate, 10% zinc phosphate, and 10% vermiculate are seen in Figure 6.14.

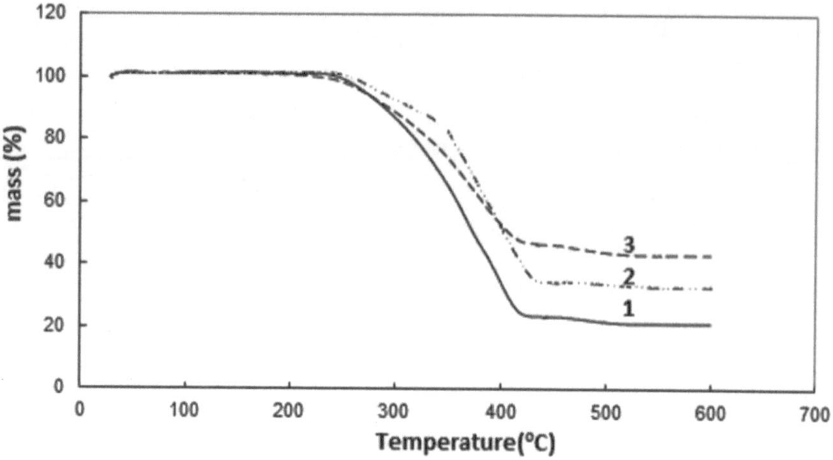

FIGURE 6.14 TG curves of composites with (1) 20% zinc phosphate, (2) 10% zinc phosphate and 10% vermiculite, and (3) 20% vermiculite.

The important points in TG curves of the composites are reported in Table 6.2. The onset of mass loss for composites with 20% vermiculate, 10% vermiculate+10%, zinc phosphate, and 20% zinc phosphate were 235°C, 260°C, and 261°C, respectively. The composites with 20% vermiculate and 20% hopeite lost their 20% of mass at 330°C and 325°C, respectively. 20% mass loss occurred at much higher temperature, 352°C for 10% zinc phosphate and 10% vermiculate composite indicating the synergistic effect on thermal degradation of polypropylene. The remaining mass values at 600°C are 22%, 33%, and 45% for 20% zinc phosphate, 10% zinc phosphate, and 10% vermiculate and 20% vermiculate, respectively. The remaining mass consists of filler and carbonized polypropylene. Since the composites had 20% filler, 2%, 13%, and 25% mass of the composite are carbonized as solid and was not gasified. 78%, 67%, and 55% of the composites were volatilized as smoke. Vermiculate had the highest smoke suppression capacity since only 55% of the composite was gasified.

6.3.3.2 ACTIVATION ENERGY OF PP THERMAL DEGRADATION

The effect of fillers on the rate of degradation of polypropylene could be determined by considering the activation energy (E). The activation energy of polypropylene degradation in polypropylene, masterbatch, and the composites prepared in the present study was determined by using Broido's method. The $\ln\ln(1/\alpha)$ versus $1/T$ was plotted using TG data given in Figures 6.6 and 6.14.

α is the fractional decomposition at any time and is given by:

$$\alpha = (M_t - M_\infty) / (M_o - M_\infty), \quad (6.1)$$

where M_o, M_t, and M_∞ represent the initial mass, residual mass at time t, and the residual mass at the end of degradation respectively. The slope of the linear plots equals E/R according to Broido's equation[32] as given below:

$$\ln\ln(1/\alpha) = -E/RT + \text{constant}, \quad (6.2)$$

where E is the activation energy and T is the absolute temperature.

The activation energy values reported in Table 6.2 were calculated from the slopes of the lines obtained by linear regression with r^2 values of at least 0.98 in Figure 6.15. Polypropylene had the highest activation energy 72.2 kJ/mol for the initial degradation temperature range (265–350°C). The masterbatch also had a similar activation energy (73.2 kJ/mol) with polypropylene. The presence of vermiculate, zinc phosphate, or both in polypropylene did not change significantly activation energy for decomposition of polypropylene, however, they lead the formation of more solid products from polypropylene. The masterbatch does not lead to the formation of carbonized degradation products as the EDX analysis showed presence of very small amount of carbon in the ash of the masterbatch.

6.3.4 LOI VALUES OF POLYPROPYLENE, MASTERBATCH, AND COMPOSITES

LOI values were measured for all the samples in the present study. The commercial masterbatch had LOI value as 22% oxygen, indicating that it will not burn in ambient air having 21% oxygen. On the other hand, polypropylene which has LOI value of 19% oxygen will burn in ambient air. Adding solid fillers in the present study did not have a significant effect on LOI value. The composites with 20% vermiculate, and 10%

vermiculate and 10% zinc phosphate had the same LOI value with polypropylene, 19% and composite with 20% zinc phosphate had a lower LOI value than polypropylene, 18%. The additives vermiculate and zinc phosphate at this level did not change the flammability of polypropylene, however composite with 20% vermiculate, and 10% vermiculate and 10% zinc phosphate are smoke suppressants. They caused more carbonization of polypropylene than producing gaseous products.

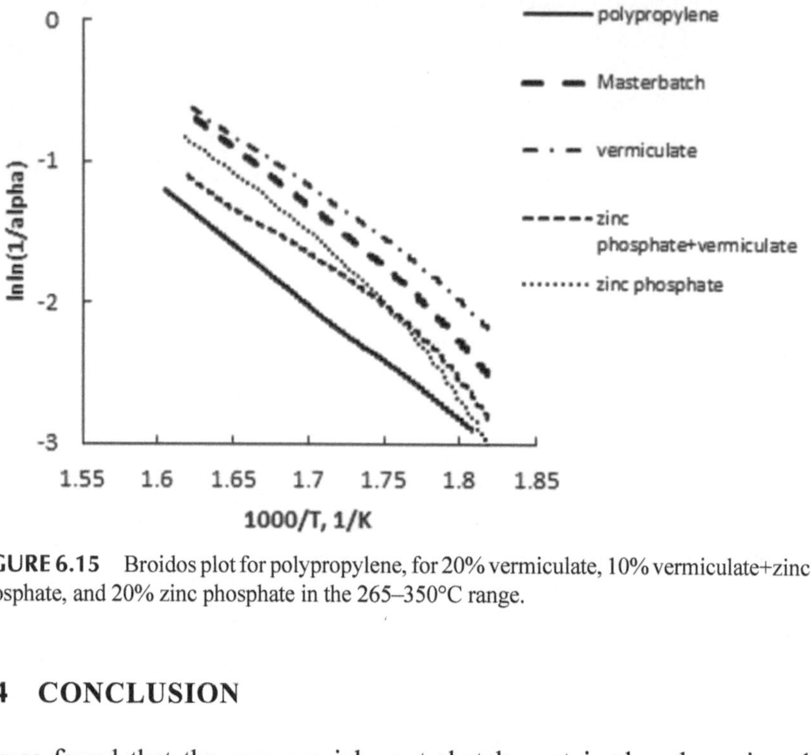

FIGURE 6.15 Broidos plot for polypropylene, for 20% vermiculate, 10% vermiculate+zinc phosphate, and 20% zinc phosphate in the 265–350°C range.

6.4 CONCLUSION

It was found that the commercial masterbatch contained a clay mineral most probably montmorrilonite and a phosphorous compound. Thus a clay, vermiculite and a phosphorous compound zinc phosphate were coated with stearic acid and were added to polypropylene. Composites without any additives, with 20% vermiculite, 10% vermiculite and 10% zinc phosphate and 20% zinc phosphate were prepared. They were mixed at 180°C in a Haake Rheomixer and pressed in Carver press into plates. The thermal stability of the composites was determined by TGA measurement and their flammability and combustion characteristics were examined by LOI.

It was observed that the composite having 20% zinc phosphate had the highest degradation onset temperature (261°C). The mass loss at 600°C was 79% for commercial masterbatch, 78% for 20% zinc phosphate, 67% for 10% zinc phosphate and 10% vermiculite and 67% for 20% vermiculite composite. It was understood that 20% vermiculite increased the thermal stability to highest extent and formed less gaseous and more solid degradation products. The LOI was 22% for flame retardent masterbatch, 19% for polypropylene and composites having vermiculite and 18% for composite with zinc phosphate. In conclusion the composites prepared were not as succesfull as commercial masterbatch in flame redardency, however the composite with 20% vermiculite had higher smoke suppression capability and the composite with 10% vermiculite and 10% zinc phosphate had the highest temperature(352°C) for 20% decomposition.

Flame retardant formulation we developed by simulating the inorganic content of the masterbatch was not successful in the present work. Using much lower nanosized inorganic content with an IFR composition with APP and pentaerytritol would be more effective. Since the same polypropylene was made to have 42% LOI by using 1–10% surface modified natural zeolite in intumescent formulation,[12] further work using vermiculate and/ or zinc phosphate as alternative inorganic additive in intumescent system would be necessary.

KEYWORDS

- **master batch**
- **polypopylene**
- **flame retardant**
- **vermiculite**
- **zinc phosphate**

REFERENCES

1. Mishra, S.; Sonawane, S. H.; Singh R. P.; Bendde A.; Patil, K. Effect of Nano-$Mg(OH)_2$ on the Mechanical and Flame-retarding Properties of Polypropylene Composites. *J. Appl. Polym. Sci.* **2004,** *94* (1), 116–122.

2. Ma, Z. L., Fan, C. R.; Lu, G. Y.; Liu, X. Y.; Zhang, H. Synergy of Magnesium and Calcium oxides in Intumescent Flame-retarded Polypropylene. *J. Appl. Polym. Sci.* **2012,** *125* (5), 3567–3574.
3. Shen, L.; Chen, Y. H. Synergistic Catalysis Effects of Lanthanum oxide in Polypropylene/Magnesium Hydroxide Flame Retarded System. *Comp. Part a Appl. Sci. Manuf.* **2012,** *43* (8), 1177–1186.
4. Dogan, M.; Yilmaz. A.; Bayramli, E. Synergistic Effect of Boron Containing Substances on Flame Retardancy and Thermal Stability of Intumescent Polypropylene Composites. *Polym. Degr. Stab.* **2010,** *95* (12), 2584–2588.
5. Yi, D. Q.; Yang, R. J. Ammonium Polyphosphate/Montmorillonite Nanocompounds in Polypropylene. *J. Appl. Polym. Sci.* **2010,** *118* (2), 834–840.
6. Du, B. X.; Guo, Z. H., Song, P.; Liu, H.; Fang, Z. Flame Retardant Mechanism of Organo-bentonite in Polypropylene. *Appl. Clay Sci.* **2009,** *45* (3), 178–184.
7. Wang, J. J.; Wang, L.; Xiao, A. Recent Research Progress on the Flame-Retardant Mechanism of Halogen-Free Flame Retardant Polypropylene. *Polym. Plast. Technol. Eng.* **2009,** *48* (3), 297–302.
8. Zhang, S.; A. Richard Horrocks, A. R.; Hull, R.; Kandola, B. K. Flammability, Degradation and Structural Characterization of Fibre-forming Polypropylene Containing Nanoclay–flame Retardant Combinations. *Polym. Degr. Stab.* **2006,** *91* (4), 719–725.
9. Qin, H. L., Zhang, S. M.; Zhao, C.; Hu, G.; Yang, M. Flame Retardant Mechanism of Polymer/Clay Nanocomposites Based on Polypropylene. *Polymer* **2005,** *46* (19), 8386–8395.
10. Courtat, J.; Flavien Melis, F.; Taulemesse, J. M.; Bounor-Legare, V.; Sonnier, R.; Ferry, L.; Cassagnau, P. Effect of Phosphorous-modified Silica on the Flame Retardancy of Polypropylene Based Nanocomposites. *Polym. Degr. Stab.* **2015,** *119*, 260–274.
11. Pappalardo, S.; Russo, P.; Acierno, D.; Rabe, S.; Schartel, B. The Snergistic Effect of Organically Modified Sepiolite in Intumescent Flame Retardant Polypropylene. *Eur. Polym. J.* **2016,** *76*, 196–207.
12. Demir, H.; Arkis, E.; Balkose, D.; Ulku, S. Synergistic Effect of Natural Zeolites on Flame Retardant Additives. *Polym. Degrad. Stab.* **2005,** *89* (3), 478–483.
13. Gomes, E. V. D.; Visconte, L. L. Y.; Pacheco, E. B. A. V. Morphological. Thermal and Mechanical Properties of Polypropylene and Vermiculite Blends. *Int. J. Polym. Mater.* **2008,** *57* (10), 957–968.
14. Gomes, E. V. D.; Visconte, L. L. Y.; Pacheco, E. B. A. V. Thermal Characterization of Polypropylene/Vermiculite Composites. *J. Therm. Anal. Calor.* **2009,** *97* (2), 571–575.
15. Tjong, S. C.; Meng, S. C. Y. Z. Meng; Hay, A.S. Novel Preparation and Properties of Polypropylene-Vermiculite Nanocomposites. *Chem. Mater.* **2002,** *14* (1), 44–51.
16. Chen, S. H.; Wang, B.; Chen, S.; Wang, B.; Jıan Kang, J.; Chen, J.; Gai, J.; Yang, L.; Cao, Y. Synergistic Effect of Organic Vermiculite on the Flame Retardancy and Thermal Stability of Intumescent Polypropylene Composites. *J. Macromol. Sci. Part B Phys.* **2013,** *52* (9), 1212–1225.
17. Ren, Q. A.; Zhang, Y.; Li, J.; Jin Chun Li, J. C. Synergistic Effect of Vermiculite on the Intumescent Flame Retardance of Polypropylene. *J. Appl. Polym. Sci.* **2011,** *120* (2),1225–1233.

18. Nevare, M. R.; Gite, V. V., Pramod, P.; Mahulikar, P. P.; Ahamad, A.; Rajput, S. D. Synergism Between LDH and Nano-Zinc Phosphate on the Flammability and Mechanical Properties of Polypropylene. *Polym. Plast. Technol. Eng.* **2014,** *53* (5), 429–434.
19. Erdogdu, C. A.; Atakul, S.; Balkose, D.; Ulku, S. Development of Synergistic Heat Stabilizers for PVC from Zinc Borate-Zinc Phosphate. *Chem. Eng. Commun.* **2009,** *196* (1–2), 148–160.
20. Ramos, J. H.; Ploux, L.; Anselme K.; Balan, L.; Simon-Masseron, A. Hydrothermal Synthesis and Characterization of Bio-Sourced Macroporous Zinc Phosphates Prepared with Casein Protein. *Cryst. Growth Des.* **2016,** *16* (9), 4897–4904.
21. Onoda, H.; Haruki, M.; Toyama T. Preparation and Powder Properties of Zinc Phosphates with Additives. *Cer. Int.* **2014,** *40* (2), 3433–3438.
22. Onoda, H.; Haruki, M. Influence of Phosphate Source on Preparation of Zinc Phosphate White Pigments. *Int. J. Ind. Chem.* **2016,** *7* (3), 309–314.
23. Türkçü, H. N. Investigation of the Crystallinity and Orientation of Polypropylene with Respect to Temperature Changes Using FT-IR, XRD, and Raman Techniques, Master Thesis, Bilkent University, Ankara, Turkey, August 2004.
24. Ulku, S.; Balkose, D.; Arkis, E. A Study of Chemical and Physical Changes During Biaxially Oriented Polypropylene Film Production. *J. Polym. Eng.* **2003,** *23* (6), 437–456.
25. Marcos, C.; I. Rodriguez, I. Structural Changes on Vermiculite Treated with Methanol and Ethanol and Subsequent Microwave İrradiation. *Appl. Clay Sci.* **2016,** *123*, 304–314.
26. Frost, R. L. An İnfrared and Raman Spectroscopic Study of Natural Zinc Phosphates. *Spectrochim. Acta Part A Mol. Biomol. Spectrosc.* **2004,** *60* (7), 1439–1445.
27. Hillier, S.; Marwa, E. M. M.; Rice, C. M. On the Mechanism of Exfoliation of 'Vermiculite'. *Clay Miner.* **2013,** *48* (4), 563–582.
28. de la Calle, C.; Pezerat, H.; Gasperin, M. Problemes d'ordre-desordre dans les vermiculites structure du mineral calcique hydrate a 2 couches. *J. Phys.* **1977,** *38*, 128–133.
29. https://www.mindat.org/min-1999.html. (accessed Feb 13, 2018).
30. Herschke, L.; Enkelmann, V.; Lieberwirth, I.; Wenger, G. The Role of Hydrogen Bonding in the Crystal Structures of Zinc Phosphate Hydrates. Note: Beta-Hopeite. *Chem. Eur. J.* **2004,** *10*, 2795–2803.
31. https://www.unamur.be/services/microscopie/sme-documents/Energy-20table-20for-20EDS-20analysis-1.pdf. (accessed Feb 13, 2018).
32. Broido, A. A Simple, Sensitive Graphical Method of Treating Thermogravimetric Analysis Data. *J. Polym. Sci. Part A-2* **1969,** *7*, 1761–1773.

CHAPTER 7

INFLUENCE OF FRUCTOSE IN THE DIFFUSION OF SODIUM BORATE IN AQUEOUS SOLUTIONS AT 298.15 K

LUÍS M. P. VERÍSSIMO[1], ANA C. F. RIBEIRO[1,*],
DANIELA F. S. L. RODRIGUES[1], ANA M. T. D. P. V. CABRAL[2],
and M. A. ESTESO[3,*]

[1]*Department of Chemistry, Coimbra Chemistry Centre, University of Coimbra, 3004-535 Coimbra, Portugal*

[2]*Faculty of Pharmacy, University of Coimbra, 3000-295 Coimbra, Portugal*

[3]*U.D. Química Física, Universidad de Alcalá, 28871 Alcalá de Henares, Spain*

[*]*Corresponding author. E-mail: anacfrib@ci.uc.pt and miguel.esteso@uah.es*

ABSTRACT

Mutual diffusion coefficients have been measured at 298.15 K for aqueous systems containing sodium borate at concentration 0.100 mol dm^{-3} and fructose at various concentrations (from 0.001 to 0.100 mol dm^{-3}), by using a conductimetric Lobo's cell. From these data, we can say that the presence of fructose affects the diffusion of sodium borate. Two effects can explain this behavior: the ionic association and the salting-out effect.

7.1 INTRODUCTION

Fructose is a monosaccharide found in nature in fruits and honey. However, the exponential growth in the use of high fructose corn syrup (HFCS) as a sweetener in processed foods, appeared on the focus of recent scientific literature as an established cause of the obesity epidemics and a growing number of fructose-correlated diseases and conditions.[1]

Recent research shows that borate complexed with fructose provides an alternative, time-saving, and specific method for serum fructose determination.[2] However, studies are still needed to better model the ionic behavior in the presence of fructose. In order to establish new mobility data, mutual diffusion coefficients for sodium borate in aqueous solutions (0.100 mol dm^{-3}) containing fructose at various concentrations (from 0.001 to 0.100 mol dm^{-3}) have been measured at 298.15 K, by using a conductimetric Lobo's cell coupled to an automatic data acquisition system to follow the diffusion.[3–18]

This apparatus proved itself to be very useful when studying interactions between an electrolyte and non-electrolytes, as it can measure the whole system transient response using a pseudo-binary approach while maintaining the concentration changes of the non-electrolyte component inconsequent to the conductimetric measure.

The analysis of the diffusion behavior of the system fructose-sodium borate should prove to be useful in modeling metabolic syndrome mechanisms and other fructose-related diseases and conditions recently reported.

7.2 DETERMINATION OF MUTUAL DIFFUSION COEFFICIENT BY CONDUCTIMETRIC METHOD

Being the main goal of this work the determination of mutual diffusion coefficients for aqueous systems, and from some confusion in the literature about this concept, it is important to refer some notes having in mind its clarification. That is, there are two processes of diffusion very distinguished, that is, self-diffusion D^* (also named as intradiffusion, tracer diffusion, single ion diffusion, or ionic diffusion) and mutual diffusion D (also known as interdiffusion, concentration diffusion, or salt diffusion).[19–22] In this work, we refer to the mutual diffusion D. This phenomenon, denominated by isothermal diffusion, is an irreversible process results of the gradient of concentration inside a solution (without convection or migration) at constant temperature, producing a flow of matter in the opposite direction, which arises from random fluctuations in the positions of molecules in the space.

The diffusion coefficient, *D,* in a binary system (i.e., with two independent components) may be defined in terms of the concentration gradient by phenomenological equations, known as Fick's first and second laws (eqs 7.1 and 7.2).[19–22]

$$J_i = -D_F\left(\frac{\partial c_i}{\partial x}\right) \tag{7.1}$$

$$\frac{\partial c}{\partial t} = \frac{\partial}{\partial x}\left(D_F \frac{\partial c}{\partial x}\right), \tag{7.2}$$

where J represents the flow of matter of component i across a suitably chosen reference plane, per unit area and per unit time, in a one-dimensional system, c is the concentration of the solute in moles per unit volume at the point considered, and D_F is the Fikian coefficient diffusion.

Really, the gradient of chemical potential in the real solution must be considered as the true virtual force producing diffusion and not the concentration gradient

$$J_i = -D_T\left(\frac{\partial \mu_i}{\partial X}\right), \tag{7.3}$$

where μ_1 represents the chemical potential of the i component defined by eq 7.4.

$$\mu_i = \mu_i^0 + \mathrm{RT}\ln f_i c_i, \tag{7.4}$$

μ_i^0 and f_i being the chemical potential at zero concentration and the activity coefficient of component i, respectively.

Combining eqs 7.3 and 7.4, and comparing the final result with the eq 7.1, we obtain the relation between D_F and $D_{T,}$ given by eq 7.5.

$$D_F = D_T\left(1 + \frac{\partial \ln f_i}{\partial \ln ci}\right) \tag{7.5}$$

This parameter, D_F, is not a pure kinetic parameter, because it depends on two contributions: a kinetic (D_T) and a thermodynamic ($\partial\mu/\partial c$).[19–22] In other words, two different effects can control the diffusion process: the ionic mobility and the gradient of free energy.

However, as an approach, we can assume that the variation of the activity coefficient is not significant for the difference of concentration responsible for the diffusion (eq 7.5), and, consequently, that for all practical purposes, D is a constant (the thermodynamic diffusion coefficient equals the Fikian diffusion coefficient) (eq 7.6).

$$D_F = D_T. \quad (7.6)$$

Thus, it is possible to describe the isothermal diffusion from two different approaches: the thermodynamics of irreversible processes and the Fick's laws (eqs 7.1 and 7.2).[19–22]

Based on previous studies,[12,14,15,17] our group has now focused the study of the diffusion behavior of this system (Borate, Fructose, and water) as a pseudo-binary one. Pseudo-binary systems are actually ternary systems for which the cross-diffusion coefficients, D_{12} and D_{21}, are negligible and can be disregarded. As a consequence, it is only necessary to measure the main diffusion coefficient D_{11} (which can be considered as binary diffusion coefficients, D) to characterize the diffusion process of it. Thus, from the experimental conditions, we may consider all those systems as pseudo-binary ones, and, consequently, take the measured parameters as the corresponding binary diffusion coefficients, D.

7.3 EXPERIMENTAL

7.3.1 REAGENTS

Fructose Tetraborate ($Na_2B_4O_7$ (*Riedel-de Häen*, > 99%) and D(-)fructose (*Baker*, 99%) were used as received without further purification. Work solutions were prepared in calibrated volumetric flasks using ultrapure water (*Millipore*, 18.2 Mohm/cm at 298.15 K). They were freshly prepared and de-aerated for about 30 min before each set of runs.

7.3.2 SUMMARY DESCRIPTION OF THE CONDUCTIMETRIC TECHNIQUE

In order to establish new mobility data, mutual diffusion coefficients for sodium borate in aqueous solutions (0.100 mol dm^{-3}) containing fructose at various concentrations (from 0.001 to 0.100 mol dm^{-3}) were measured at 298.15 K, by using a conductimetric cell (Lobo's cell) coupled to an automatic data acquisition system to follow the diffusion (Fig. 7.1).

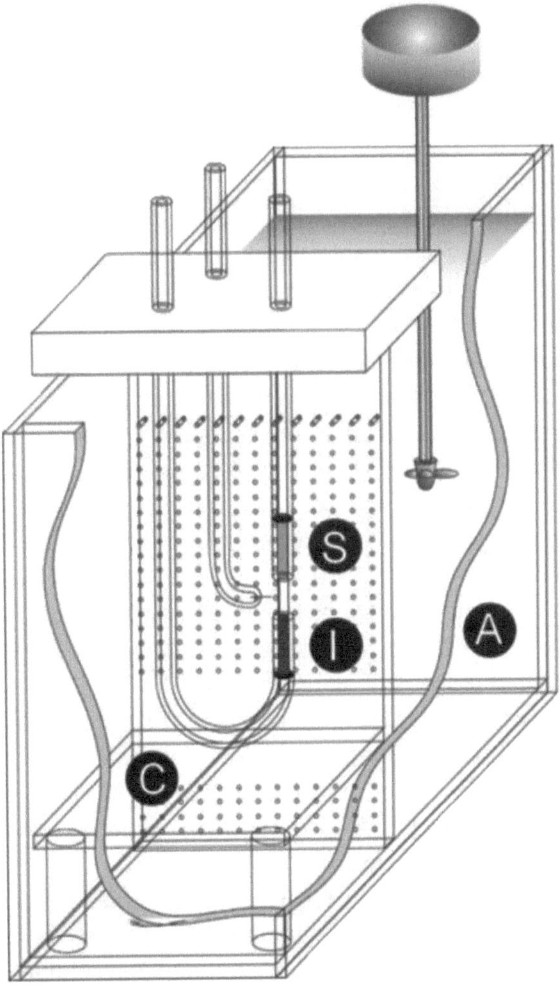

FIGURE 7.1 Lobo's conductimetric cell layout.[3,4] A—Glass tank with stirrer with "bulk" solution (Fructose [*c*] + 0.1 mol.dm^{-3} $Na_2B_4O_7$); C—perspex fitted flow grids; S—"Top" capillary with (Fructose [c] + 0.075 mol.dm^{-3} $Na_2B_4O_7$), and I—"Bottom" capillary with (Fructose[*c*] + 0.125 mol.dm^{-3} $Na_2B_4O_7$).

This apparatus has proved itself to be very useful. When studying interactions of an electrolyte with a non-electrolyte, the whole system transient response can be measured by using a pseudo-binary approach, while maintaining the non-electrolyte component inconsequent to the conductimetric measure itself.[3,4]

The cell has two vertical capillaries, each closed at one end by a platinum electrode and positioned one above the other with the open ends faced and separated by a distance of about 14 mm.

The upper and lower tubes, initially filled with solutions of concentrations 0.75 *c* and 1.25 *c*, respectively, were surrounded by a solution of *c* concentration. This ambient solution was contained in a glass tank (200 × 140 × 60) mm immersed in a thermostat at 298.15 K. The tank was divided internally by Perspex sheets and a glass stirrer created a slow lateral flow of ambient solution across the open ends of the capillaries. Experimental conditions were such that the concentration at each of the open ends was equal to the ambient solution value *c*, that is the physical length of the capillary tube coincided with the diffusion path, such that the boundary conditions described in the literature[3,4] to solve Fick's second law of diffusion are applicable. As a consequence, the so-called Δl-effect is reduced to negligible proportions. In contrast to a manual apparatus, where diffusion is followed by measuring the ratio of resistances of the top and bottom tubes, $w = Rt/Rb$, by an alternating current transformer bridge, in our automatic apparatus w was measured by a Solartron digital voltmeter (DVM) 7061 with 6 1/2 digits. A Bradley Electronics Model 232 power source supplied a 30 V sinusoidal signal of 4 kHz (stable up to 0.1 mV) to a potential divider that applied a 250 mV signal to the platinum electrodes at the top and bottom capillaries. By rapidly (< 1 s) measuring the voltages V′ and V″ from top and bottom electrodes with respect to the central electrode at ground potential, the value $w = Rt/Rb$ was then calculated from the DVM reading.

To measure the differential diffusion coefficient, *D*, at a given concentration *c*, a "top" solution of concentration 0.75 *c* and a "bottom" solution of concentration 1.25 *c* were prepared, each in a 2 L volumetric flask. The "bulk" solution of concentration *c* was produced by mixing accurately measured volumes of 1 L of "top" solution with 1 L of "bottom" solution. The glass tank and the two capillaries were filled with solution *c*, immersed in the thermostat, and were allowed to come to thermal equilibrium. The quantity $TR_{inf} = 10^4/(1 + w)$ was now measured very accurately. Here $w = Rt/Rb$ is the electrical resistance (R) ratio for solutions of concentration *c* of the top (t) and bottom (b) diffusion capillaries at infinite time. $TR = 10^4/(1 + w)$ is the equivalent at any time *t*.

The capillaries were then filled with "top" and "bottom" solutions, which were allowed to diffuse into the "bulk" solution. Resistance ratio readings were taken at recorded times, beginning 1000 min after the start of each experiment. The diffusion coefficient was evaluated using a linear least-squares procedure to fit the data, followed by an iterative process which uses 20 terms of the expansion series of the solution of Fick's second law for the present boundary conditions. The theory developed for this cell has been described previously.[3,4]

7.4 RESULTS AND DISCUSSION

Table 7.1 shows the diffusion coefficients, *D*, for each concentration, of the mutual diffusion coefficients, *D*, for 0.100 mol.dm^{-3} $Na_2B_4O_7$ aqueous solutions and fructose at various concentrations (c) at 298.15 K, as well as their standard deviation values. These *D* values are the mean ones of, at least, three independent measurements. On the basis of previous papers reporting data obtained with this technique[3–17] we believe that our uncertainty is not larger than (1–3%).

TABLE 7.1 Mutual Diffusion Coefficients, *D*, for Aqueous Solutions of 0.100 mol dm^{-3} $Na_2B_4O_7$ (1) and Fructose (2) at Various Concentrations (*C*), at $T = 298.15$ K and $P = 101.3$ kPa.

c_2 (mol kg^{-1})	$(D \pm S_D)^a$ (10^{-9} m^2 s^{-1})	$\Delta D/D$ %^{b}
0.0000	1.133 ± 0.011	–
0.0010	0.810 ± 0.048	−28.6
0.0100	0.842 ± 0.009	−25.7
0.0200	0.863 ± 0.019	−23.8
0.0350	0.860 ± 0.005	−24.1
0.0500	0.796 ± 0.037	−29.7
0.0750	0.858 ± 0.012	−24.3
0.1000	1.083 ± 0.013	−4.4

Standard uncertainties, u, are: $u(c) = 0.03$ mol dm^{-3}, $u(T) = 0.01$ K, and $u(p) = 2.03$ kPa.

[a]*D* is the diffusion coefficient obtained as the mean value of, at least, three independent experiments; S_D is the standard deviation of that mean.

[b]Deviations between our *D* values, here indicated, and the experimental diffusion coefficient ones of borate in aqueous solutions at $c = 0.100$ mol dm^{-3}.

Figure 7.2 shows the distribution results which suggests the parted model here proposed.

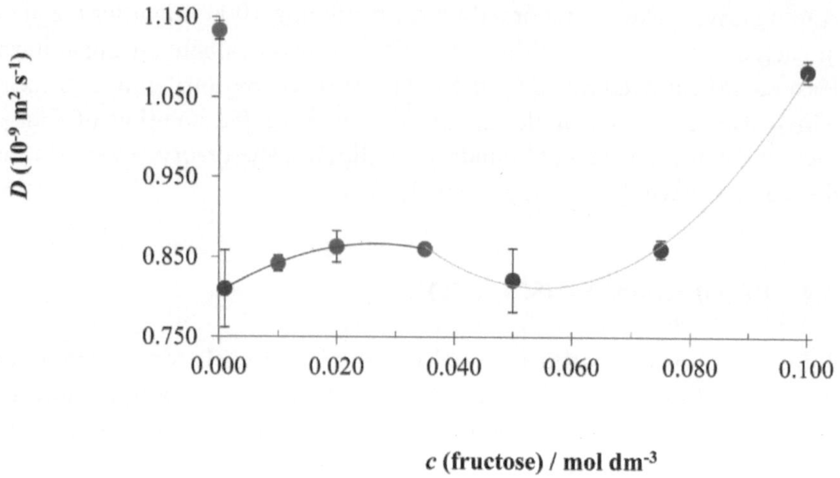

FIGURE 7.2 Behavior of the diffusion coefficients, *D*, for aqueous solutions of 0.100 mol.dm^{-3} $Na_2B_4O_7$ (1) and fructose (2) at various concentrations (*c*) at 298.15 K.

Under the present experimental conditions, that is, $[Na_2B_4O_7]$/[fructose] ratio ≥ 1 and dilute solutions, the motion of the solvent and the change with concentration of parameters such as viscosity, dielectric constant, and hydration degree can be neglected.

For [fructose] ≤ 0.035 as well as for [fructose] > 0.035 mol.dm^{-3}, the following polynomial equations in *c* (eqs 7.7 and 7.8, respectively) were used to fit the data by a least squares method. That is,

$$D\,(10^{-9}\ \text{m}^2\ \text{s}^{-1}) = -8.623 \times 10^{-4}\, c^2 + 4.590 \times 10^{-5}\, c + 8.054 \times 10^{-6} \quad (7.7)$$

$(R^2 = 0.999)$ and

$$D\,(10^{-9}\ \text{m}^2\ \text{s}^{-1}) = 1.3248 \times 10^{-3}\, c^2 + 1.451 \times 10^{-4}\, c + 1.209 \times 10^{-5} \quad (7.8)$$

$(R^2 = 0.9981)$.

The goodness of the fit (obtained with a confidence interval of 98%) can be assessed by the excellent correlation coefficients, R^2, and the low standard deviation (< 1%) found.

By comparison of these results (Table 7.1) with the value obtained for solutions of 0.100 mol dm^{-3} $Na_2B_4O_7$ at the same temperature and with the same technique, it is evident that the diffusion behavior of sodium borate in aqueous solutions at 298.15 K is affected by the presence of the fructose molecules. In general, these values are lower than the value obtained for 0.100 mol dm^{-3} $Na_2B_4O_7$ aqueous solutions (deviations between 24% and 29%; Table 7.1). This behavior may be explained by the eventual presence of new different species resulting from the short-range electrostatic interactions between $Na_2B_4O_7$ and fructose which, due to their size, have a lower mobility than both the cation Na^+ and the anion $B_4O_7^{2-}$.[19,23,24] From our data and considering the eqs 7.5 and 7.6, we can say that the gradient of chemical potential also decreases.

However, when the concentration of fructose increases, approaching the concentration of $Na_2B_4O_7$ (that is, c = 0.100 mol dm^{-3}), it is observed a less pronounced deviations between the diffusion coefficients of the aqueous systems studied (fructose + 0.100 mol dm^{-3} $Na_2B_4O_7$) and the diffusion coefficient for 0.100 mol dm^{-3} $Na_2B_4O_7$ aqueous solutions. This fact can be interpreted on the basis of a salting-out effect. This effect plays an important role, as it is shown by the higher diffusion coefficient values observed, and leads to conclude that the presence of fructose contributes to what the interactions between sodium borate and water molecules are not too much favored.

At $c_{fructose}$ = 0.100 mol dm^{-3}, we can say that fructose does not affect the diffusion of the $Na_2B_4O_7$, once their diffusion coefficient values are practically equals.

7.5 CONCLUSIONS

Based on these measurements of diffusion coefficients of systems containing sodium borate and fructose in aqueous solutions, and assuming that in the present experimental conditions this system can be considered as pseudo-binary, we conclude that the diffusion of this electrolyte is strongly affected by the presence of fructose, probably due to the appearance of new different species resulting from various equilibria (e.g., aggregation). In fact, the pseudo diffusion coefficients, D, are not identical to the binary diffusion coefficients of aqueous sodium borate, and, in these circumstances, this fact suggests that there are interacting solutes. However, when the added fructose increases, these sodium borate pseudo-diffusion coefficients also increase, which leads to conclude that these facts are due to a salting-out effect.

The analysis of the fructose-shaped sodium borate diffusion coefficient values obtained should be useful in modeling metabolic mechanisms in fructose-related diseases and conditions recently reported.

ACKNOWLEDGMENTS

The authors are grateful for funding from "The Coimbra Chemistry Centre" which is supported by the Fundação para a Ciência e a Tecnologia (FCT), Portuguese Agency for Scientific Research, through the projects UID/QUI/UI0313/2013 and COMPETE.

KEYWORDS

- **sodium borate**
- **fructose**
- **diffusion**
- **transport properties**
- **solutions**
- **taylor technique**

REFERENCES

1. Mastrocola, R.; Nigro, D.; Cento, A. S.; Chiazza, F.; Collino, M.; Aragno, M. High-Fructose Intake as Risk Factor for Neurodegeneration *Neurobiol. Dis. Elsevier* **2016**. DOI: 10.1016/j.nbd.2016.02.005.
2. Cheng, C. Y.; Liao, C. I,; Lin, S. F. Borate–fructose Complex: A Novel Mediator for Laccase and its New Function for Fructose Determination. *FEBS Lett.* **2015,** *589*, 3107–3112. DOI:10.1016/j.febslet.2015.08.032.
3. Agar, J. N.; Lobo, V. M. M. Measurement of Diffusion Coefficients of Electrolytes by a Modified Open-Ended Capillary Method. *J. Chem. Soc. Faraday Trans.* **1975,** *1* (71), 1659–167013.
4. Lobo, V. M. M. Mutual Diffusion Coefficients in Aqueous Electrolyte Solutions. *Pure Appl. Chem.* **1993,** *65*, 2613–2640.
5. Lobo, V. M. M.; Ribeiro, A. C. F.; Verissimo, L. M. P. Diffusion Coefficients in Aqueous Solutions of Magnesium Nitrate at 298 K. *Ber Bunsenges Phys. Chem.* **1994,** *98*, 205–208.
6. Lobo, V. M. M.; Ribeiro, A. C. F.; Verissimo, L. M. P. Diffusion Coefficients in Aqueous Solutions of Beryllium Sulphate at 298 K. *J. Chem. Eng. Data* **1994,** *39*, 726–728.
7. Lobo, V. M. M.; Ribeiro, A. C. F.; Valente, A. J. M. Célula de difusão condutimétrica de capilares abertos - Uma análise do método. *Corros. Prot. Mat.* **1995,** *14*, 14–21.

8. Lobo, V. M. M.; Ribeiro, A. C. F.; Andrade, S. G. C. S. Diffusion Coefficients in Aqueous Solutions of Divalent Electrolytes. *Ber. Buns. Phys. Chem.* **1995,** *99,* 713–720.
9. Lobo, V. M. M.; Ribeiro, A. C. F.; Verissimo, L. M. P. Diffusion Coefficients in Aqueous Solutions of Potassium Chloride at High and Low Concentrations. *J. Mol. Liq.* **1998,** *78,* 139–149.
10. Lobo, V. M. M.; Ribeiro, A. C. F.; Natividade, J. J. S. Diffusion Coefficients in Aqueous Solutions of Potassium Thiocyanate at 298.15 K. *J. Mol. Liq.* **2001,** *94,* 61–66.
11. Ribeiro, A. C. F.; Lobo, V. M. M.; Azevedo, E. F. G. Diffusion Coefficients of Ammonium Monovanadate in Aqueous Solutions at 25 °C. *J. Solut. Chem.* **2001,** *30,* 1111–1115.
12. Lobo, V. M. M.; Ribeiro, A. C. F.; Azevedo, E.F.G.; Burrows, M. M. G. H. D. Diffusion Coefficients of Sodium Dodecylsulfate in Aqueous Solutions and in Aqueous Solutions of Sucrose. *J. Mol. Liq.* **2001,** *94,* 193–201.
13. Ribeiro, A. C. F.; Lobo, V. M. M.; Natividade, J. S. S. Diffusion Coefficients in Aqueous Solutions of Cobalt Chloride at 298.15 K. *J. Chem. Eng. Data* **2002,** *47,* 539–541.
14. Ribeiro, A. C. F.; Lobo, V. M. M.; Azevedo, E. F. G.; Miguel, M. da G.; Burrows, H. D. Diffusion Coefficients of Sodium Dodecylsulfate in Aqueous Solutions and in Aqueous Solutions of β-cyclodextrin. *J. Mol. Liq.* **2003,** *102,* 285–292.
15. Ribeiro, A. C. F.; Valente, A. J. M.; Lobo, V. M. M.; Azevedo, E. F. G.; Amado, A. M.; da Costa, A. M. A.; Ramos, M. L.; Burrows, H. D. Interactions of Vanadates with Carbohydrates in Aqueous Solutions. *J. Mol. Struct.* **2004,** *703,* 93–101.
16. Ribeiro, A. C. F.; Lobo, V. M. M.; Oliveira, L. R. C.; Burrows, H. D.; Azevedo, E. F. G.; Fangaia, S. I. G.; Nicolau, P. M. G.; Guerra, F. A. D. R. A. Diffusion Coefficients of Chromium Chloride in Aqueous Solutions at 298.15 K and 303.15 K. *J. Chem. Eng. Data* **2005,** *50,* 1014–1017.
17. Ribeiro, A. C. F.; Esteso, M. A.; Lobo, V. M. M.; Valente, A. J. M.; Simões, S. M. N.; Sobral, A. J. F. N.; Burrows, H. D. Interactions of Copper (II) Chloride with Sucrose, Glucose and Fructose in Aqueous Solutions. *J. Mol.* ***Struct.*** **2007,** *826,* 113–119.
18. Lobo, V. M. M. *Handbook of Electrolyte Solutions*; Elsevier: Amsterdam, 1990.
19. Robinson, R. A.; Stokes, R. H. *Electrolyte Solutions*, 2nd Ed; Butterworths: London, 1959.
20. Harned, H. S.; Owen, B. B. *The Physical Chemistry of Electrolytic Solutions*, 3rd Ed; Reinhold Pub. Corp: New York, 1964.
21. Tyrrell, H. J. V.; Harris, K. R. *Diffusion in Liquids: a Theoretical and Experimental Study,* Butterworths: London, 1984.
22. Bockris, J. O.; Reddy, A. K. N. *Modern Electrochemistry. An Introduction to an Interdisciplinary Area*, Vol 1; 6th printing, Plenum Press: New York, 1977.
23. Burguess. J. *Metal Ions in Solution*; John Wiley & Sons, Chichester: Sussex, England, 1978.
24. Lobo, V. M. M.; Ribeiro, A. C. F. Ionic Association: Ion Pairs. *Port. Electrochim. Acta* **1994,** *12,* 29–41.

PART II

New Insights and Achievements

CHAPTER 8

HEMOLYTIC ASSAY OF BIOCOMPATIBLE NANOMATERIALS IN DRUG DELIVERY SYSTEMS

POONAM KHULLAR*, LAVANYA TANDON, RAJPREET KAUR, and DIVYA MANDIAL

Department of Chemistry, BBK DAV College for Women, Amritsar 143001, Punjab, India

Corresponding author. E-mail: virgo16sep2005@gmail.com

ABSTRACT

The toxicity of nanomaterials in red blood cells (RBCs) is of great interest, as RBCs are important in transporting oxygen in blood circulation. Investigations are being done for the size-dependent toxicity of well-known semiconductor quantum dots (QDs) and have revealed the exact toxic mechanism at the molecular level by confocal microscopy and Fourier-transform infrared spectroscopy techniques. The QDs bind to the RBCs membranes and cause the structural changes of lipid and protein in RBCs. But only the red-emitting QDs cause the breakage of the phosphodiester bond, which may cause the heavy hemolysis. The cell membrane-coated nanoparticle (NP) provides biomimetic platform which consist of a nanoparticulate core coated with membrane, which are derived from a cell, such as a RBC, platelet, or cancer cell. The cell membrane allows the particles to be perceived by the body as the source cell through interacting them with its surroundings using the translocate surface membrane components. The newly bestowed characteristics of the membrane-coated NP and can be utilized for biological interfacing in the body, providing natural solutions to many biomedical issues.

8.1 INTRODUCTION

Nanotechnology is opening new ways to solve the problems occurring in the biomedical field because of properties of nanoparticles (NPs) like high drug loading and controlled release,[1] delivery of hydrophobic drugs,[2] access to multiple endocytic routes,[3] and passive accumulation in certain organs on the basis of their size, shape, and surface charge.[4,5] Due to these characteristics, NPs are being used in various applications like drug delivery, [6,7] vaccination, [8–10] gene delivery,[11] and antimicrobial purposes.[12] NPs can be externally decorated and thus bioavailability of the encapsulated drugs is increased without modifying them. Because of their surface functionalization, the particles are being specifically targeted to the disease sites like tumor drug delivery,[21,22] toxin removal, [23–25] and vaccination.[26–28]

Biomimetic NPs are being studied so as to facilitate the development of therapeutics for understanding biologically complex applications. Surface functionalization is a bottom-up technique in which moieties are incorporated into the particle surface through the chemical conjugation or noncovalent binding.[29] These are singular strategies which exploit biological interactions. Challenging applications include highly specific targeted drug delivery, secreting unknown toxins, autoimmune diseases, and cancer immunotherapy. Biomimetic nanoengineering particles deal with cell membrane-coated nanoparticle, consisting of a core material coated with membrane which is derived from a source cell.[30] Red blood cells (RBCs) are the first cell membrane-coated nanoparticle which is being used as the source cell. Through hypnotic treatment, the RBC membrane is derived and is made to coat onto the negatively charged polymeric NPs by extrusion. Wide expansion has occurred in field of cell membrane coating of nucleated cells, which can be separated from nuclear and mitochondrial components by using the a sucrose gradient[31] or by differential centrifugation,[32] sonication method[33] is also used to coat particles. This method is used to fully coat the particles and it is believed to occur through the asymmetric charge of the cell membranes due to which the particles get coated in such a manner that the charge repulsions are minimized. Cell membrane coating is a top-down method in which the cell membrane coating is able to retain the cell surface on the nanoparticle. These particles function well in the body and appear as the source cell by using the multiple interactions occurring between the cell membrane and its substrates. The cell membrane-coated NPs are being used for the following:

i) delivery of drug
ii) detoxification
iii) modulation of immune

It is shown in Figure 8.1.

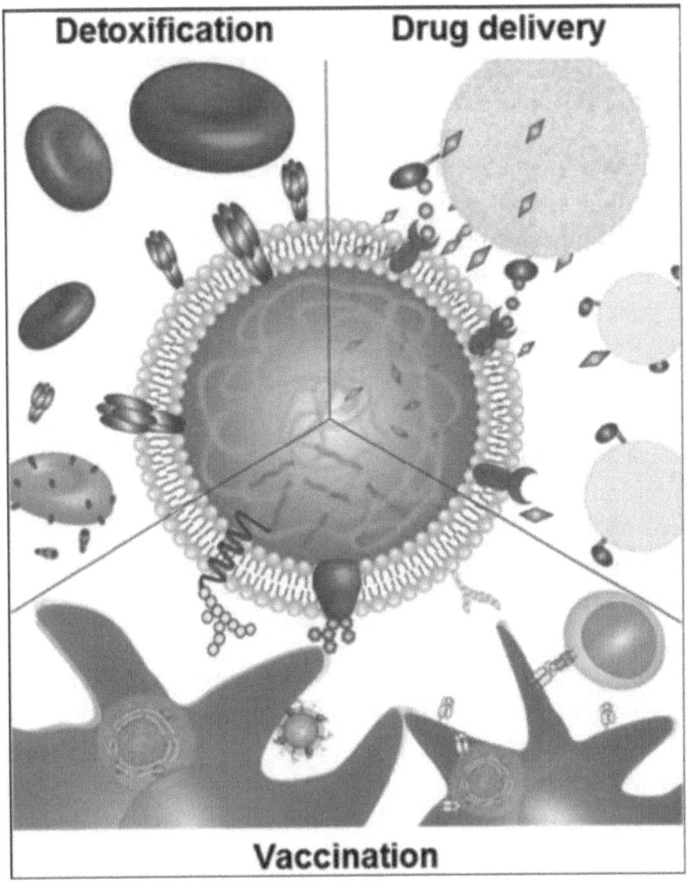

FIGURE 8.1 (See color insert.) Illustration showing three applications of cell membrane-coated NPs. The particles can be used for targeted drug delivery to tumors, sites of inflammation, or pathogens via translocate surface markers (top right). They can also act as a decoy for toxins that damage cells through membrane interactions, safely detaining them and sparing their intended targets (top left). Finally, cell membrane-coated NPs faithfully present antigens from their source cells and can be used for improved anticancer or antibacterial vaccination (bottom).

The foreign substances which are introduced into the body are captured and excreted by the immune system and the body is further trained for the rapid clearance of the foreign material on their reintroduction.

Great efforts have been made to develop polymeric NPs as desirable drug delivery systems (DDSs) for their attractive characteristics like longer circulation time, targeted drug delivery, protection from enzymatic degradation, and reduced drug toxicity or side effects.[34]

8.2 NONSPECIFIC DRUG DELIVERY AND STEALTH COATING

One of the major problems for in vivo applications of NPs is the clearance of rapid immune system clearance. The immune system has been trained to capture the foreign substance and excrete them from the body and further the body is trained for the quick clearance of impurities from the body. The PEGylated NPs are able to escape uptake by phagocytic cells without the adsorption of opsonins on the surface. RBC membrane-coated NPs are used to improve the circulation half-life of NPs. Naturally, RBCs have a long circulation life in the body approximately 100–120 days before the clearance of immune system. The RBC membrane-coated NPs show an elimination half-life of 39.6 h and PEGylated NPs show an elimination half-life of 15.8 h. The RBC membrane coating has extended the circulation of various nanostructures. Rao et al. has reported that RBC membrane-coated NPs reduce the reticuloendothelial system and no in vivo toxicity is observed. Expansion has been made in coating the cell membrane by using other cell types. Parodi et al. has developed the leukocyte membrane-coated porous silica particles in which the membrane of leukocyte has the retention of N-acetylglucosamine glycans, which are important for cellular self-recognition that serve to reduce the binding to similar immune cells. The stealth properties are useful for nonspecific cancer drug delivery. Angiogenesis characterizes the tumor growth, angiogenesis is the rapid growth of new blood vessels which provides sufficient oxygen and nutrients to the growing cancer cells. The new vessels which have been formed were abnormal and leaky and the NPs were taken up into the tumor site through extraversion and have been found to stay at the site of tumor for a long period of time. The circulation time has been improved by stealth coating and thus increases the chance of the particulate drug getting into the tumor through the EPR effect. Due to this reason, RBC membrane-coated NPs get accumulated in tumors and the delivery of chemotherapeutics is enhanced. Inside the cell, via natural diffusion, the cargo is slowed down through the membrane as the polymer matrix degrades and is released through the ultraviolet-triggered membrane degradation.

8.3 DRUG DELIVERY TO TUMORS

Tumor growth is also dependent on interactions between individual cancer cells. When cancer cells adhere to one another, then homotypic binding occurs due to which tumor masses grow. Fang et al. coated the polymeric NPs with membrane which have been derived from the breast cancer cells. The cancer cell membrane-coated NPs possess a 20-fold increase as compared with RBC membrane-coated NPs. It has been found that the NPs coated with the membrane of breast cancer cells retains the other adhesion molecules like Thomsen−Friedenreich antigen, E-cadherin, CD44, and CD326, and the delivery of paclitaxel to primary and metastatic tumors is facilitated. Same results have been interpreted with cancer cell membrane-coated, DOX-loaded magnetic NPs. In case of solid tumors, there is a requirement of an increased recruitment of various cell types so as to promote the rapid cell growth. Due to the increased demand of connective stromal cells in the area, stem cells are often recruited. The mesenchymal stem cell membrane coats the DOX-loaded gelatin NPs and gets accumulated in tumor sites and with the release of DOX, tumor destruction is enhanced. It has been observed that the DOX-loaded NPs coated with membrane derived from these cells reduce the tumor growth. In circulation of the cancer cells, platelets serve the special functions. Thrombus is formed, thus local platelets are attracted and a shield is formed around the cancer cells and immune invasion is invaded and extravasation is enabled. The platelet membrane-coated nanovehicles which are functionalized with tumor necrosis and loaded with DOX cargo havè been used to treat primary tumors and have been shown to kill circulating tumor cells. The particles are found to get accumulated at the site of tumor after the intravenous administration, and the primary tumor growth is inhibited in the breast cancer model. It has been found that the silica particles coated with the platelet membrane has been functionalized with TRAIL possess efficacy in reducing lung metastases formation after being administered.

8.4 DRUG DELIVERY TO BACTERIA

This area is of keen interest due to the growing concern of antibiotic-resistant bacteria, such as methicillin-resistant staphylococcus aureus (MRSA). The platelet membrane-coated NPs can also target to opportunistic bacteria, this is so because bacteria exploit platelets as a way to shield themselves from the immune system and localize to certain vulnerable tissues. The binding between platelets and bacteria is complex and occurs through the direct adhesion through bacterial surface protein. The platelet membrane-coated NPs (PNPs) have been developed by Hu et al. which are capable of multiple biological interactions. This binding has improved the bacteria killing efficacy and decreases the overall bacterial load.

8.5 DRUG DELIVERY TO INFLAMED INJURIES

Delivery of drugs to sites of inflammation is important for p wound healing. Platelets and leukocytes are found at sites of injury and inflammation and bleeding is clotted and extracellular matrices are formed and cells become the natural choice of the coating of membrane. Leukocyte-coated NPs can traverse the endothelium and enabled the NPs like porous silica to traffic through inflamed endothelium due to the retention for drug delivery. But the platelets get bound to collagen in the subendothelium and are exposed when upper endothelium layer is damaged.

These biomimetic nanocarriers are also immunocompatible and can improve treatment safety. Luk et al. found that RBC MNPs delivering DOX to established lymphoma tumors could double survival time of mice compared with control mice, with no systemic release of inflammatory cytokines and no myelosuppressing indications, such as the low white blood cell count that is typical of free DOX administration.

8.6 DRUG DELIVERY USING GOLD NPs

Gold NPs (AuNPs) and their surface functionalized variants are also being used for various diagnostic and therapeutic applications.[35,36] Due to physicochemical and surface characteristics, these behave as ideal candidates for developing diverse prognostic, diagnostic, and therapeutic modalities. AuNPs are also used in biomedical fields like biosensors, bioimaging, drug delivery, photo thermal therapy, radiofrequency ablative therapy, and photodynamic therapy.[37–40] It has been suggested that upon entry into the bloodstream, NPs are exposed to a highly complex biological environment and acquire an interfacial robust layer of biomolecules known as the "protein corona." The protein corona is a major determining factor for the successful delivery of drugs to targeted sites.[41]

In medicinal chemistry, development of alternate drug delivery system is an ongoing task.[42–44] The recent burst of research involves the use of AuNPs as transfection vectors, DNA-binding agents, and protein inhibitors. This demonstrates the versatility of these systems in the biological applications. AuNPs are well suited for biomedical applications which include straightforward synthesis, stability, and the facile ability to incorporate secondary tags such as peptides targeted to specific cell types to afford selectivity. The catatonically functionalized mixed monolayer protected gold clusters (MMPCs) can meditate DNA translocation across the cell membrane in mammalian cells at levels much higher than polyethylene mine (PEI), a transfection vector. Due to cytotoxicity, the transfection efficiency decreases at higher concentration.

To determine the toxicity effect, cationic and anionic NPs are synthesized as shown in given Figure 8.2.

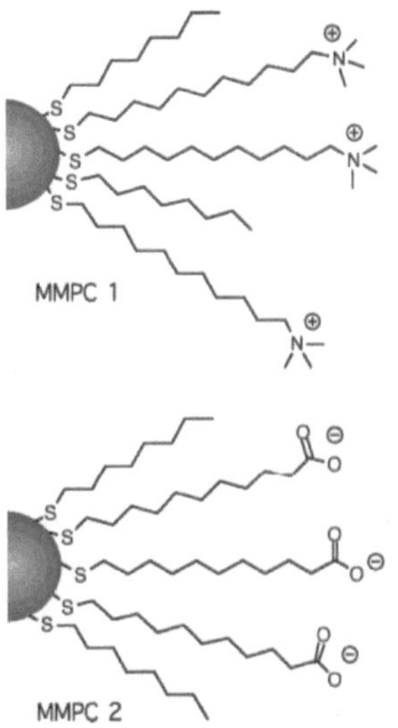

FIGURE 8.2 Schematic of the structures of MMPCs 1 and 2. Each monolayer consists of approximately 70 charged thiols and 30 unsubstituted thiols.

MMPCs 1 and 2 have been tested for toxicity in RBCs. MMPC 1 is also known as quaternary ammonium functionalized nanoparticle and it is composed of monolayer components same as to those which have been demonstrated which causes cytotoxicity in the mammalian cell cultures. The surface of MMPC 2 consist of the negative substituents which does not participate to interact with the cell membrane, and thus the bilayer of the lipid gains overall negative charge.[45] The structure of thiol components and gold core are similar to those of MMPC 1. RBCs, Cos-1 cells, and bacterial cultures are being used to probe the effect of NPs on the viability of cells. Dye release studies using lipid vesicles have been used to understand the origin of toxicity. Dye released from the lipid vesicles can be correlated with the ability of amphiphilic polymers to lyse bacterial cells. It indicated that membrane disruption is an important factor in the mechanism of toxicity. These studies are used to understand the origin of observed toxicity and allow the designing of new particles with an improved toxicity profile. MMPCs 1 and 2 have been tested for toxicity in RBCs, Cos-1 cells, and bacteria.

The toxicity profiles of the mammalian cells are virtually indistinguishable, and a 2- to 3-fold increase is required as a necessary concentration for the bacterial cultures. This increase is due to the nature of the bacteria. Usage of a low-density cross-linked agar matrix decreases the mobility of both the cells and NPs, and potentially alters the way in which the two components interact. In a passive way, NPs interact with cells because the receptors and other molecules are active in energy-dependent processes. The observed difference is due to the increased protection which has been provided by the outer membrane and also by the wall which is surrounding the bacteria, all this requires the high concentration of nanoparticle so as to fully rupture the bacteria. It has been studied that the NPs interact with the cells in a passive mode. This is because receptors and other related molecules which are active in energy-dependent processes are different for the separate species.

It has also been found that the endocytic pathway has been actively utilized by Cos-1 cells, but not by erythrocytes, and thus it has been suggested that the toxicity is not caused by cellular uptake through endocytosis. Membrane adhesion or cell lysis by the NPs is one of the mechanisms that have been anticipated to be similar in comparing the prokaryotic and eukaryotic samples. The specific lipids occurring in the bilayer for each cell type possess different identity or percentage composition but all cell types have an overall negative charge with amphiphilic and with highly charged substituents. MMPC 1 has been found to get attracted toward the negative membrane but such strong attractions have not been found in the case of MMPC 2. On binding with the cell, the amphiphilicity of the mixed monolayer of MMPC 1 induced the variety of further interactions.

A vesicle disruption assay has been performed to analyze the above potential mode of action. The molecules which were capable of interrupting the lipid bilayer release the fluorescent dye. An increase in fluorescence has been observed and it indicated that the designed molecule can be antibacterial due to the leakage of membrane, the bilayer represents the bacterial cell membrane.

Two vesicle systems have been examined, one was composed of phosphotidylcholine (SOPC) and phosphotidylserine (SOPS), it possesses net negative charge, and the other was composed of SOPC only, no net charge. Due to the electrostatic complementarity of the cationic substituents and the SOPS lipids, it has been observed that MMPC 1 lysed the SOPC/SOPS vesicles more efficiently than MMPC 2. In the neutral SOPC vesicles, high level of fluorophore release has been observed in MMPC 2 than MMPC 1 although the intensity of MMPC 2 was less than that of MMPC 1. It has been depicted in Figure 8.3.

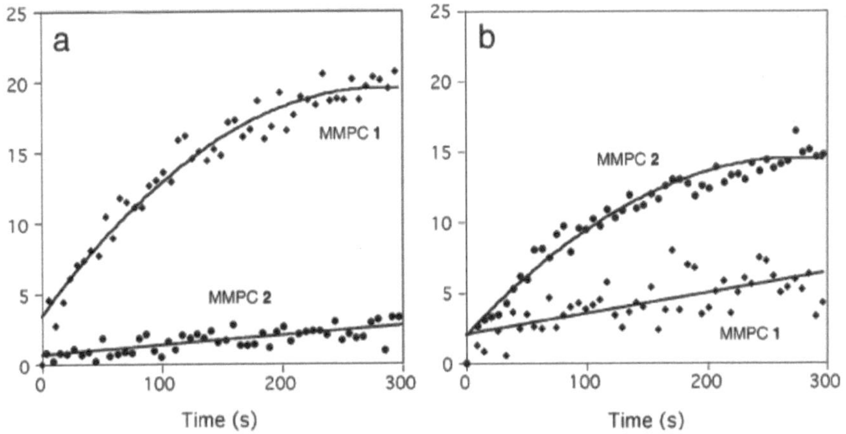

FIGURE 8.3 Comparison of MMPC **1** and **2** in disrupting vesicles with an overall negative (SOPC/SOPS; panel a) or neutral (SOPC only; panel b) charge. Nanoparticle concentration is 220 nM. Curves are meant to lead the eye.

This result has demonstrated that the specific charge pairing of the NPs and lipid bilayers has mediated membrane interaction and lysis for releasing the fluorophore. This is the concentration-dependent process as it has been observed that at higher concentration of MMPC1, the rate of lysis increases. Thus, it is said that toxicity of the gold NPs is related to their interactions with the cell membrane.

AuNPs have been used as key materials for biomedical applications like cell labeling, imaging, biosensing, and gene and drug delivery due to their unique structure, tunable optical properties, chemical inertness, and biocompatibility. The size, shape, surface charge, and functionality of NPs have played a significant role in determining the intracellular uptake and localization of the NPs. The most important factor is the surface charge of the nanomaterial which helps in determining the molecular interactions, cellular uptake, and cytotoxicity of NPs. In addition to this, NP surface functionality is also involved in the process of eliciting cellular responses and cellular uptakes. Stellacci et al. have reported the effect of surface properties of NPs on the cell membrane which has been negatively charged. The Rotello group have demonstrated the behavior of AuNPs with different hydrophobicity and also determined the surface functionality effect on hemolysis. Linear hemolytic behavior has been observed as hydrophobicity is increased in the absence of serum media. It has been found that the positively charged AuNPs have been more internalized by cells than neutral or negatively charged NPs. We all know that cancer has been a major threat to public health and its mortality rates are higher worldwide. Conventionally practiced therapy has failed to cure most cancer patients due to the presence of resistance to anticancer agents. It is being expected that nanomaterials will revolutionize cancer diagnosis and therapy. Many studies have shown that AuNPs serve as carriers for biomolecules against some cancer cells. AuNPs can be envisioned as anticancer agents through the generation of reactive oxygen species (ROS) in cells. It has been reported by Zhao et al that the AuNPs generated more intracellular ROS in lung cancer cells and Morgan et al. have observed that ROS causes cellular damage and leads to cell death, including apoptosis and necrosis at the higher concentration. It has been observed that hydrophobicity plays an important role in the generation of ROS, DNA damage, and hence potential genotoxicity. A hemocompatible and in vitro anticancer activity against A549 lung cancer cell lines have been reported using three bile acid-based amphiphile. In vitro hemotoxicity studies have been done. It is known that the hemolysis assay is routinely procedure, followed by hospital-based clinical and research laboratories. The toxicity of NPs is highly dependent on its physicochemical and surface properties and it has been essential to evaluate the blood compatibility of AuNPs for application of biomedical purpose. The results of hemolytic assay have shown that all the tested amphiphile-stabilized AuNPs exhibited concentration-dependent hemolysis with the very lowest activity. It can be seen in given Figure 8.4.

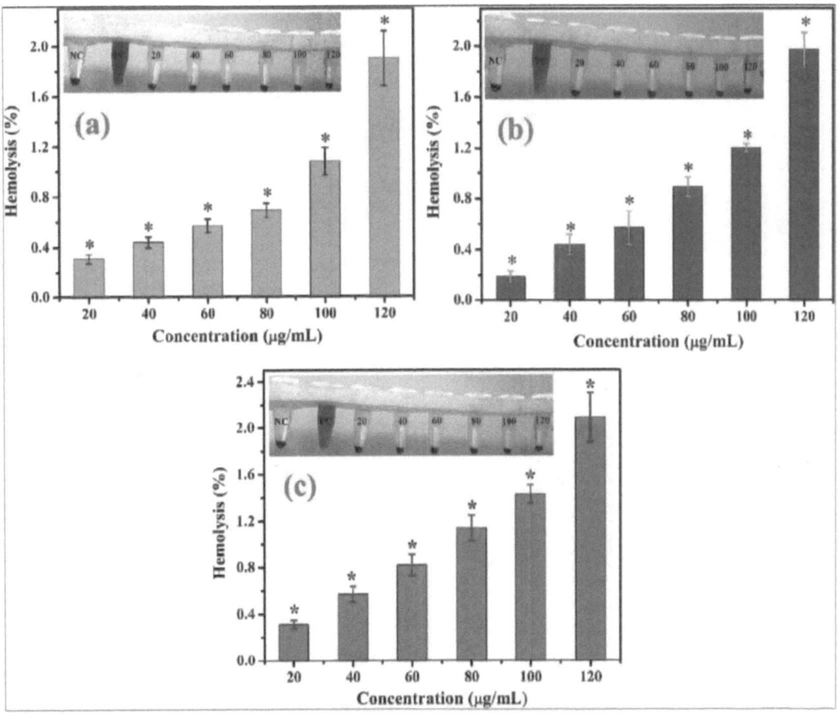

FIGURE 8.4 (See color insert.) Hemolytic activity of dicationic amphiphile-stabilized AuNPs on human RBC at various concentrations ranging from 20 to 120 μg/mL (a) DCaC-AuNPs, (b) DCaDC-AuNPs, and (c) DCaLC-AuNPs. The inset shows the photographs of the corresponding solution. Here, (*) represents a significant difference ($p < 0.05$) compared with that of positive control.

At the highest concentration, the hemolytic activities of different samples, that is, DCaC-, DCaDC-, and DCaLC-stabilized AuNPs were found to be 1.90, 1.97, and 2.09%, respectively. It has been reported that the tested samples with less than 5% hemolytic activity could be considered hemocompatible. In this study, it has been observed that these amphiphile-stabilized AuNPs possess good hemocompatibility upto the concentration of 120 µg/mL.

The result of hemocompatibility has been further supported by the observations of the light microscopy of RBCs treated with dicationic amphiphile-stabilized AuNPs.

It is shown in Figure 8.5.

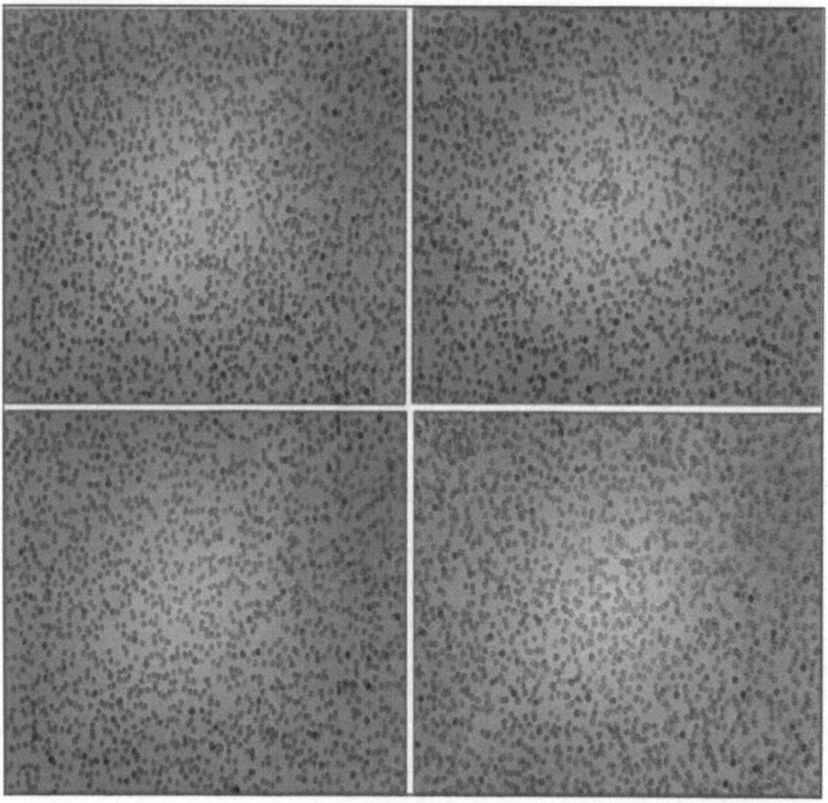

FIGURE 8.5 (See color insert.) Photomicrographs of dicationic amphiphile-stabilized AuNP treated human RBC by light microscopy: (a) control RBC, (b) RBC treated with DCaC-AuNPs, (c) DCaDC-AuNPs, and (d) DCaLC-AuNPs at the highest concentration of 120 µg/mL.

More than 90% of the erythrocytes have been found to retain their original oval shape without any aggregation, and these could be used for future biological and biomedical applications.

8.7 BIOCOMPATIBLE NANOMATERIALS IN DRUG DELIVERY

Biocompatible nanomaterials are being developed for cell imaging, tracking, and drug delivery.[46-49] The field of nanotheranostics has emerged due to the surge of the application of nanomaterials for the diagnosis and study of the development of cancer. There is problem in the potential toxicity of the carriers due to poor metabolism and elimination, and to solve this problem, Ying's group has developed nanocomplexes, which consist of green tea catechin derivatives and protein drugs and thus demonstrates high amount of safety and good antitumor effects.[12] The drug delivery systems comprises active components, and the absence of carrier materials leads to safe clinical applications.

Some artificial materials like natural polysaccharides, for example hyaluronic acid, chitosan, and heparin, have inspired the development of NPs due to their nontoxic and nonimmunogenic properties. One of the most unique materials is unfractionated heparin (UFH). It possesses anticoagulant activity which is not possessed by other polysaccharides and it also has inhibitory effects on tumor angiogenesis and metastasis. It has some side effects like thrombocytopenia and hemorrhagic complications and due to this its clinical usage is limited. To solve this issue, low molecular weight heparin (LMWH) has been used as a good alternate for UFH. This LMWH is obtained from the depolymerization of UFH. The treatment should provide the inhibitory effect on the tumor cells and the cytotoxicity in drug delivery system (DDS). DDS prevents the tumor growth and the metastasis. The tumor acidic environment-responsive amphiphilic NPs based on LMWH and DOX are demonstrated in Figure 8.6.

Thus, low molecular weight heparin–doxorubicin (LH–DOX) has been hypothesized. It is used as a carrier, anticancer, and antimetastatic. The drug is gathered in the hydrophobic core and hydrophilic LMWH protects it. In the blood, it reduces opsonization and lengthens the circulation time. After its accumulation in tumor tissues, drug is released in an acidic microenvironment and tumor cells are killed effectively.

Self-assembled NPs are formed by the conjugates in the aqueous solution, and are confirmed by DLS and TEM which can be seen in Figure 8.7,

TEM images have revealed the spherical shape of the NPs.

For drug delivery or gene therapy, a variety of inorganic core nanomaterials which are surrounded by a biocompatible ligand shell have been synthesized using diagnostic imaging tools and transport vessels.[50] A

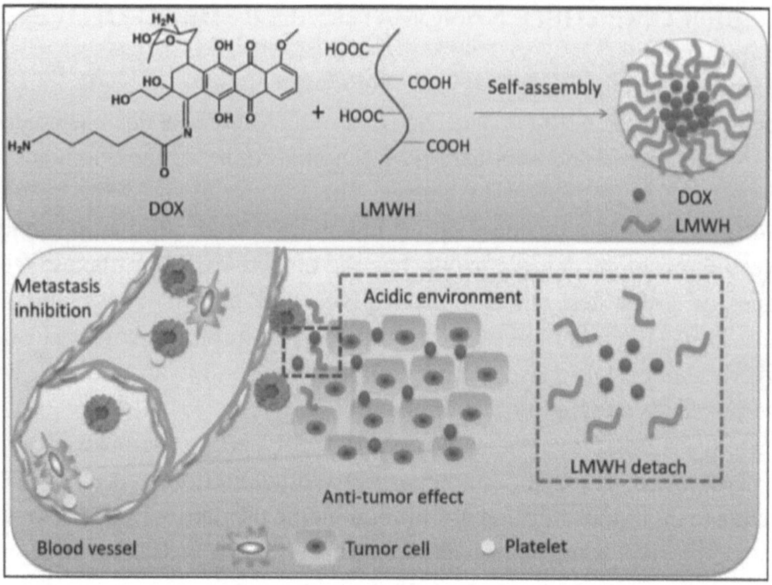

FIGURE 8.6 **(See color insert.)** Schematic design of the multifunctional polymer–drug conjugate.

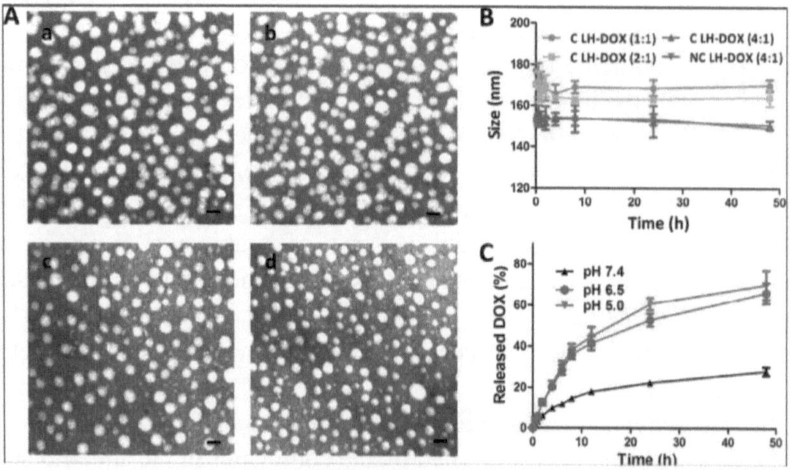

FIGURE 8.7 **(See color insert.)** (A) Representative transmission electron microscopy images of NPs: (a) C (cleavable) LH–DOX 1:1, (b) 2:1, (c) 4:1, (d) NC (noncleavable) LH–DOX. The scale bar represents 150 nm. (B) Size changes of NPs in 50% FBS for 48 h. (C) Drug release profiles of C LH–DOX (4:1) under various pH conditions (pH 5.0, pH 6.5, and pH 7.4) for 48 h. Errors are standard errors of the mean (SEM) for N = independent experiments.

bright fluorescence signals for noninvasive live cellular imaging and bio distribution tracking is required for these applications. Various types of nanomaterials like quantum dots (QDs), which are inherently fluorescent, are not used in applications much due to their toxicity and difficulty in surface functionalization. Issues of biocompatibility has been overcomed through the use of fluorescent core–shell silica NPs (NPs) but the ease of functionalization remain unsuitable for some applications due to their large size, hemolytic activity, and surface chemistry. There is a large variety of nanotheranostic tools which are inherently nonfluorescent and fluorescent. Dyes are employed as noncovalently bound labels at their surface. In the plasmonic particles, the fluorescence of the dye is quenched by the core material, for example, gold. The fuorescence is regained when the dye is released from the surface of the nanomaterial.[51] The functionalization scheme of dye-labeled nanomaterials depends upon the chemistry of the core material, and it also involves the use of biofunctional molecule, known as a ligand. Dye molecules can be prethiolated and exchanged onto the surface of nanoparticle in case of the AuNPs. Ligands are moderately stable, the collection and characterization of the nanomaterial is necessary to ensure dye viability which is impractical and sometimes impossible restriction. It is necessary to develop a control experiment so as to detect molecules free of dye in the presence of dye functionalized nanotheranostic carriers as there is an issue which deals with contamination of a sample with free dye from the labeling procedure. The bound dye–particle assemblies are purified from unreacted free dye by using centrifugation. But this method is not trustworthy that the free dye has been truly removed.

The experiment should be made to run in parallel with in vitro experiments on other types of cells on the dye-labeled sample which are in use. RBCs are also known as erythrocytes, these are being used for delivering oxygen to the tissues in vertebrates. The RBCs of mature mammalian lack nuclei and mitochondria, thus distinguishing them from most other mammalian eukaryotic cells. Their surface does not possess phagocytic receptors and thus cannot employ endocytosis that is uptake of foreign material or proteins at the cell surface. For nonphagocytic/nonendocytic cells, RBCs are commonly used model. It is known that small molecules like fluorescent dyes can partition into the erythrocyte membrane.

Compounds containing fluorescent dyes have been developed for imaging application because of their environment-independent fluorescent yields, high brightness, and sharp excitation and emission peaks.

These are readily inserted at trace levels into the RBC plasma membrane. Therefore, erythrocytes are considered as a unique system which takes up the nanomaterials and also the amphiphilic molecules into their plasma membranes. Efforts are being made to develop RBCs as a quick screen so as to detect free dye in a dye-functionalized nanomaterial. The interaction of nanomaterials with RBCs has led into the field of applications which includes passive membrane penetration mechanisms, nanopore generation, hemocompatibility, and enhanced-contrast blood flow imaging.

In vivo circulation time can be improved by adhering the NPs on to the outer surface of RBCs. Interaction of RBCs and nanotheranostics are used for the separation and use of RBC ghost cells. After the osmotic hemolysis of RBCs in diluted, the naked RBC membranes can be isolated. Under the hypertonic conditions, the swelled membranes become shrink and crenate and empty membranes, or ghost cells are left behind. The erythrocyte inspired delivery is an inspired field which is primarily utilizing RBC ghosts so as to encapsulate nanomaterials, cancer chemotherapeutics, genetic material, proteins, and is increasing circulation in the body and bioavailability.

These types of cells can be easily identified by confocal microscopy and are compared with normal RBCs and these are excluded from quantification because of complexities which arise from interactions of NP–ghost cell. For the study of cell penetration, the method based on sulfonate-terminated NPs is used. These NPs can be functionalized with dye through place-exchange procedures and these NPs are also soluble in water. No qualitative changes in the RBC morphology are observed after the coincubation of NPs with RBCs. The fluorescence of the RBC membrane has been measured by the confocal laser scanning microscopy (CLSM) and flow cytometry (FC). No fluorescence takes place for the particles that do not enter RBCs. The fluorescence signal from the RBCs is detected when free dye is present. A rapid detection method for detecting the free dye in nanoparticle samples is the NPs' RBC incubation.

The fluorescent dye place-exchange reaction is shown in Figure 8.8.

The samples used as nanotheranostic carriers are sulfonate-terminated AuNPs or heteroligand sulfonate-methyl terminated NPs and these are used to label the dyes. By the place exchange procedures, these NPs are used to incorporate the secondary ligands like thiolated dyes into the protecting shell.

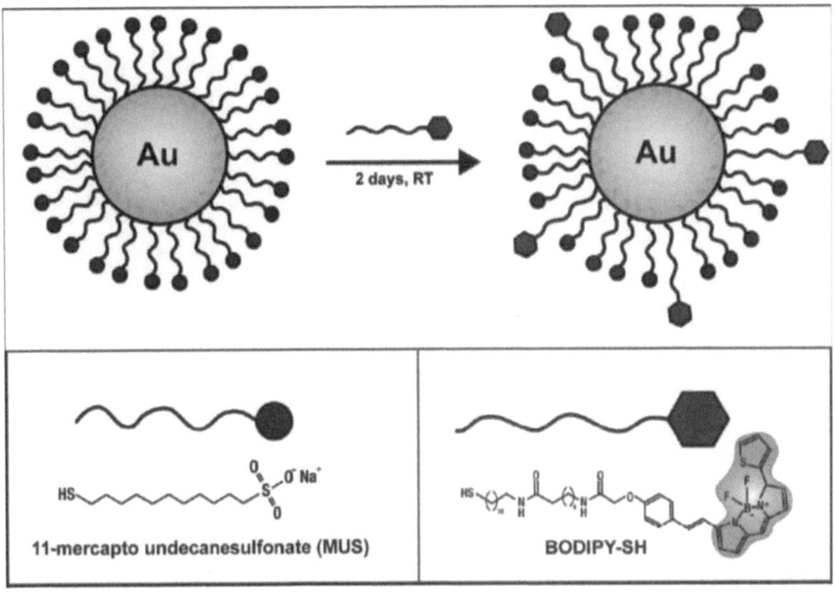

FIGURE 8.8 AuNP structure during place-exchange reaction. The fluorescent BODIPY-SH molecules are mixed with monolayer protected NPs in molar excess for 2 days at room temperature.

A washing experiment has been designed. Thiolated free dye is made to mix with the NPs and a typical place-exchange procedure is followed. An equilibrium is established between the dye molecules that remain in solution and the dye molecules that attach to the surface of the nanoparticle. Through centrifugation, the dye-labeled NPs are pelleted and in the supernatant the free dye is left behind. After a spin, the NPs, which were pelleted, were made to divide from the supernatant, which contains the remaining free dye and after this the separate incubation with RBCs was provided.

In the given Figure 8.9, the results of this experiment can be seen,

Figure 8.9a shows that the RBCs alone are the control.

Figure 8.9d shows the unwashed NP/dye mixed solution.

Figure 8.9b shows that after the first wash the supernatant contains all of the free dye. Figure 8.9e shows that the pelleted NPs are free of any unbound dye.

Figure 8.9c shows that the washing step is repeated again and again.

Figure 8.9f shows that NP pellet was incubated separately with RBCs. Both of these are free from unbound dyes.

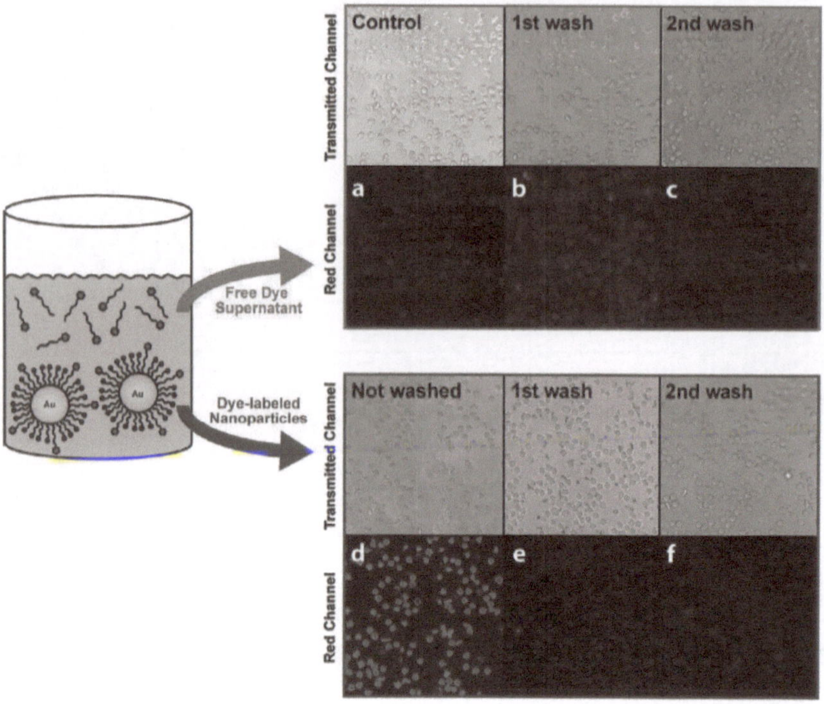

FIGURE 8.9 **(See color insert.)** Dye-labeled NP incubations with RBCs. Following the dye place-exchange reaction, the supernatants (a–c) and dye-labeled NP pellets (d–f) of successive washes were incubated with RBCs. All of the unbound dye is removed in the first wash.

Numerous efforts have been made for the development of safe and effective DDS for selective delivery of toxic drugs to tumors.[52,53] But the targeted DDS have some limitations, involving the release of premature drug before reaching the target site. The ideal DDS deals with the delivery of toxic drug to specific target sites without the leakage of drug on its way. There are two important keys to increase the therapeutic efficacy and reduce the side-effects of anticancer drugs, which are targeting and stimuli-responsive drug release.

A "zero premature release" has been developed based on the conjugation of drugs to vehicles. A redox-responsive bond that is disulphide bond is cleaved under a high concentration of glutathione (GSH), this covalent bond has an advantage that it can be used in dealing with the stability of extracellular fluids and plasma. An easy rupture in intracellular fluid occurs due to a difference in the concentration of GSH between extracellular fluids and the intracellular fluids. Inorganic counterparts exhibit particular properties such as inertness and chemical stability as compared with traditional lipid-based NPs or organic polymers. Mesoporous silica NPs (MSNs) possess high surface area, tunable pore size, ease of surface functionalization, and excellent biocompatibility and thus can be used as ideal candidates for the development of stimuli-responsive DDS. A redox-responsive disulfide bond is used to connect a mercapto-containing drug to mercapto-functionalized silica. Anticancer drugs, such as doxorubicin and cisplatin consist of mercapto group and can accept a mercapto group by a simple modified grafting process, and grafted to silica by disulfide bonds covalently.

Hyaluronan also known as hyaluronic acid (HA) possess large surface area, is anionic, biodegradable, linear polysaccharide composed of 1000–25,000 repeated units of glucuronic acid and nacetylglucosamine. And for DDS, HA is being widely used. This is due to the specific interaction with the receptors which are overexpressed on various cancer cells. HA-modified MSNs are used for the tumor targeting delivery and dispersity is also improved but the macromolecule HA also has some limitations as a drug carrier. It is rapidly cleared and destructed by the liver which is due to the strategy of HA-modified MSNs for the tumor targeting delivery and improved dispersity as reported previously.

Degradation of macromolecule HA leads to the oligosaccharides (fragments) of HA (oHA) with an Mw $< 10,000$ through hyaluronidase. oHA provides a longer blood circulation time and is slowly removed from the liver. The surface of silica was modified by cleavable of disulfide bonds rather than the amido bond. The cellular uptake of DDS can be improved by using targeting ligand. The surface targeting group retards the release of drugs after the DDS is taken by tumor cells, by the cleavage of the disulphide bond the surface of silica can be modified.

Figure 8.10 shows the cellular uptake and GSH triggered drug release.

The drug-loaded oHA NPs has shown good dispersibility and stability in PBS containing blood cells at high concentrations upto 1500 $\mu g\ mL^{-1}$ and the percentage hemolysis of RBCs could be calculated.

Figure 8.11 shows the percentage hemolysis of drug is related to the concentration range of 20–1500 $\mu g\ mL^{-1}$. When the concentration of drug is 500, 1000, and 1500 $\mu g\ mL^{-1}$, the percentage hemolysis was 21.3%, 57.6%, and 66.5%, respectively. The high degree of hemolysis raises serious safety concerns regarding the application of NPs for drug delivery.

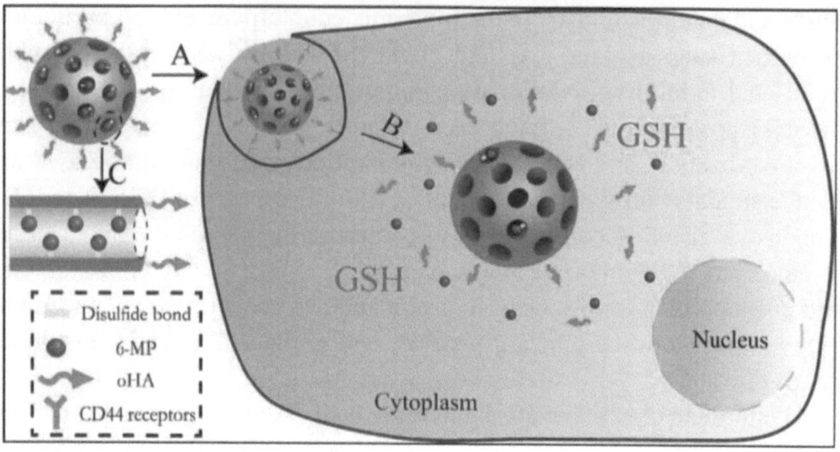

FIGURE 8.10 (See color insert.) (A) Cellular uptake through oHA-mediated CD44 interaction, (B) GSH-triggered drug release inside the cell, (C) magnified image of pore structure of CMS-SS-MP/oHA.

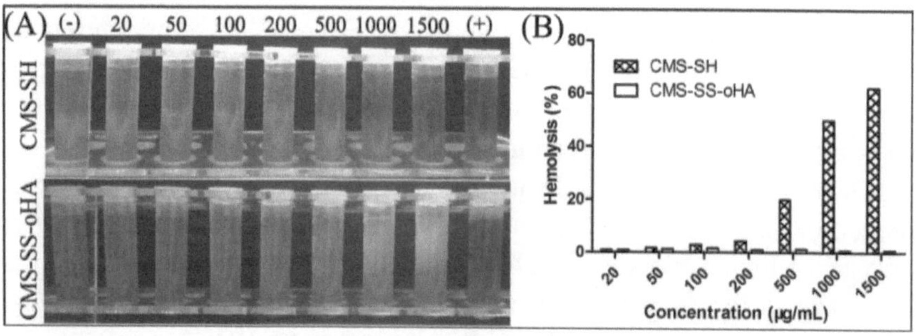

FIGURE 8.11 (See color insert.) (A) Hemolytic photographs and (B) hemolysis percentages of drug and drug-loaded oHA at different concentrations (μg mL^{-1}).

Engineered NPs are being used in different areas like that of industry, technology, and medicine. The NPs can enter into the body, and intravenous injection is the most common way of entrance; the respiratory system, skin, and gastrointestinal tract are also the routes that NPs enter into the body. The NPs reach the blood stream and interact with blood components. Interaction occurs between nanomaterials and serum proteins and due to this the protein conformation, orientation is altered, and it leads to functional disturbing.[54] The surface of NPs is modified through the adsorbed protein. RBC hemolysis is caused due to the interaction of NPs with RBCs. The responses of human body defense system is changed by the RBCs. RBCs have significance in the design and engineering of NPs. Nanotoxicity can be reduced by the evaluation of blood compatibility. MSNs are used in biomedical applications like bioimaging, biosensing, biocatalysis, and drug delivery. These possess properties like easy synthesis, controllable particles scale, and different morphologies ranging from spheres to rods, uniform cylindrical mesopores, high surface areas, and available surface modification. MSNs are thus used in various biomedical applications. The synthesized MSNs are used as intravascular drug carriers and these have low hemolytic activity than the nonporous counterparts of similar size. It has been found that the small MSNs with a diameter of 100 nm reduce the hemolytic activity than big MSNs. It has been reported by Madhura Jogkekar et al. that spherical NPs exhibit low hemolytic activity 20. Serum proteins are used onto the MSNs due to their wide range of applications in nanomedicine.

Hydroxyapatite ($Ca_{10}(PO_4)_6(OH)_2$, HAP) is an artificial bone substitute which possesses good biocompatibility, bioactivity, osteoconductivity, nontoxicity, noninflammatory behavior, and nonimmunogenicity. HAP can improve the fracture toughness. HAP NPs are used as carriers for delivering genes, drugs, and/or proteins.

The light microscopy images of RBCs are shown in Figure 8.12.

TEM image of HAP is shown below in Figure 8.13.

HAP NPs exhibits different size and surface effects on the aggregation of RBCs. The HAP NPs lead to the aggregation of the RBCs. And, the aggregation of the RBCs can be lowered by the use of surface modification of the HAP NPs with negatively charged groups.

There are two exclusive "models" for aggregation of RBCs—"bridging" and "depletion" models in the macromolecules solution.

In bridging model, macromolecules are adsorbed onto adjacent cell surfaces which result in the bridging force. When these attractive bridging forces exceed the disaggregating forces, then the aggregation of the RBCs occurs.

In the depletion model, the depletion layer of the macromolecules occurs near the cell surface because of lower concentration of macromolecules in the fluid than that in the bulk solution. The merging of the depletion layers results in an attractive force. The aggregation of the RBCs appears when the attractive force is sufficient to overcome the disaggregating forces.

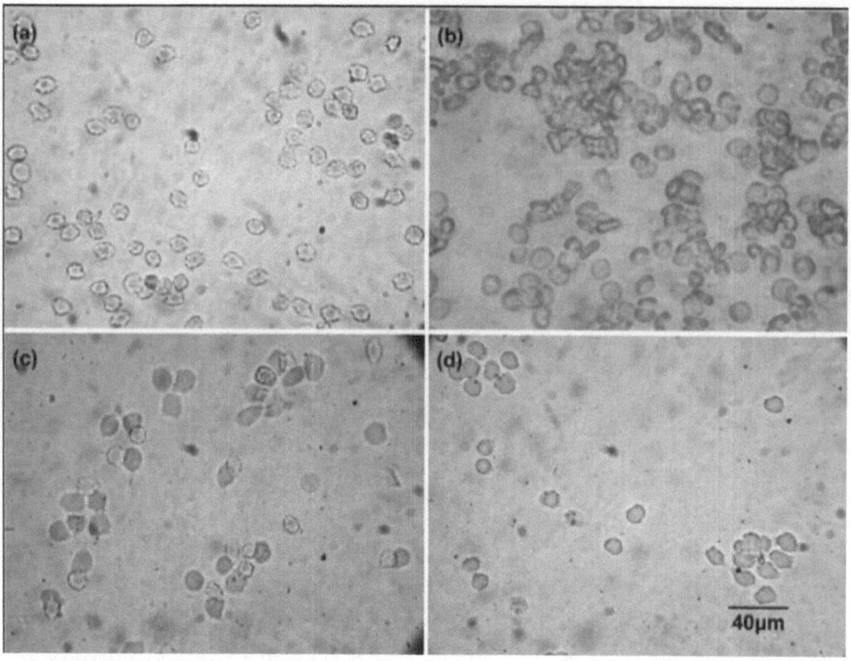

FIGURE 8.12 Light microscopy images of RBCs after 1 h incubation (a) without HAP particles and with (b) HAP NPs, (c) HAP microparticles, and (d) heparin-modified HAP NPs. The concentration was 140 μg/mL for each of the HAP particle groups.

Nanomaterials are also used in the field of biomedicine. Due to their unique morphological traits and physicochemical properties, silica-based nanomaterials (SiNMs) are also used in biocatalysis, bioseparation, and bio-optics, and drug delivery. It is due to the good biocompatibility, diversity, and tunability of their morphology, size, and surface chemistry. The mesoporous structure provides a large space for the loading of different kinds of biomolecules and drugs, which are helpful for the therapeutics delivery; MCM-41, MCM-48, and SBA-15 are the most common mesoporous silica NPs and these possess different pore sizes and structural characteristics. The mesopores are used as an intracellular delivery nanocarrier and is used for loading cargo molecules and the external surface offers a site for chemical functionalization so as to enable cellular targeting and bimolecular interaction. The shape, size, surface chemistry, and mesoporosity have significantly influenced interaction with cargo molecules and also has biological influences on cells and tissues. Mesoporous silica materials are used as a carrier for drug delivery. This is achieved by tailoring the internal mesopore geometry and by functionalizing the silanol-containing surface to control the drug diffusion kinetics.

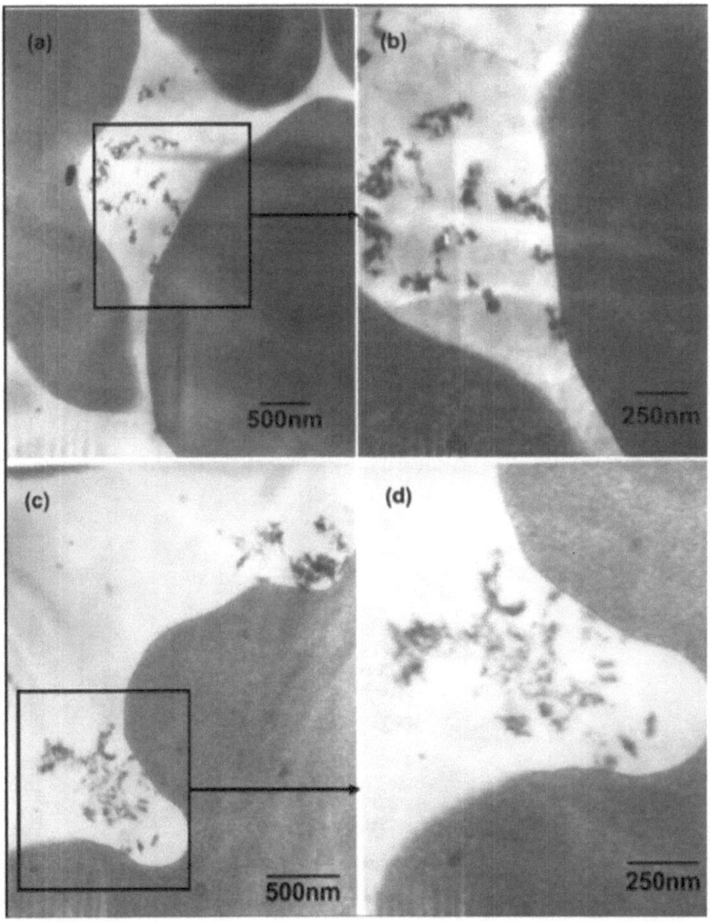

FIGURE 8.13 TEM images of RBCs incubated with HAP NPs for different times. (a), (b): 15 min; (c), (d): 1 h.

Controlled release of drugs is maintained by tailoring the mesopores with the sensitive materials sensitive to an external stimulus. Mesopores can be capped with stimuli-responsive groups like inorganic NPs, chemical moieties, and proteins. These provide the on-demand release of therapeutic drugs by external stimulus such as pH, temperature, redox potential, light and enzymatic reactions additional functional properties like magnetic or luminescent properties to improve the potential for targeting and tracking of the delivered drugs. The method of soft templating are used to synthesize SiNMs. It involves two steps hydrolysis and condensation precursors of silica in the presence of micelle templates and through calcinations or by solvent extraction so as to form ordered mesopores within the network of silica. The pore size can be controlled by using different sized templates. The precursor used for silica can be TEOS and nanostructures are formed by the use of cationic quaternary ammonium salts and cetyltrimethyl ammonium bromide (CTAB). By using organosilanes with silica, precursor's mesopore surface can be functionalized. Organosilanes induces unique morphological transition rod-like particles which were obtained when mercapto propyl triethoxy silane is used, mesopore structural orientation by addition of different amount of urea-containing organosilane (UDPTMS).

The mechanism of formation of SiNMs with different particle morphologies and mesochannel orientations are shown below in Figure 8.14.

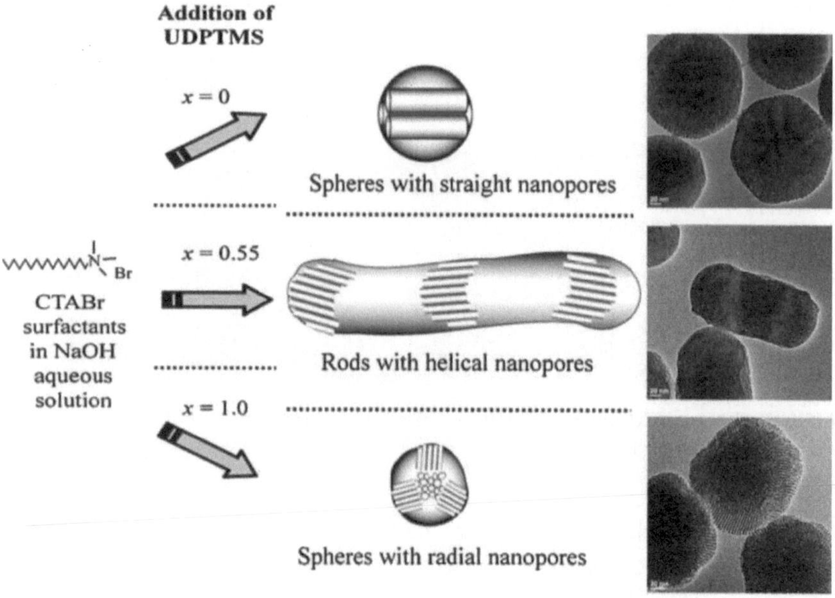

FIGURE 8.14 The formation mechanism of SiNMs with various particle morphologies and mesochannel orientations.

SiNMs can be synthesized with different features which include particle size, shape, surface area, pore volume, pore size, and surface-functional groups on their internal and external surface. The physical and chemical properties of the SiNMs are determined by the change in functional groups present on the surface. SiNMs can become more biocompatible by controlling the surface properties. Surface-functionalized SiNMs are used in biological applications as the functional groups can improve the adsorption capacity so that the bioactive molecules increase the binding ability to target cell. SiNMs possess a high density of surface. The functional groups play an important role due to the following reasons:

i) Tendency to control the surface charge of SiNMs.
ii) Linking functional molecules chemically to the pores.
iii) Controlling the size of the pore so that the size of the cargo pore can be modified.

Figure 8.15 depicts the mesoporous vehicle with a functionalized surface to serve different roles.

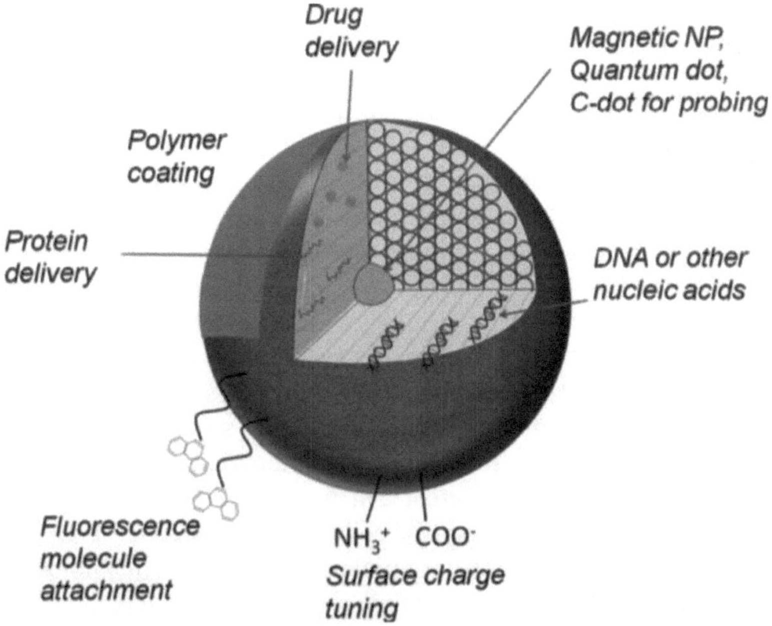

FIGURE 8.15 Surface tailoring and functional modification of SiNMs for specific purposes.

Small molecule drugs, proteins, or genes are loaded onto themesopores of SiNMs through adsorption or linking to pore walls. Molecules are introduced on the surface target cell receptors or intracellular organelles. The surface properties, such as charge, are also changed through the functionalization. For bioimaging purposes, the surface is further functionalized with probes, through combining with optical dyes or NPs, magnetic NPs, QDs, and carbon dots. The physicochemical properties of NPs influence the cell internalization of nanocarriers. Surface charge is one of the determinant factors among various physicochemical parameters.

From accidental exposure, potential hazards can originate and to prevent them, their toxicological effects have been studied from toxicological, environmental, health, and scientific perspectives. Crystalline silica dust is used in mining operations, foundry work, mineral processing, and construction sites. Chronic obstructive pulmonary disease, silicosis, or even lung cancer is induced through inhalation of crystalline silica. Amorphous fumed or precipitated silica is considered safe and is also used as a food or animal feed ingredient. It has been found that inhalation of amorphous silica possess minimal or no health-related risks. Amorphous silica is used in nanobiotechnology which has covered vast areas including diagnostics, drug delivery, bioanalysis and imaging, and gene transfer; therefore, it is conceivable that amorphous silica can be incorporated into the human body by oral ingestion, inhalation, intravenous injection, and transdermal delivery. In future, silica can be used as the biomaterial, therefore, information regarding its biodistribution, retention, absorption, degradation, clearance, and safety of silica is of vital importance. Through Sto¨ber (sol-gel) process, silica is prepared in the solution by the hydrolysis and polycondensation of silicon alkoxide.

Chitosan, a naturally derived polysaccharide, is capable of forming composite NPs with silica. The toxicity of silica and chitosan is known well but the cytotoxicity of silica NPs modified with chitosan has not been reported. Therefore, an interest in the investigation to determine the cytotoxicity of chitosan modification on these NPs in contact with silica. In vitro studies are made to examine the cytotoxicity of composite and amorphous silica NPs to different cell lines. Cell lines include dermal fibroblast cells, normal pulmonary cells, and tumor cells of the colon, gastric system, and lung. By examining the response of multiple cell lines to silica NPs, information dealing with the cytotoxicity of silica can become available.

Scanning electron microscope (SEM) is used to determine the average size of NPs by measuring the size of about 100 particles in the micrograph. The scanning electron micrographs of silica are shown in Figure 8.16.

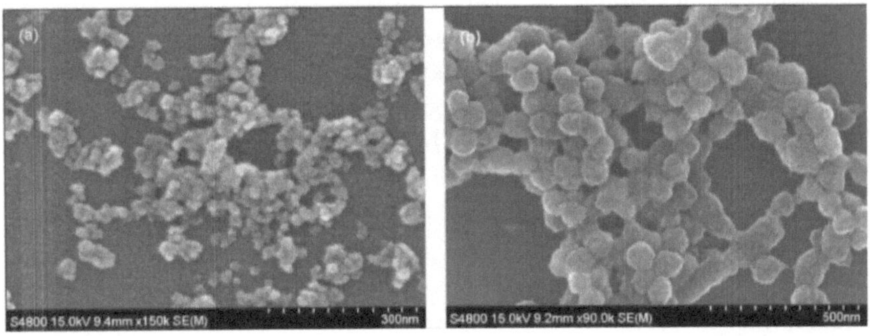

FIGURE 8.16 SEM picture of silica NPs (a) from sodium silicate and (b) from TEOS.

The figure showed that freeze-dried silica NPs are aggregated and spherical. The aggregation is due to the freeze-drying and sample treatment procedures. The size of the individual silica particles prepared from prepared from sodium silicate is 21.58 (4.36 nm) and size of silica particles prepared from TEOS is 80.21 (14.43 nm). An increase can be made in the particle size of the silica–chitosan composite NPs from 10 to 14 nm with a reaction time of 15–360 min. It can be seen in Figure 8.17.

It can be concluded that at high concentrations, the silica NPs becomes highly toxic to the human cells. It has been found that high dosage of silica nanoparticles are more toxic to the fibroblast of human cells than to the cancer cells and amorphous silica at high dosage can retard the cell proliferation and damage the cell membrane and also induces the necrosis. By synthesizing the nanoparticles with chitosan, the cytotoxic effects of silica nanoparticles were significantly reduced. The silica nanoparticles can be used in future applications in drug delivery.

Hemolysis of RBCs takes place as RBCs are exposed to nanomaterials, the toxicity of nanomaterials in RBCs has been widely investigated. There are various factors which influence the hemolytic activity. These include particle size, geometry, concentration, composition, surface modifications, and the surrounding environment. One of the key factors which determine that whether the particles can be recognized and cleared away by mononuclear phagocytic system is the particle size. For the application of

NPs in nanomedicine, the study of size-dependent hemolysis is extremely important. It has been reported that small NPs with a larger surface area are more hemolytic than larger ones, and it has also been found that larger NPs exhibits greater hemolytic activity than smaller ones. The NPs induced morphological changes of RBCs. It have been investigated by some researchers that the damage to cell membranes, hemolysis was induced. There are some factors which control the shape of RBCs, one of which is the content of ATP and the other is the structure of phospholipid bilayer and spectrin in membranes. To study the mechanism of hemolysis in detail, a study of nanoparticle-induced structural changes of RBCs is necessary and equally important is the observation of morphological transformation, FT-IR spectroscopy has been used to study the molecular conformational order of biological membranes. FT-IR spectroscopy has a great ability to detect global biochemical information on the intact cells, it is also used in distinguishing differences among cell populations. Semi-conductor QDs have been widely used in biomedical applications in vivo and in vitro imaging including vascular imaging.

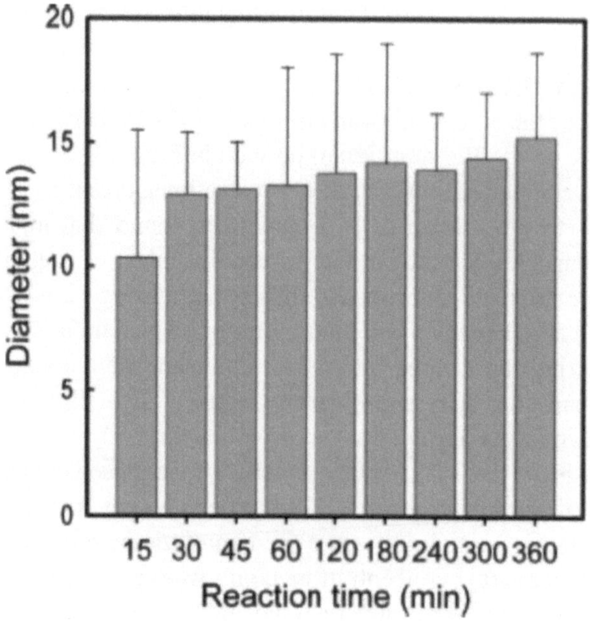

FIGURE 8.17 Average diameters of silica–chitosan composite NPs after different reaction times.

Various biological barriers can be overcome through the small size and the living organisms can be invaded. Semiconductors QD also possess nontoxicity. In many works, the toxicity of QDs has been studied. MSAQDs are used as a model to study the size-dependent toxicity of QDs in RBCs via hemolysis testing.

The confocal microscopy revealed that the particles of different sizes cause different changes in the shape of RBCs, and FT-IR is used to analyze structural changes of RBCs membranes. Based on all this, it has been analyzed that the phosphodiester bond may be the key factor that induced hemolysis.

The confocal images of RBCs treated with different MSA-QDs are shown below in Figure 8.18.

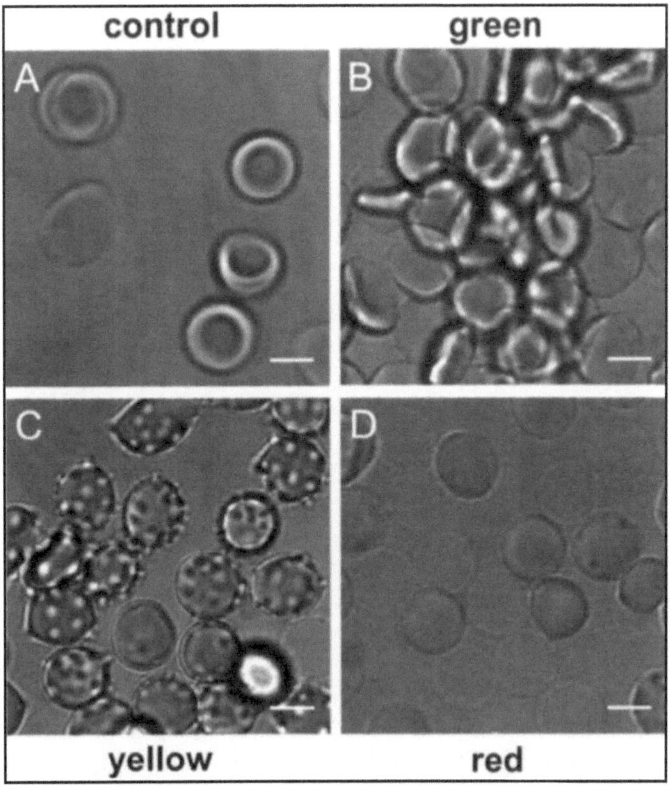

FIGURE 8.18 Confocal images of RBCs incubated with (A) PBS, (B) the green-emitting, (C) yellow-emitting, and (D) red-emitting MSA-QDs at a concentration of 500 μg mL^{-1} for 3 h. Scale bar = 10 μm.

Figure 8.19 shows the interaction of QDs with RBCs, the hydrogen bond between the free COO^- groups at the surface of QDs and the lipid membrane may be formed, irrespective of their size. During the approach of the RBCs, many of these have a strong interaction by forming hydrogen bonds, and part of them may enter the gaps between the long biopolymer chains and interact with outer membrane proteins. They have the strongest ability to form hydrogen bonds with lipids and cause a bigger disturbance in the conformers of lipids. Besides, they break the phosphate ester bond to form a terminal phosphoryl group and a phosphate–cadmium complex, which may cause the pore formation in lipid membrane of RBCs and subsequent hemolysis. The difference in the interaction models for smaller and bigger QDs is due to the difference in their surface energy regarding size.

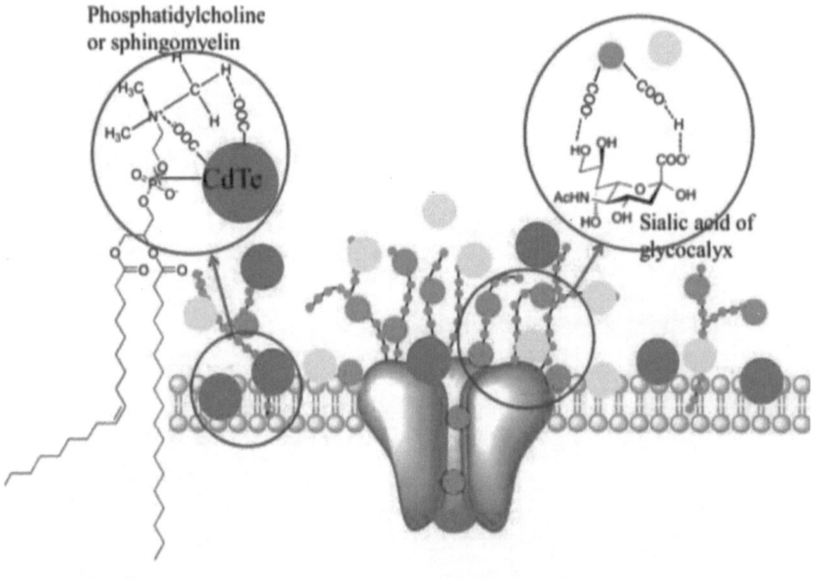

FIGURE 8.19 **(See color insert.)** Possible interaction of MSA-QDs with RBC membrane.

Indocyanine green (ICG) is a fluorescence dye which is used for near-infrared imaging. There is a limited application of this dye due to its various disadvantageous properties in aqueous solution, which also includes concentration-dependent aggregation, poor aqueous stability in vitro and low quantum yield. ICG is highly bound to nonspecific plasma proteins, which leads to rapid elimination from the body with a half-life of 3–4 min. ICG was made to encapsulate within various micellar systems so as to overcome these limitations. The micellar systems have been characterized by their optical properties, particle size distribution, zeta potential, and hemolytic activity. Encapsulation efficiency was determined using analytical ultracentrifugation.

ICG is a water-soluble, amphiphilic tricarbocyanine dye which provides a wide range of applications because of its low toxicity and capacity to absorb and emit in the near-infrared spectral range. ICG is used for many therapeutic and diagnostic applications such as cardiac output. ICG's optical activity depends upon the concentration of dye and the solvent used. ICG forms dimmers and oligomers in aqueous solutions at dye concentrations over 3.9 mg/L and this leads to nanoapplicability of Beer–Lambert's law. Large J-aggregates are formed at concentration greater than 103 mg/L. Self-quenching of the dye occurs due to the presence of aggregates and the fluorescence is reduced. Also the degradation takes place in aqueous media which results in simultaneous loss of absorption and fluorescence. It is due to saturation of the double bonds in the conjugated chain in the case of aqueous solutions solvent radicals and ions occur and the degradation process gets activated. ICG possess amphiphilic character and due to this it interact with both the lipophilic and hydrophilic molecules in plasma, ICG binds with the major plasma proteins such as albumin, globulins, and lipoproteins.

And after the binding, ICG is excreted by means of the liver. The uptake of ICG from the plasma is ambiguous. After the dissociation of the protein ICG complex, the ICG molecule is carried across the sinusoidal membrane via active carrier-mediated transport. Through the hepatocyte, ICG is vesicular transported and is secreted. In peripheral tissue, there is negligible uptake of the dye and excretion by the kidney does not occur. Thus, rapid elimination from the bloodstream occurs through ICG, with an initial half-life of 3–4 min to overcome the high degradation rate and the short plasma half-life. Various nanosized carriers for ICG have been investigated. Blood half-life and absorption capacity can be increased by oil-in-water emulsions.

8.8 CONCLUSION

The structure and properties of various NPs make them useful for a large variety of biological application. Toxicity is being observed at high concentrations using these systems. Hemolysis and bacterial viability assays have been used to explore differential toxicity among the various cell types. These studies show that cationic particles are moderately toxic, whereas anionic particles are quite nontoxic. Cell membrane-coated NPs by using the surface coating interacts with the pathogenic materials due to which these are used for various therapeutic procedures. Concentration-dependent lyses which were mediated by initial electrostatic binding were observed in case of dye release studies. In case of lipid bilayer, it prevents the damage to the outer surface of RBC. A redox-responsive drug delivery system has been developed. The drug and the targeting ligand have been conjugated in the interior and exterior through the cleavable disulphide bond. The oHA increases the stability and biocompatibility under physiological conditions. The drug and oHA exhibited highly redox-responsive release. The multivalence interactions allow the particles to interface with the diseases and exhibited the binding sites and trained the immune system against antigens.

KEYWORDS

- **biocompatibility**
- **cytotoxicity**
- **drug delivery**
- **hemolysis**
- **nanotechnology**

REFERENCES

1. Kamaly, N.; Yameen, B.; Wu, J.; Farokhzad, O. C. Degradable Controlled-release Polymers and Polymeric Nanoparticles: Mechanisms of Controlling Drug Release. *Chem. Rev.* **2016,** *116*, 2602–2663.
2. Snipstad, S.; Westrom, S.; Morch, Y.; Afadzi, M.; Aslund, A. K.; de Lange Davies, C. Contact-mediated Intracellular Delivery of Hydrophobic Drugs from Polymeric Nanoparticles. *Cancer Nanotechnol.* **2014,** *5*, 8.
3. Zhang, S.; Gao, H.; Bao, G. Physical Principles of Nanoparticle Cellular Endocytosis. *ACS Nano.* **2015,** *9*, 8655–8671.

4. Blanco, E.; Shen, H.; Ferrari, M. Principles of Nanoparticle Design for Overcoming Biological Barriers to Drug Delivery. *Nat. Biotechnol.* **2015,** *33*, 941–951.
5. Bazak, R.; Houri, M.; El Achy, S.; Hussein, W.; Refaat, T. Passive Targeting of Nanoparticles to Cancer: A Comprehensive Review of the Literature. *Mol. Clin. Oncol.* **2014,** *2*, 904–908.
6. Assanhou, A. G.; Li, W.; Zhang, L.; Xue, L.; Kong, L.; Sun, H.; Mo, R.; Zhang, C. Reversal of Multidrug Resistance by Co-delivery of Paclitaxel and Lonidamine Using a TPGS and Hyaluronic Acid Dual-functionalized Liposome for Cancer Treatment. *Biomaterials* **2015,** *73*, 284–295.
7. Zhou, H.; Fan, Z.; Deng, J.; Lemons, P. K.; Arhontoulis, D. C.; Bowne, W. B.; Cheng, H. Hyaluronidase Embedded in Nanocarrier PEG Shell for Enhanced Tumor Penetration and Highly Efficient Antitumor Efficacy. *Nano Lett.* **2016,** *16*, 3268–3277.
8. Kranz, L. M.; Diken, M.; Haas, H.; Kreiter, S.; Loquai, C.; Reuter, K. C.; Meng, M.; Fritz, D.; Vascotto, F.; Hefesha, H., et al. Systemic RNA Delivery to Dendritic Cells Exploits Antiviral Defence for Cancer Immunotherapy. *Nature* **2016,** *534*, 396–401.
9. Siefert, A. L.; Caplan, M. J.; Fahmy, T. M. Artificial Bacterial Biomimetic Nanoparticles Synergize Pathogen-associated Molecular Patterns for Vaccine Efficacy. *Biomaterials* **2016,** *97*, 85–96.
10. Carrillo-Conde, B.; Song, E. H.; Chavez-Santoscoy, A.; Phanse, Y.; Ramer-Tait, A. E.; Pohl, N. L.; Wannemuehler, M. J.; Bellaire, B. H.; Narasimhan, B. Mannose-functionalized "Pathogen-like" Polyanhydride Nanoparticles Target C-type Lectin Receptors on Dendritic Cells. *Mol. Pharm.* **2011,** *8*, 1877–1886.
11. Giljohann, D. A.; Seferos, D. S.; Prigodich, A. E.; Patel, P. C.; Mirkin, C. A. Gene Regulation with Polyvalent siRNANanoparticle Conjugates. *J. Am. Chem. Soc.* **2009,** *131*, 2072–2073.
12. Le Ouay, B.; Stellacci, F. Antibacterial Activity of Silver Nanoparticles: A Surface Science Insight. *Nano Today* **2015,** *10*, 339–354.
13. Steichen, S. D.; Caldorera-Moore, M.; Peppas, N. A. A Review of Current Nanoparticle and Targeting Moieties for the Delivery of Cancer Therapeutics. *Eur. J. Pharm. Sci.* **2013,** *48*, 416–427.
14. Jia, H.; Titmuss, S. Polymer-functionalized Nanoparticles: From Stealth Viruses to Biocompatible Quantum Dots. *Nanomedicine* **2009,** *4*, 951–966.
15. Yen, S. K.; Janczewski, D.; Lakshmi, J. L.; Dolmanan, S. B.; Tripathy, S.; Ho, V. H.; Vijayaragavan, V.; Hariharan, A.; Padmanabhan, P.; Bhakoo, K. K., et al. Design and Synthesis of Polymer-functionalized NIR Fluorescent Dyes–Magnetic Nanoparticles for Bioimaging. *ACS Nano.* **2013,** *7*, 6796–6805.
16. Vertegel, A. A.; Reukov, V.; Maximov, V. Enzyme-nanoparticle Conjugates for Biomedical Applications. *Methods Mol. Biol.* **2011,** *679*, 165–182.
17. Arruebo, M.; Valladares, M.; Gonzalez-Fernandez, A. Antibody-conjugated Nanoparticles for Biomedical Applications. *J. Nanomater.* **2009,** 439389.
18. Mout, R.; Moyano, D. F.; Rana, S.; Rotello, V. M. Surface Functionalization of Nanoparticles for Nanomedicine. *Chem Soc. Rev.* **2012,** *41*, 2539–2544.
19. Chan, J. M.; Monaco, C.; Wylezinska-Arridge, M.; Tremoleda, J. L.; Gibbs, R. G. Imaging of the Vulnerable Carotid Plaque: Biological Targeting of Inflammation in

Atherosclerosis Using Iron Oxide Particles and MRI. *Eur. J. Vasc. Endovasc. Surg.* **2014,** *47*, 462–469.

20. Kwon, H. J.; Cha, M. Y.; Kim, D.; Kim, D. K.; Soh, M.; Shin, K.; Hyeon, T.; Mook-Jung, I. Mitochondria-targeting Ceria Nanoparticles as Antioxidants for Alzheimer's Disease. *ACS Nano*. **2016,** *10*, 2860–2870.
21. Yeh, C. Y.; Hsiao, J. K.; Wang, Y. P.; Lan, C. H.; Wu, H. C. Peptide-conjugated Nanoparticles for Targeted Imaging and Therapy of Prostate Cancer. *Biomaterials* **2016,** *99*, 1–15.
22. Hu, C. M.; Kaushal, S.; Cao, H. S. T.; Aryal, S.; Sartor, M.; Esener, S.; Bouvet, M.; Zhang, L. Half-antibody Functionalized Lipid-polymer Hybrid Nanoparticles for Targeted Drug Delivery to Carcinoembryonic Antigen Presenting Pancreatic Cancer Cells. *Mol. Pharm.* **2010,** *7*, 914–920.
23. Herrmann, I. K.; Urner, M.; Graf, S.; Schumacher, C. M.; Roth- Z'graggen, B.; Hasler, M.; Stark, W. J.; Beck-Schimmer, B. Endotoxin Removal by Magnetic Separation-based Blood Purification. *Adv. Healthcare Mater*. **2013,** *2*, 829–835.
24. Herrmann, I. K.; Urner, M.; Koehler, F. M.; Hasler, M.; Roth- Z'graggen, B.; Grass, R. N.; Ziegler, U.; Beck-Schimmer, B.; Stark, W. J. Blood Purification Using Functionalized Core/Shell Nanomagnets. *Small* **2010,** *6*, 1388–1392.
25. Hoshino, Y.; Koide, H.; Urakami, T.; Kanazawa, H.; Kodama, T.; Oku, N.; Shea, K. J. Recognition, Neutralization, and Clearance of Target Peptides in the Bloodstream of Living Mice by Molecularly Imprinted Polymer Nanoparticles: A Plastic Antibody. *J. Am. Chem. Soc*. **2010,** *132*, 6644–6645.
26. Li, H.; Li, Y.; Jiao, J.; Hu, H. M. Alpha-alumina Nanoparticles Induce Efficient Autophagy-dependent Cross-presentation and Potent Antitumour Response. *Nat. Nanotechnol.* **2011,** *6*, 645–650.
27. Lin, A. Y.; Lunsford, J.; Bear, A. S.; Young, J. K.; Eckels, P.; Luo, L.; Foster, A. E.; Drezek, R. A. High-density Sub-100-nm Peptide-gold Nanoparticle Complexes Improve Vaccine Presentation by Dendritic Cells In Vitro. *Nanoscale Res. Lett.* **2013,** *8*, 72.
28. Clemente-Casares, X.; Blanco, J.; Ambalavanan, P.; Yamanouchi, J.; Singha, S.; Fandos, C.; Tsai, S.; Wang, J.; Garabatos, N.; Izquierdo, C., et al. Expanding Antigen-specific Regulatory Networks to Treat Autoimmunity. *Nature* **2016,** *530*, 434–440.
29. Pandey, A.; Roy, M. K.; Pandey, A.; Zanella, M.; Sperling, R. A.; Parak, W. J.; Samaddar, A. B.; Verma, H. C. Chloroformand Water-soluble Sol-Gel Derived Eu+++/Y2o3 (Red) and Tb++ +/Y2o3 (Green) Nanophosphors: Synthesis, Characterization, and Surface Modification. *I.E.E.E. Trans. Nanobiosci.* **2009,** *8*, 43–50.
30. Hu, C. M.; Zhang, L.; Aryal, S.; Cheung, C.; Fang, R. H.; Zhang, L. Erythrocyte Membrane-camouflaged Polymeric Nanoparticles as a Biomimetic Delivery Platform. *Proc. Natl. Acad. Sci. U. S. A*. **2011,** *108*, 10980–10985.
31. Boone, C. W.; Ford, L. E.; Bond, H. E.; Stuart, D. C.; Lorenz, D. Isolation of Plasma Membrane Fragments from HeLa Cells. *J. Cell Biol*. **1969,** *41*, 378–392.
32. Suski, J. M.; Lebiedzinska, M.; Wojtala, A.; Duszynski, J.; Giorgi, C.; Pinton, P.; Wieckowski, M. R. Isolation of Plasma Membrane-associated Membranes from Rat Liver. *Nat. Protoc*. **2014,** *9*, 312–322.
33. Pang, Z.; Hu, C. M.; Fang, R. H.; Luk, B. T.; Gao, W.; Wang, F.; Chuluun, E.; Angsantikul, P.; Thamphiwatana, S.; Lu, W., et al. Detoxification of Organophosphate Poisoning Using Nanoparticle Bioscavengers. *ACS Nano.* **2015,** *9*, 6450–6458.

34. Kumaresh, S. S.; Tejraj, M. A.; Anandrao, R. K.; Walter, E. R. J. *Controlled* Release **2001,** *70*, 1–20.
35. Sasidharan, A.; Monteiro-Riviere, N. A. Biomedical Applications of Gold Nanomaterials: Opportunities and Challenges. *Wiley Interdiscip. Rev. Nanomed. Nanobiotechnol.* **2015,** *7* (6), 779–796.
36. Lin, Z.; Monteiro-Riviere, N. A.; Riviere, J. E. Pharmacokinetics of Metallic Nanoparticles. *Wiley Interdiscip. Rev. Nanomed. Nanobiotechnol.* **2015,** *7* (2), 189–217.
37. Cai, W.; Gao, T.; Hong, H.; Sun, J. Applications of Gold Nanoparticles in Cancer Nanotechnology. *Nanotechnol. Sci. Appl.* **2008,** *1* (1).
38. Giljohann, D. A.; Seferos, D. S.; Daniel, W. L.; Massich, M. D.; Patel, P. C.; Mirkin, C. A. Gold Nanoparticles for Biology and Medicine. *Angew. Chem. Int. Ed.* **2010,** *49* (19), 3280–3294.
39. Thakor, A. S.; Jokerst, J.; Zavaleta, C.; Massoud, T. F.; Gambhir, S. S. Gold Nanoparticles: A Revival in Precious Metal Administration to Patients. *Nano Lett.* **2011,** *11* (10), 4029–4036.
40. Dykman, L.; Khlebtsov, N. Gold Nanoparticles in Biomedical Applications: Recent Advances and Perspectives. *Chem. Soc. Rev*. **2012,** 2256–2282.
41. Docter, D.; Strieth, S.; Westmeier, D.; Hayden, O.; Gao, M.; Knauer, S. K.; Stauber, R. H. No King Without a Crown–Impact of the Nanomaterial-protein Corona on Nanobiomedicine. *Nanomedicine* **2015,** *10* (3), 503–519.
42. Hruby, V. J. Designing Peptide Receptor Agonists and Antagonists. *Nat. Rev. Drug Discov.* **2002,** *1*, 847–858.
43. Opalinska, J. B.; Gewirtz, A. M. Nucleic-acid Therapeutics: Basic Principles and Recent Applications. *Nat. Rev. Drug Discov.* **2002,** *1*, 503–514.
44. LaVan, D. A.; Lynn, D. M.; Langer, R. Moving Smaller in Drug Discovery and Delivery. *Nat. Rev. Drug Discov.* **2002,** *1*, 77–84.
45. Voet, D.; Voet, J. G. *Biochemistry*; John Wiley and Sons, Inc.: New York, 1995.
46. Ferrari, M. Cancer Nanotechnology: Opportunities and Challenges. *Nat. Rev. Cancer* **2005,** *5*, 161–171.
47. Ghosh, P.; Han, G.; De, M.; Kim, C.; Rotello, V. Gold Nanoparticles in Delivery Applications. *Adv. Drug Delivery Rev*. **2008,** *60*, 1307–1315.
48. Farokhzad, O. C.; Langer, R. Impact of Nanotechnology on Drug Delivery. *ACS Nano*. **2009,** *3*, 16–20.
49. Youan, B. B. C. Impact of Nanoscience and Nanotechnology on Controlled Drug Delivery. *Nanomedicine* **2008,** *3*, 401–406.
50. Xie, J.; Lee, S.; Chen, X. Y. Nanoparticle-based Theranostic Agents. *Adv. Drug Delivery Rev.* **2010,** *62*, 1064–1079.
51. Hong, R.; Han, G.; Fernandez, J. M.; Kim, B. J.; Forbes, N. S.; Rotello, V. M. Glutathione-mediated Delivery and Release Using Monolayer Protected Nanoparticle Carriers. *J. Am. Chem. Soc.* **2006,** *128*, 1078–1079.
52. Langer, R. Drug Delivery and Targeting. *Nature* **1998,** *392*, 5–10.
53. Soppimath, K. S.; Aminabhavi, T. M.; Kulkarni, A. R.; Rudzinski, W. E. Biodegradable Polymeric Nanoparticles as Drug Delivery Devices. *J. Controlled Release* **2001,** *70*, 1–20.
54. Cedervall, T.; Lynch, I.; Lindman, S.; Berggård, T.; Thulin, E.; Nilsson, H.; Dawson, K. A.; Linse, S. *Proc. Natl. Acad. Sci. U. S. A.* **2007,** *104*, 2050–2055.

CHAPTER 9

MAGNETIC SEPARATION: A NANOTECHNOLOGY APPROACH FOR BIOLOGICAL MOLECULES PURIFICATION

ANA KARINA PÉREZ-GUZMÁN[1], ARIEL GARCÍA-CRUZ[1], ANNA ILYINA[1,*], RODOLFO RAMOS-GONZÁLEZ[2], MÓNICA L. CHÁVEZ-GONZÁLEZ[1], ARTURO I. MARTÍNEZ-ENRÍQUEZ[3], JUAN ALBERTO ASCACIO-VALDES[4], ALEJANDRO ZUGASTI CRUZ[4], MARÍA LOURDES VIRGINIA DÍAZ-JIMÉNEZ[3], ELDA PATRICIA SEGURA CENICEROS[1], JOSÉ LUIS MARTÍNEZ HERNÁNDEZ[1], and CRISTÓBAL NOÉ AGUILAR[4]

[1]*Research Group of NanoBioscience, Chemistry School, Autonomous University of Coahuila, Saltillo 25280, Coahuila, México*

[2]*CONACYT, Autonomous University of Coahuila, Saltillo 25280, Coahuila, México*

[3]*Center for Research and Advanced Studies of the National Polytechnic Institute, CINVESTAV, Saltillo 25900, Ramos Arizpe, Coahuila, Mexico*

[4]*Research Group of Science Food Department, Chemistry School, Autonomous University of Coahuila, Bvd. V. Carranza e Ing. J. Cáredenas V., Col. República, Saltillo 25280, Coahuila, México*

[*]*Corresponding author. E-mail: annailina@uadec.edu.mx*

ABSTRACT

Extraction and purification are the procedures widely applied in almost all areas of science, biosciences, and biotechnology. Recently, magnetic separation techniques based on magnetic nanoparticles application have been used to achieve this goal. The present review describes some aspects related to properties of magnetic nanosystems, functional groups, and examples of their use for protein and antioxidant extraction. The purpose of this paper is to summarize various strategies and materials which can be used for the isolation and purification of biochemical compounds with the help of magnetic field. It is hoped that the use of magnetic nanosystems will lead to the development of more efficient, economical, and environmentally-friendly technological processes applied to the food and biotechnology industries.

9.1 INTRODUCTION

Isolation and separation of biochemical molecules are used in all areas of biological sciences and biotechnology. It is known that the cost of purification is that defines the price of the final product. Various procedures can be used to achieve this goal. However, recently, more attention has been given to the development and application of magnetic separation techniques where magnetic nanoparticles (MNPs) are used.[2,3]

The MNPs possess attractive properties such as catalysts of chemical reactions, magnetic resonance imaging, biomedicine, sensors, magnetic recording, nanofluids, environment remediation, etc. Most applications are based on possibility of developing magnetic separation in several processes. Some biocatalysts (enzymes and microbial cells) applied in food industry can be immobilized on magnetic nanocarriers to be reused in repeated cycles of catalysis.[12,33]

Purification by means magnetic separation is one of the most versatile processes used in biotechnology. By means of this method, cells, viruses, proteins, and nucleic acids can be purified directly from the crude samples. The rapid process in combination with its easy development and automation provides unique advantages to perform the separation technique.[2,32,37] In general, magnetic affinity separations can be performed in two different modes. In the direct method, an appropriate affinity ligand is directly coupled to the magnetic particles or biopolymer exhibiting the affinity toward target compound(s) is used in the course of preparation of magnetic affinity particles. These particles are added to the sample and target compounds then bind to them. In the indirect method, the free affinity ligand (in most cases an appropriate antibody) is added to the solution or suspension to enable the interaction with the target compound. The resulting complex is then captured by appropriate magnetic particles with affined ligand. The two methods perform equally well, but, in general, the direct technique is more controllable. The indirect procedure may perform better if affinity ligands have poor affinity for the target compound.[31]

The present review briefly describes some aspects of enzymes and antioxidants purification by means of MNP use, mentioning some of their important properties and technique for their obtaining.

9.2 MNPs AND THEIR FUNCTIONALIZATION

Comprehension of nanoparticles magnetic properties is an important subject for magnetic nanomaterials development. MNPs themselves are employed as the components of different systems: recording tapes, ferrofluids, flexible disk recording media, as well as catalysts and biomedical materials, etc. It is the reason why MNPs are the object of studies in several fields of research from chemistry and physics to biotechnology and medicine.[17]

At small dimension of magnetic particles, in a range less than a certain critical diameter, the total magnetic energy is so small that there is no magnetic wall within the particle, and a case of single magnetic domain particle is considered. This critical diameter is approximately, $2A^{1/2}/M_S$, where A = exchange constant, M_S = moment per unit volume.[9] The comportment of single domain particles can be described by supposing that all atomic moments are strictly aligned as a single spin that is used as the essence of the superparamagnetism theory.[17] Superparamagnetism is frequently presented in nanomagnetic materials: nanoparticles or magnetic fine particles remaining in the metal. Superparamagnetic substances would not show magnetism without external magnetic field because the magnetic moment easily changes the direction with thermal vibration.[12] This characteristic can be useful to develop the processes involving magnetic separation of nanoparticles with affinity groups to some biochemical compounds. This constitutes the key point for magnetic separation application in biotechnology.

To simplify the recovery process, MNPs have been modified with compounds, which contain the chemical groups to modify them by chemical reactions.[23] Moreover, coating leads to stabilize magnetic systems and prevent nanoparticle agglomeration. Table 9.1 describes some magnetic system coated with biopolymers and some of their applications.

TABLE 9.1 Some Magnetic Nanosystems Contained Biopolymer and Their Applications.

Type of synthesized nanostructure	Used biopolymer	Application	References
Iron loaded alginate	Alginate	An efficient delivery system for Fe^{2+}	[14]
Polypyrrole/Fe_3O_4/ alginate bead	Alginate	It is used as magnetic solid-phase extraction sorbent for the extraction and enrichment of estriol, β-estradiol and bisphenol A in water samples	[4]
Fe_3O_4-k-carrageenan	Carrageenan	Pullulanase inmobilized onto Fe_3O_4-k-carrageenan nanoparticles could be used to continuous starch processing applications in the food industry	[18]
Magnetite-maghemite coated with chitosan in one coprecipitation step	Chitosan	Bivalent ion adsorbtion Carrier for lectin and enzyme immobilization	[11, 25, 26]

MNPs have been applied to selectively attach, manipulate, and transport targeted species to a desired location under the application of an external magnetic field. Owing to their size, MNPs feature a superparamagnetic behavior, offering an immense potential in a diversity of applications in their bare form or through attachment on their surface of a molecule or functional group for a specific application. Nanoparticles, used in many technological applications, can be synthesized by ball milling, coprecipitation, autocombustion, sol–gel, ultrasonic cavitation, and other methods.[15]

In analogy with affinity chromatography, the groups bound to the MNPs can serve as ligands for selective adsorption of the compounds. The incubation of these nanosystems with the mixtures containing the biochemical substances of interest leads to the formation of complexes. Later application of the magnetic field allows to separate these complexes in a precipitate, of which the biochemical compounds can be eliminated with selecting and application of the conditions that alter the interaction with ligand. This principle is very similar to affinity chromatography and the findings of this method can be useful to develop new systems of magnetic purification of compounds.

9.2.1 USE OF MNPs FOR PROTEIN EXTRACTION

MNPs can be used as a simple, fast method for the separation and purification of proteins and peptides[5,24,31] In recent years, magnetic and nanostructured materials have attracted great attention due to their unique physical, chemical, and structural properties. In these materials, there are magnetic fields of atomic origin, produced by a dipole or magnetic moment, which is united by a cohesive force known as "exchange energy" that allows the separation of nanocomposites. The efficiency of the magnetic separation depends on the quality of the MNPs and the stability of the bond between the magnetic support and "nonmagnetic" components.[26] Safarik and Safarikova[31] described the advantages of this technique and tools applied for its performance. The equipment applied for laboratory experiments is very simple. MNPs coated with polymer or biopolymer contained ligand immobilized on them to functionalize the surface exhibiting affinity for the target compound(s) are usually used. Magnetic separators of different types[31] can be used for magnetic extractions, but frequently cheap strong permanent magnets are equally efficient. Flow-through magnetic separators are usually more expensive, and high-gradient magnetic separators (HGMS) are the typical examples. Table 9.2 shows some examples of magnetic nanosystems applied for protein and peptide purification.

The magnetic nature of such adsorbent particles allows their selective manipulation and separation from suspended solids, colloids, or solutions. Therefore, it becomes possible to magnetically separate the selected target (biochemical molecules from cells, plasma, serum, and plant extracts, or complete cells or spores) simply by their linking the magnetic adsorbents before the application of a magnetic field.[27,28]

The use of MNPs for the adsorption or immobilization of biomolecules such as proteins, peptides, and others have the following advantages: (1) specific surface area that can be attached to large amounts of protein samples; (2) the selective orientation of biomolecules with other solids in suspension, even for a small volume of sample.[10] The application of functionalized magnetic adsorbent particles in combination with separation techniques has received considerable attention in recent years. The magnetically sensitive nature of such adsorbent particles allows their selective manipulation.[8] Magnetic separation techniques have several advantages compared with standard protein separation procedures. This process is usually very simple and with only a few steps of handling (Table 9.2). Biomolecules isolated using magnetic techniques have to be usually eluted from the magnetic separation materials. In most cases, bound proteins and peptides can be submitted to standard elution methods such as the change of pH, change of ionic strength, use of polarity reducing agents (e.g., dioxane or ethyleneglycol), or the use of deforming eluents containing chaotropic salts. Affinity elution (e.g., elution of glycoproteins from lectin-coated MNPs by the addition of free sugar) may be both a very efficient and gentle procedure.

TABLE 9.2 Example of Proteins Isolated with MNP.

Isolated protein	Applied nanoparticles	Attached ligand	References
Moringa oleifera coagulant protein	Iron oxide nanoparticles obtained from *water-in-oil* microemulsion systems (ME-MIONs)	Oil molecules	[24]
His-tagged human superoxide dismutase 1	Fe_3O_4/PMG/ IDA-Ni^{2+} nanoparticles	Ni(II)	[38]
Proteases secreted by *Bacillus licheniformis*	Iron oxide ferrite	Bacitracin	[16]
Aprotinin (trypsin inhibitor)	MNP of chitosan	Trypsin	[1]
Trypsin	Silanized magnetite particles	p-Aminobenzamidine	[31]
SA-α-2,6-Gal receptors affined to influenza virus	Magnetite/maghemite coated with chitosan	*Sambucus nigra* lectin	[11]
SA-α-2,3-Gal receptors affined to influenza virus	Magnetite/maghemite coated with chitosan	*Maackia amurensis* lectin	[6]

So, the magnetic separation applied for compounds purification is based on the principles similar to affinity chromatography, that is, as method that depends essentially on the interaction between the molecule to be purified and a solid phase with specific ligand that will allow the separation of pollutants.[29] This point is essential for the development of new systems for the magnetic separation applied for purification.

Both methods have in common "bioselective adsorption," a term which is used to refer to an adsorption based on an affinity between the desired biological compound and a molecule in an interaction with it.[19] For example, it may be the biological affinity between an enzyme and its substrate, inhibitor and/or other small ligand, usually in the active site of the enzyme. These ligands are classified according to their chemical nature or their selectivity for the retention of analytes, the latter is classified into specific ligands (antibodies) and general (such as lecithins).[19] Conventional affinity ligands have originated from natural sources such as peptides, oligonucleotides, antibodies, substrates, or other receptors binding proteins.[7] These ligands are usually extremely specific, but, at the same time, they are expensive and sometimes difficult to immobilize and preserve their biological activity.[11]

MNPs coated with polymers or biopolymers (Table 9.1) may be a good support for affinity ligands binding. They should be chemically inert or have a minimum of interaction with other molecules, and provide numerous free hydroxyl, carboxyl, amine, or other groups which is the most used for ligand union.[33]

During the development of magnetic separation methods, one of the key applications for this method is its use in the selective purification of biochemical products. The variety of ligands and supports that is now available has made this technique valuable for both small- and large-scale purification methods. The purity of a biochemical compound is an important requirement for its biological or industrial application and depends on their use. For example, for therapeutic purposes, very high purity (> 99.9%) is mandatory to minimize the danger of side effects or immunogenic response, whereas for industrial applications, such as in the food industry, a lower purity is sufficient. Magnetic separation may be useful to allow both levels of purity.

As a result of the many current applications for this method, affinity magnetic separation is expected to continue to grow and develop in the future as a vital tool in areas ranging from the production of biopharmaceuticals to clinical trials, environmental testing, pharmaceutical testing, biomedical and food chemistry researches.

9.2.2 USE OF MNPs FOR ANTIOXIDANTS EXTRACTION

The motivating application of magnetic separation refers to the extraction of antioxidants which will bring extraordinary prospects to the food industry.

Aroma, taste, texture, consistency, and appearance can be altered in the presence of free radicals which produce food deterioration. Oxidation-causing factors may be avoided if substances to reduce the effect of free radicals are used. These substances are named antioxidants.[20] Antioxidants delay or prevent the substrate oxidation. They are used to keep the nutritional value of foods protecting them and consumers against reactive oxygen species (ROS), those of nitrogen (RNS) and chlorine (RCS) species[35] when those species are in excess in human body. ROS contribute in the development of different diseases such as cancer, aging, and cardiovascular and neurodegenerative diseases. The consumption of fresh fruits, vegetables, or teas offer natural antioxidants to combat those diseases.[22] Recently, antioxidants extractions from food residues or raw materials have been developed.

Classic and not so classic methods have been used to extract antioxidant from natural sources, such as conventional methods using solvents, ultrasonic assisted, pulse electric field, enzyme-assisted, microwave, pressurized low-polarity water, and others.[34] Magnetic extraction using nanosystems is a novel method to extract antioxidants. This technique has been still described poorly in the literature. Wu et al.[36] reported the use of a composite of graphene oxide (GO) and Fe_3O_4 nanoparticles to extract flavonoids from tea, wine, and urine samples. In this case, authors used the advantage of the high superficial area of GO and the magnetic properties of magnetite to extract quercetin, kaempferol, and luteolin. As authors point out, magnetic extraction has excellent adsorption efficiency and rapid separation from the mixture which contains to the antioxidant substances. Using an external magnetic field, the use of a lot of solvents applied in classical methods is avoided. Moreover, MNPs application allows to reduce the operation time of antioxidant extraction process and to adsorb chemical substances from large volume of samples.[36]

Qing et al. (2011) extracted eight flavonoids from *Rosa chinensis* using baicalin-functionalized magnetite. After extraction, flavonoids were recovered from MNPs using methanol as solvent. Baicalin, a flavone, has a high electronic conjugation system which stimulates the antioxidants adsorption. The functionalization of MNPs with baicalin allows the use of magnetic nanosystem for selective extraction of flavonoids.[30]

Nanoparticles can increase the antioxidant activity. Kanagesan et al.[13] reported synthesis of ferromagnetic nanoparticles of $ZnFe_2O_4$ and $CuFe_2O_4$ ferrite using a sol–gel self-combustion technique. Both synthesized ferrite nanoparticles were equally effective in scavenging

2,2-diphenyl-1-picrylhydrazyl hydrate (DPPH) free radicals: $ZnFe_2O_4$ and $CuFe_2O_4$ nanoparticles showed 30.57 ± 1.0% and 28.69 ± 1.14% scavenging activity at 125 μg/mL concentrations, respectively. In vitro cytotoxicity study performed with MCF-7 cells revealed that at low concentrations (<125 μg/mL) $ZnFe_2O_4$ and $CuFe_2O_4$ were nontoxic, suggesting their biocompatibility.

Antioxidants may be applied to synthetize nanoparticles. For example, an extract from *Sambucus nigra* L. fruits contained polyphenols was used to synthesized silver nanoparticles with diameters of around 26 nm.[21] These nanoparticles demonstrated antioxidant activity in ABTS test. The authors suggested that these nanoparticles could be used as pharmaceutical products against oxidative tissue injuries.[21]

9.3 CONCLUSIONS

The present analysis demonstrates that MNPs are widely applied for magnetic extraction of biochemical compounds. Magnetic nanosystems are an evolving innovative field which synergistically assimilates nanotechnological and biotechnological developments. Actually, real advances are achieved to improve their properties and selectivity, as well as stability and magnetic properties to be applied in different bioprocesses. Functional magnetic nanomaterials can be used as affinity nano-carriers to extract substances by the application of an external magnetic field. Excellent results are revealed for proteins, peptide, and antioxidants extractions. It is hoped that the use of magnetic nanosystems will lead to the development of more efficient, economical, and environmentally-friendly technological processes applied to the food and biotechnology industries.

ACKNOWLEDGMENT

The authors would thank the Mexican Council of Science and Technology (CONACyT) for its financial support by grant No. 213844 (PDCPN2013-01 CONACyT-Mexico). Also, they acknowledge CONACyT for the undergraduate and graduate scholarships, and for the financial support under the program "Cátedras-CONACyT" (Researcher No. 2498).

KEYWORDS

- **affinity ligand**
- **magnetic nanoparticles**
- **magnetic extraction**
- **nanosystems**
- **separation**

REFERENCES

1. An, X.; Su, Z.; Zeng, H. Preparation of Highly Magnetic Chitosan Particles and their Use for Affinity Purification of Enzymes. *J. Chem. Technol. Biotechnol.* **2003,** *78*, 596–600.
2. Borlido, L.; Azevedo, A.; Roque A.; Aire-Barros M. R. Magnetic Separations in Biotechnology. *Biotechnol. Adv.* **2013,** *31*, 1374–1385.
3. Bucak, S.; Jones, D. A.; Laibinis, P. E.; Hatton, T. A. Protein Separations Using Colloidal Magnetic Nanoparticles. *Biotechnol. Progr.* **2003,** *19*, 477–484.
4. Bunkoed, O.; Nurerk, P.; Wannapob, R.; Kanatharana, P. Polypyrrole-Coated Alginate/Magnetite Nanoparticles Composite Sorbent for the Extraction of Endocrine-disrupting Compounds. *J. Separation Sci.* **2016,** *39* (18), 3602–3609.
5. Cao, M.; Li, Z.; Wang, J.; Yu, W. W. Food Related Applications of Magnetic Iron Oxide Nanoparticles: Enzyme Immobilization, Protein Purification, and Food Analysis. *Trends Food Sci. Technol.* **2012,** *27*, 47–56.
6. Carrizales Álvarez, S. A.; Ilyina, A.; Gregorio Jáuregui, K. M.; Martínez Hernández, J. L.; Vazquez Gutiérrez, B. B.; Segura Ceniceros, E. P.; Zugasti Cruz, A.; Saade Caballero, H.; López Campos, R. C. Extraction and Immobilization of N-Sa-Alpha-2,3-Gal Receptors Using Magnetic Nanoparticles Coated with Chitosan and *Maackia amurensis* Lectin. *Appl. Biochem. Biotechnol.* **2014,** *174*, 1945–1958. DOI 10.1007/s12010-014-1178-6.
7. Clonis, Y. D. Affinity Chromatography Matures as Bioinformatic and Combinatorial Tools Develop. *J. Chromatography A* **2006,** *1101*, 1–24.
8. Franzreb, M.; Siemann-Herzberg, M.; Hobley, T. J.; Thomas, O. Protein Purification Using Magnetic Adsorbent Particles. *Appl. Microbiol. Biotechnol.* **2006,** *70*, 505–516.
9. Frei, E. H.; Shtrikman, S.; Treves, D. Critical Size and Nucleation Field of Ideal Ferromagnetic Particles. *Phys. Rev.* **1957,** *106*, 446–455. DOI:10.1103/PhysRev.106.446.
10. García, P. F.; Brammen, M.; Wolf, M.; Reinlein, S.; Freiherr von Roman, M.; Berensmeier, S. High-gradient Magnetic Separation for Technical Scale Protein Recovery Using Low Cost Magnetic Nanoparticles. *Separation and Purification Technol.* **2015,** *150*, 29–36. https://doi.org/10.1016/j.seppur.2015.06.024.
11. Gregorio-Jauregui, K. M.; Carrizalez-Alvarez, S. A.; Rivera-Salinas, J. E.; Saade, H.; Martinez, J. L.; López, R. G.; Segura, E. P.; Ilyina, A. Extraction and Immobilization

of SA-α-2,6-Gal Receptors on Magnetic Nanoparticles to Study Receptor Stability and Interaction with *Sambucus nigra* Lectin. *Appl. Biochem. Biotechnol.* **2014,** *172,* 3721–3735. DOI 10.1007/s12010-014-0801-x. ISSN: 0273-2289.

12. Ilyina, A.; Ramos-González, R.; Vargas-Segura, A.; Sánchez-Ramírez, J.; Palacios-Ponce, S. A.; Martínez-Hernández, J. L.; Segura-Ceniceros, E. P.; Contreras-Esquivel, J. C.; Aguilar-González, C. N. Magnetic Separation of Nanobio-structured Systems for Innovation of Biocatalytic Processes in Food Industry. In: *Novel Approaches of Nanotechnology in Food*; Grumezescu, A. M., Ed.; Elsevier, 2016; Vol. 1, pp 67–92. ISBN: 978-0-12-804308-0.
13. Kanagesan, S.; Hashim, M.; Aziz, A. A. B.; Ismail, I.; Tamilselvan, S.; Alitheen, N. B.; Swamy, M. K.; Rao, B. P. C. Evaluation of Antioxidant and Cytotoxicity Activities of Copper Ferrite ($Cufe_2o_4$) and Zinc Ferrite ($Znfe_2o_4$) Nanoparticles Synthesized by Sol-Gel Self-combustion Method. *Appl. Sci.* **2016,** *6* (184), 2–13. Available at: http://www.mdpi.com/2076-3417/6/9/184.
14. Katuwavila, N.; Perera, C.; Dahanayake, D.; Karunaratne, V.; Amaratunga, G.; Karunaratne, N. Alginate Nanoparticles Protect Ferrous from Oxidation: Potential Iron Delivery System. *Int. J. Pharm.* **2016,** *513* (1–2), 404–409. Available at: http://dx.doi.org/10.1016/j.ijpharm.2016.09.053.
15. Kharisov, B. I.; Rasika Dias, H. V.; Kharissova, O. V. Mini-review: Ferrite Nanoparticles in the Catalysis. *Arabian J. Chem.* **2014.** DOI: 10.1016/j.arabjc.2014.10.049.
16. Käppler, T.; Cerff, M.; Ottow, K.; Hobley, T.; Posten, C. *In situ* Magnetic Separation for Extracellular Protein Production. *Biotechnol. Bioeng.* **2009,** *102*, 535–545.
17. Kodama, R. H. Magnetic Nanoparticles. *J. Magnetism Magnetic Mater.* **1999,** *200*, 359–372. Doi:10.1016/S0304-8853(99)00347-9.
18. Long, J.; Xu, E.; Li, X.; Wu, Z.; Wang, F.; Xu, X.; Jin, Z.; Jiao, A.; Zhan, X. Effect of Chitosan Molecular Weight on the Formation of Chitosan-pullulanase Soluble Complexes and their Application in the Immobilization of Pullulanase onto Fe_3O_4-K-Carrageenan Nanoparticles. *Food Chem.* **2016,** *202*, 49–58. Available at: http://dx.doi.org/10.1016/j.foodchem.2016.01.119.
19. Lowe, C. R. Combinatorial Approaches to Affinity Chromatography. *Curr. Opin. Chem. Biol.* **2001,** *5*, 248–256.
20. Mohamed, R.; Pineda, M.; Aguilar, M. Antioxidant Capacity of Extracts from Wild and Crop Plants of the Mediterranean Region. *J. Food Sci.* **2007,** *72* (1), 59–63.
21. Moldovan, B.; Luminiţa, D.; Achim, M.; Clichici, S.; Filip, A. A Green Approach to Phytomediated Synthesis of Silver Nanoparticles Using *Sambucus Nigra* L. Fruits Extract and their Antioxidant Activity. *J. Mol. Liquids* **2016,** *221*, 271–278.
22. Moo-Huchin, V. M.; Moo-Huchin, M. I.; Estrada-León, R. J.; Cuevas-Glory, L.; Estrada-Mota, I. A.; Ortíz-Vázquez, E.; Betancur-Ancona, D.; Sauri-Duch, E. Antioxidant Compounds, Antioxidant Activity and Phenolic Content in Peel from Three Tropical Fruits from Yucatan, Mexico. *Food Chem.* **2015,** *166*, 17–22.
23. Netto, C. G. C. M.; Toma, H. E.; Andrade, L. H. Superparamagnetic Nanoparticles as Versatile Carriers and Supporting Materials for Enzymes. *J. Mol. Catalysis B Enzymatic* **2013,** *85–86*,71–92. Doi:10.1016/j.molcatb.2012.08.010.
24. Okoli, C.; Boutonnet, M.; Mariey, L.; Järås, S.; Rajarao, G. Application of Magnetic Iron Oxide Nanoparticles Prepared from Microemulsions for Protein Purification. *J. Chem. Technol. Biotechnol.* **2011,** *86*, 1386–1393.

25. Osuna, Y.; Gregorio-Jauregui, K. M.; Gaona-Lozano, J. G.; De La Garza-Rodríguez, I. M.; Ilyna, A.; Barriga-Castro, E. D.; Saade, H.; López, R. G. Chitosan-coated Magnetic Nanoparticles with Low Chitosan Content Prepared in One-step. *J. Nanomater.* **2012,** 1–7. DOI:10.1155/2012/327562.
26. Osuna, Y.; Sandoval, J.; Saade, H.; López, R. G.; Martinez, J. L.; Colunga, E. M.; Segura, E. P.; Arévalo, F. J.; Zon, M. A.; Fernández, H.; Ilyina, A. Immobilization of *Aspergillus niger* Lipase on Chitosan-coated Magnetic Nanoparticles Using Two Covalent-binding Methods. *Bioproc. Biosys. Eng.* **2015,** *38*, 1437–1445. DOI: 10.1007/s00449-015-1385-8.
27. Palacios Ponce, A. S.; Pérez Guzmán, A. K.; Olivares Tobanche, D. L.; Cortés Arganda, J. F.; Iliná, A.; Ramos González, R.; Ruiz Leza, H. A.; Martínez Hernández, J. L.; Segura Ceniceros, E. P.; Aguilar González, C. N.; Hernández Flores, H.; Aguilar González, M. A.; Martínez Enríquez, A. I. Solicitud de Patente: Sistema Superparamagnético Nanoestructurado Recubierto de Quitosán Para Inmovilización de Microorganismos, su Eliminación de Medio Líquido o Aplicación en Procesos de Fermentación. Datos de aplicación: Expediente MX/a/2016/008117, Folio MX/E/2016/041879.
28. Palacios-Ponce, S.; Ramos-González, R.; Ruiz, H. A.; Aguilar, M. A.; Martínez-Hernández, J. L.; Segura-Ceniceros, E. P.; Aguilar, C. N.; Michelena, G.; Ilyina, A. *Trichoderma* sp. spores and *Kluyveromyces marxianus* Cells Magnetic Separation: Immobilization on Chitosan-coated Magnetic Nanoparticles. *Preparative Biochem. Biotechnol.* **2017,** *47* (6), 554–561. https://doi.org/10.1080/10826068.2016.1275007.
29. Porath, J. Conditions for Biospecific Adsorption. *Biochimie* **1973,** *55*, 943–951.
30. Qing, L.; Xiong, J.; Xue, Y.; Liu, Y.; Guang, B.; Ding, L.; Liao, X. Using baicalin-functionalized magnetic nanoparticles for selectively extracting flavonoids from Rosa chinensis. J. Sep. Science 2011, 34, 3240-3245. DOI:10.1002/jssc.201100578
31. Safarik, I.; Safarikova, M. Magnetic Techniques for the Isolation and Purification of Proteins and Peptides. *BioMagnetic Res. Technol.* **2004,** *2*, 1–17.
32. Safarik, I.; Safarikova, M. Use of Magnetic Techniques for the Isolation of Cells. *J. Chromatogr. B: Biomed. Sci. Appl.* **1999,** *722*, 33–53.
33. Sánchez-Ramírez, J.; Martínez-Hernández, J. L.; Segura-Ceniceros, E. P.; Contreras-Esquivel, J. C.; Medina-Morales, M. A.; Aguilar, C. N.; Iliná, A. Artículo de Revisión: Inmovilización de Enzimas Lignocelulolíticas en Nanoparticulas Magnéticas. *Quimica Nova* **2014,** *37*, 504–512. http://dx.doi.org/10.5935/0100-4042.20140085.
34. Selvamuthukumaran, M.; Shi, J. Recent Advances in Extraction of Antioxidants from Plant By-products Processing Industries. *Food Quality and Safety* **2017,** *1*, 61–81.
35. Shahidi, F. Antioxidants in Food and Food Antioxidants. *Mol. Nutr. Food Res.* **2000,** *44* (3), 158–163.
36. Wu, J.; Xiao, D.; Zhao, H.; He, H.; Peng, J.; Wang, C.; Zhang, C.; He, J. A Nano-composite Consisting of Graphene Oxide and Fe_3O_4 Magnetic Nanoparticles for the Extraction of Flavonoids from Tea, Wine and Urine Samples. *Microchimica Acta* **2015,** *182*, 2299–2306.
37. Yavuz, C. T.; Prakash, A.; Mayo, J.; Calvin, V. L. Magnetic Separations: From Steel Plants to Biotechnology. *Chem.Eng. Sci.* **2009,** *64*, 2510–2521.
38. Zhou, Y.; Yan, D.; Yuan, S.; Chen, Y.; Fletcher, E. E.; Shi, H.; Han, B. Selective Binding, Magnetic Separation and Purification of Histidine-tagged Protein Using Biopolymer Magnetic Core-shell Nanoparticles. *Protein Expression and Purification* **2018,** *144*, 5–11.

CHAPTER 10

PECAN NUT EXTRACTS OBTAINED BY GREEN TECHNOLOGIES: ANTIMICROBIAL EFFECT AGAINST FOODBORNE PATHOGENS

JOSÉ LUIS VILLARREAL-LÓPEZ, JORGE A. AGUIRRE-JOYA, LLUVIA I LÓPEZ-LÓPEZ, RAÚL RODRÍGUEZ-HERRERA, JOSÉ L. MARTÍNEZ, JOSÉ SANDOVAL, and CRISTÓBAL NOÉ AGUILAR*

Facultad de Ciencias Químicas, Universidad Autónoma de Coahuila, Unidad Saltillo, México

**Corresponding author. E-mail: cristobal.aguilar@uadec.edu.mx*

ABSTRACT

Pecan nut (*Carya illinoensis* K.) is an economically important tree due to the fruit that it produces. It has been reported that extracts of some parts of the tree, like the leaves, husk, and shell of the fruit, has antimicrobial properties. Because of this it is possible to develop a green technology to extract natural antimicrobial agents such as polyphenols that has potential application in food industries as antimicrobials that permits substitute synthetic antimicrobials for natural ones. Leaves, husk, and shell of pecan nut were collected by a community in Parras de la Fuente, Coahuila, Mexico, and phenolic extracts were obtained by water heating, ohmic heating, and microwave at 40°C, 60°C, and 80°C, respectively, by 20 min for each sample. Obtained extracts were tested against *Escherichia coli* and *Staphylococcus aureus*. Present work demonstrates that extracts obtained by green technologies (ohmic and microwave) are more effective to inhibit growth of *E. coli* and *S. aureus*.

10.1 INTRODUCTION

The pecan nut tree *Carya illinoensis* (Koch) is an economically important tree due to the fruit that it produces. It belongs to the *Juglandaceae* family and the *Carya* gender. Its place of origin is said to be southeast of the United States and North of Mexico. The tree can rise to over 30 m in height and after 6–10 years, it starts its productive cycle, which can last for more than 50 years. Normally, it grows in sandy clay soil that is well-drained, nevertheless it also can be observed in alluvial soils of recent origin. At its original region, it develops in humid climate with annually rainfall average of 760 mm and a maximum of 2010 mm.[14] In actuality, Mexico is the largest producer around the world with a production of 125,758.45 t and a net income of 347,609,172.4 USD.

Pecan nut tree is a millennial plant originally found in the Middle East. It was brought to Europe maybe before the Roman age, and in America, near to the 17th century.[16] The appropriate climate for the development of the tree are summer temperatures ranging from 21°C to 46°C and winter temperatures, on an average, ranging from 10°C to −1°C, with extreme condition from −18°C to 29°C.[14]

The pecan nut fruit is considered a drupe, which consists of the pericarp, mesocarp, and seeds. Pericarp is a four-segmented structure that, when dehydrated, opens up to free the endocarp and the seed. This portion of endocarp and mesocarp is known as husk. Pecan nut (including endocarp and seed) normally measures from 2 to 6 cm in length and weighs from 4 to 12 g. The tree dormancy period comprises the months of December, January, and February. This time is recommended to do the corresponding agricultural management of the tree.[14]

Pecan nut has two different varieties: creole nut (natives or wild) that has small, hard shell and has high content of oil; and the fine or paper shelled-pecans that has a high percentage of nut and are thin-shelled.[7]

In Mexico, the most produced and sold pecan nuts are the improved varieties, like the Western (90%,) and Wichita (5%,).[7] The more important genders of the Juglandaceae family are: *Junglas* and *Carya*, which represents the Castilla walnut and the pecan nut tree, respectively.[7] It is estimated that 40% of the production is designated for national consumption, and the remaining 60%, is exported, mainly to the United States of America.[14]

Pecan nut tree (*Carya illinoenesis* K.) is an economically important crop in México not only because of the flavor of the fruit, but also because of the following: antioxidant, antibacterian, and antifungic of its compounds.[4]

There are some research studies which brings out the antioxidant, antimicrobial, and antifungal activities of pecan nut extracts[5] and such studies show the need to develop bioactive extracts from pecan nut through the use of green technology and use them for biological control of food pathogens.

There are diverse compounds identified in the leaves of pecan nut tree, like β-farnesene, limonene, myrcene, α and β cimene, α and β pinene and sabinene, sesquiterpenes, cariofilene y germacrene, and la juglone. Also, in pericarp, compounds found are cafeic acid, clorongenic, cumaric ferulic, and synaptic acids. The fruit contains quinonaes: jugloane and 1,4-naftaquinone. The bark contains juglones, betulin, and caffeic acid. The etanolic extract bark is effective against *Candida albicans* and *C. tropicales, Sporotrichum schenkii*. Also, aqueous extract has demonstrated activity against *Haemonchus contortis*. Salinas reports lethal doses (50) of ethanolic and aqueous extracts of leaves and husks of 1000 mg/kg in mice by intraepithelial injection.[20]

Those extracts are biological in origin and are biodegradable ones and manifest a minimum negative impact over human health and environment.

The identification of compounds present in different extracts of different parts of pecan nut will provide us the basis for their application over national food products to substitute synthetic chemicals often used to control or prevent spoilage but with certain degree of contamination or risk for the environment and human health Also, bioactive extracts will be obtained by green technologies, ensuring a sustainable product and process. This study focuses on the development of green, economic, effective, and fast technologies to obtain high quality and effective biological products to ensure food quality, with low impact to environment. Also, it is important to evaluate and compare novel extraction technologies like ohmic heating and microwave heating to ensure its extractive advantages.

The objective of this study is to obtain extracts of pecan nut leaves, husks, and shell by green technologies, like ohmic heating and microwave-assisted extraction compared with traditional extraction in hot water.

10.2 MATERIAL AND METHODS

10.2.1 SAMPLING

Sample collection of leaves, shell, and husks of pecan nut tree (*Carya illinoensis* K.) was done in Parras de la Fuente, Coahuila, during the months of September and October (commercial time of collection). The plots were characterized by type and frequency of irrigation, location, age of crop, application of pesticides, and the care of plots. Collected materials (leaves, husks, and shells) were let to dehydrate at room temperature in the shadows.

10.2.2 MAINTENANCE OF STRAINS

Strains of two representative food bacteria spoilage (*Escherichia coli* and *Staphylococcus aureus)* were donated by the Food Research Department of the Universidad Autonoma de Coahuila, Coahuila, Mexico. Bacterial strains were stored in nutritive broth and agar in tubes at 4°C. When needed, subcultures were taken.

10.2.3 EXTRACTS OF LEAVES, HUSKS, AND SHELLS

Samples of leaves, husks, and shells were dehydrated at room temperature in the shadows and pulverized in an electric mill. The obtained powder was screened in a sieve No. 40. Samples were maintained in plastic bags with hermetic sealed until further analysis. To obtain the extracts, powder (leaves, husks, and shells) wax of each of the samples were mixed at 7% in water, extracts were obtained by heating in water, ohomic heating, and microwave-assisted extraction, all of them at temperatures of 40°C, 60°C, and 80°C for 20 min.

10.2.4 EXTRACT OBTAINED BY OHMIC HEATING

The extract was prepared by 400 mL of a 7% suspension of powder of leaves, husks, and shells and were processed in the equipment of ohmic heating. The tree vegetal samples were processed individually at 40°C, 60°C, and 80°C by triplication; each one of the crude extracts was passed through muslin and filtered with a Wathman No. 5 filter. The extracts were stored in amber bottles at 4°C.

10.2.5 MICROWAVE-ASSISTED EXTRACTION

Aliquots of 120 mL of a 7% suspension of samples in water were used.. Suspension was irradiated with microwave (Microwave Laboratory

Systems, Ethos CFR, Microwave Continuous Flow Reactor and Cryo LAB) in a procedure consisting of three stages. Stage 1: it was of 40 s duration and had a potency of 800 watts with a temperature ranging from 24°C to 40°C; in stage 2, time was of 1 min and a potency of 800 watts at 40°C, 60°C, and 80°C for 20 min in each temperature by triplicated. Each one of the crude extracts was passed through muslin and filtered with a Wathman No. 5 filter. Extract was stored in amber bottles at 4°C (Proestos and Komaitis, 2008; Zhang et al., 2008).

10.2.6 DETERMINATION OF ANTIMICROBIAL EFFECT OF THE EXTRACTS

Escherichia coli and *Staphylococcus aureus* were cultured in Brain Heart Infusion and Agar Brain Heart and the number of bacteria was determined in a broth culture with 20 h incubation at 37°C by the method of counting viable plate This culture was planted by closed groove with a cotton swab in Brain Heart Agar boxes in which sterile filter paper discs were placed impregnated with each of the dilutions (1:2, 1:4, 1:8, 1:16, and 1:32) of each extract (leaves, husks, and shells) obtained at different temperatures (40°C, 60°C, and 80°C) for susceptibility testing.

10.2.7 STATISTICAL ANALYSIS

SAS statistical program "The Sistem SAS for Windows 9," Copyright 2002 by SAS Institute Inc. designed randomized complete block design with three replications and a factorial arrangement 3 × 3 with three levels (shell, leaf, and husk) was employed. A randomized complete block experimental design with factorial fix was used. Extraction process (microwave assisted extraction, ohmic heat assited extraction and reflux), kind of sample (shell, leaf, and husk) and extraction temperature (40°C, 60°C, and 80°C) were the evaluated variables. SAS statistical program "The Sistem SAS for Windows 9," Copyright 2002 by SAS Institute Inc. was used to analyze the otained results through of comparisons of mean values by multiple range test of Tukey.

10.3 RESULTS AND DISCUSSION

Drug-sensitivity spectrum of *S. aureus* and *E. coli* with WalkAway-96 equipment and standard techniques was obtained to determine the use of trimethoprim/sulfamethoxazole as a positive control. Figure 10.1 shows the results of sensitivity of both bacteria to antibiotic in the control test.

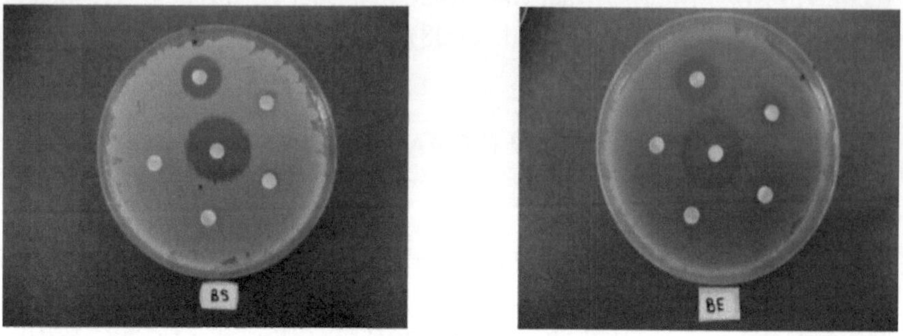

(a) (b)

FIGURE 10.1 Positive control of (a) *S. aureus* and (b) *E. coli*.

Table 10.1 shows the compounds identified in the extracts by high performance liquid chromatography (HPLC).

TABLE 10.1 Bioactive Compounds Identified in the Extracts of Pecan Nut by HPLC.

Bioactive compound	Husk	Leaf	Shell
Katequin	+	+	+
Galic acid	+	+	+
Cinamic acid/hidroxicinamic	+	+	
Resorcinol	+	+	
Quercetin	+		
Metil galate	+		
Pirogalol	+		
Clorogenic acid		+	+
Elagic acid		+	
p-coumáric acid		+	

Based on the obtained results (Fig. 10.2), it is concluded that the extracts have activity against *E. coli* and *S. aureus* and these showed significant

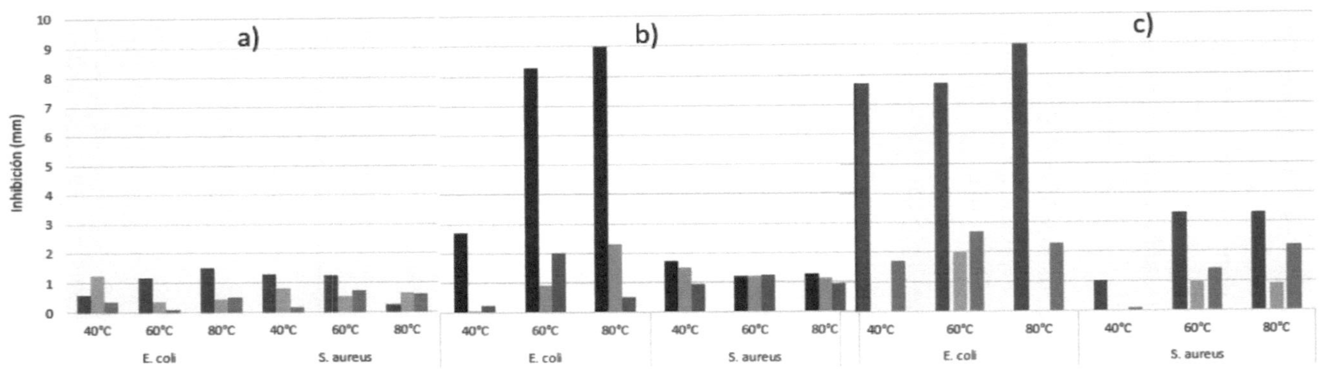

FIGURE 10.2 Antimicrobial capacity of aqueous extracts of (■) husk, (■) leaf, and (■) shell of pecan nut obtained by (a) reflux, (b) microwave, and (c) ohmic heating.

differences. The most effective was the husk, followed by leaf, and finally the shell. When extraction temperature is increased, the antimicrobial potential of samples is higher with microwave than ohmic heat. These showed significant differences, except for that samples obtained by reflux with water

The amount of total polyphenol extracted with following methods: reflux with water, microwave, and heat is highest for ohmic shell, followed by leaves, and finally, husk (*).

The extraction process of bioactive compounds should include a defined control de conditions such as temperture, time, amount and pretreatment of sample, kind of solvent, which affect the antimicrobial capacity of the extracted molecules.

The extraction method is critical to obtaining products with antibacterial activity, so these must be scaled with an appropriate methodology and should involve controlling of as many variables.

Among the factors that could explain this are: the effect of temperature, time, dilutions, the pH, the type of extraction, the sensitivity of the method, hydrolysis, or alteration of the extracts.

KEYWORDS

- **pecan nut**
- **ohmic heating**
- **antimicrobial activity**
- **microwave-assisted extraction**
- **extract**

REFERENCES

1. Albiplast-Bioplast. Dispersiones Acuosas Como Revestimientos de los Quesos. 2008. http://alinat.com.ar/download/fichas/alimold.pdf, (accessed Nov 6, 2016).
2. DL, O. B., Hernández-Rodríguez, O. A., Martínez-Téllez, J., Núñez-Barrios, A., & Perea-Portillo, E. Aplicación foliar de quelatos de zinc en nogal pecanero. Revista Chapingo. *Serie horticultura,* **2009,** *15* (2), 205–210.

3. Barrera-Necha, L.; y Bautista-Baños, S. Actividad Antifúngica de Polvos, Extractos y Fracciones de Cestrum Nocturnum L. Sobre el Crecimiento Micelial de Rhizopus Stolonifer (Ehrenb.:FR) Vuill. *Revista Mexicana de Fitopatología* **2008,** *26*, 27–31.
4. Bolling, B. W.; Chen, Diane C-Y. O.; McKay, L.; Blumberg J. B. Tree Nut Phytochemicals: Composition, Antioxidant Capacity, Bioactivity, Impact Factors. A Systematic Review of Almonds, Brazils, Cashews, Hazelnuts, Macadamias, Pecans, Pine Nuts, Pistachios and Walnuts. *Nutri. Res. Rev.* **2011,** *24* (2), 244–275.
5. De la Rosa, L. A.; Alvarez-Parrill, E.; Shahidi, F. Phenolic Compounds and Antioxidant Activity of Kernels and Shells of Mexican Pecan (*Carya illinoinensis*). *Agricult. Food Chem.* **2011,** *59* (1), 152–162.
6. Salminen, S.; Bouley, C.; Boutron-Ruault, C.; Cummings, J. H.; Franck, A.; Gibson, G. R.; Isolauri, E.; Moreau, M. C.; Roberfroid, M.; Rowland, I. Functional food science and gastrointestinal physiology and function. Brit. J. Nutr. **1998,** *80* (1), S147–S171.
7. Fundación Produce Chihuahua y JJCONSULTORES S.C. Programa Estratégico de Necesidades de la Investigación y Transferencia de Tecnología Nuez. 2003, 1–71.
8. Gimferrer, N. El Poder Antifúngico de los Aceites Esenciales de Cítricos. 2008. http://www.consumer.es/seguridad-alimentaria/sociedad-y-consumo/2008/05/14/176888.php (accessed Nov 6, 2016).
9. Gimferrer, N. La miel, posible sustituta de aditivos. 2008. http://www.consumer.es/seguridad-alimentaria/sociedad-y-consumo/2008/05/14/176888.php (accessed Nov 6, 2016).
10. Gómez, Y.; Gil, K.; González, E.**2**y Farías, L. M. Actividad Antifúngica de Extractos Orgánicos del Árbol *Fagara monophylla* (Rutaceae) en Venezuela. *Rev. Biol. Trop.* **2007,** *55* (3–4), 767–775.
11. INIFAP. Y SAGARPA. Actualización en Tecnología Para la Producción del Nogal Pecanero. 2009.
12. Jalife, C. Organización Para Fortalecer Nuestro Enfoque. "Experiencias de nogalero a nogalero" y "Huertas en proceso de cambio". XII Simposium Internacional Nogalero Nogatec, 2004.
13. PC-076-2007. Pliego de Condiciones Para el uso de la Marca Oficial México Calidad Suprema en Nuez Pecanera (Rev. 070307), 2007, 1–39.
14. Quintero-Mora, M.; Londoño-Orozco, A.; Hernández-Hernandez, F.; Manzano-Gayosso, P.; López-Martínez, R.; Soto-Zárate, C.; Carrillo-Miranda, L.; Penieres-Carrillo, G.; García-Tovar, C.; y Cruz-Sánchez, T. Efecto de Extractos de Propóleos Mexicanos de *Apis Mellifera* Sobre el Crecimiento In Vitro de *Candida albicans*. *Rev. Iberoam. Micol.* **2008,** *25*, 22–26.
15. Raissouni, T. Estudio Comparativo de la Eficacia de Varios Tratamientos Tópicos de la Estomatitis Aftosa Recurrente, Universidad Autónoma de Granada, Facultad de Odontología, 2005; pp 1–89.
16. Rios, N.; Medina, G.; Jiménez, J.; Yañez, C.; García, M.; Bernardo, M.; y Gualtieri, M. Actividad Antibacteriana y Antifúngica de Extractos de Algas Marinas Venezolanas. *Rev. Peru. Biol.* **2009,** *16* (1), 97–100.
17. Ruiz, Y.; Ramírez, V.; López, P.; Castillo, C.; Guzmán, F.; y Román, A. Buenas Prácticas de Manofactura Aplicadas al Malteado e Diferentes Variedades de Cebada (*Hordeum sativum* jess) Producidas en el Estado de Hidalgo. 2005. http://www.

respyn.vanl.mx/especiales/2005/ee-042005/cantles/normatividad.htlm (accessed Nov 6, 2016).
18. Salinas, N. Análisis de la Actividad de Extractos Crudos de *Junglans regia* L. *Juglans mollis* y *Carya illinoensis* Contra *Mycobacterium tuberculosis.* Thesis, Universidad Autónoma de Nuevo León, Facultad de Ciencias Biológicas, México, 2004. http://www.wipo.int/pctdb/en/wo.jsp?1a=es200507017&display=desc-47k-08-19 (accessed Nov 6, 2016).

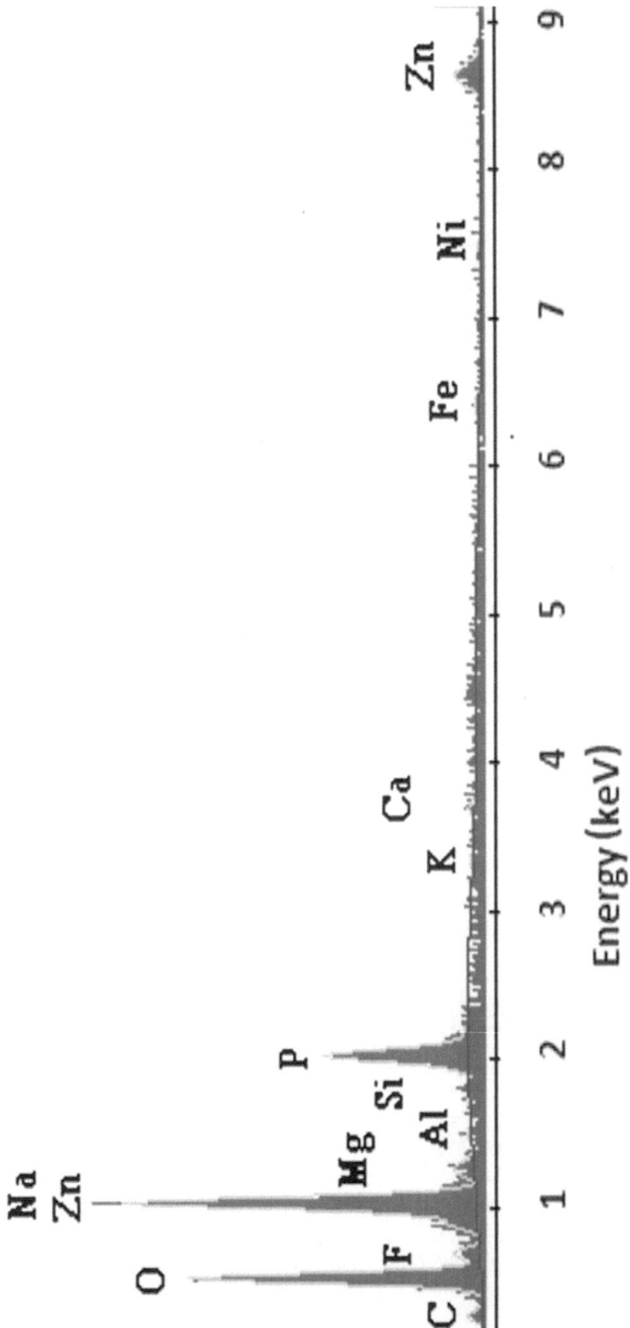

FIGURE 6.12 EDX spectrum of zinc phosphate.

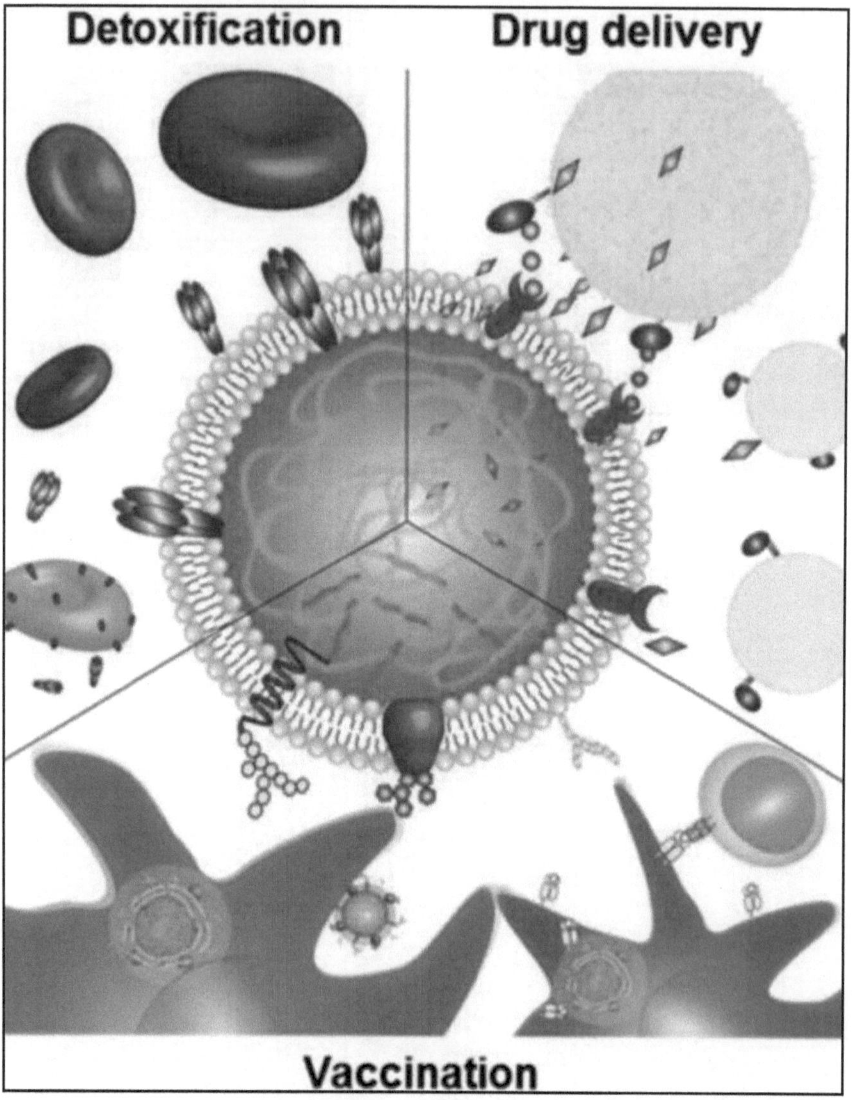

FIGURE 8.1 Illustration showing three applications of cell membrane-coated NPs. The particles can be used for targeted drug delivery to tumors, sites of inflammation, or pathogens via translocate surface markers (top right). They can also act as a decoy for toxins that damage cells through membrane interactions, safely detaining them and sparing their intended targets (top left). Finally, cell membrane-coated NPs faithfully present antigens from their source cells and can be used for improved anticancer or antibacterial vaccination (bottom).

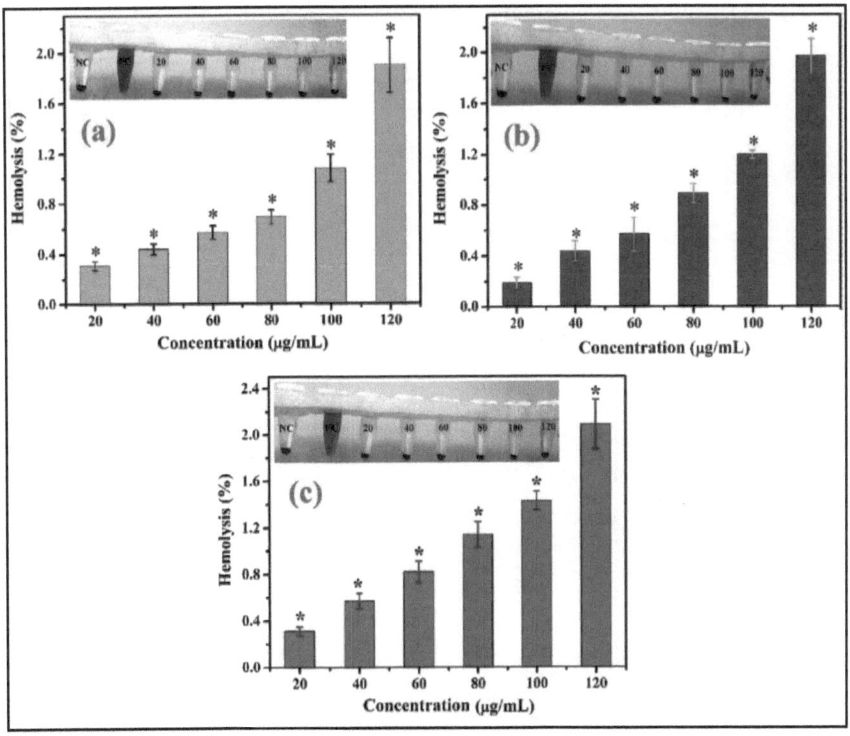

FIGURE 8.4 Hemolytic activity of dicationic amphiphile-stabilized AuNPs on human RBC at various concentrations ranging from 20 to 120 μg/mL (a) DCaC-AuNPs, (b) DCaDC-AuNPs, and (c) DCaLC-AuNPs. The inset shows the photographs of the corresponding solution. Here, (*) represents a significant difference ($p < 0.05$) compared with that of positive control.

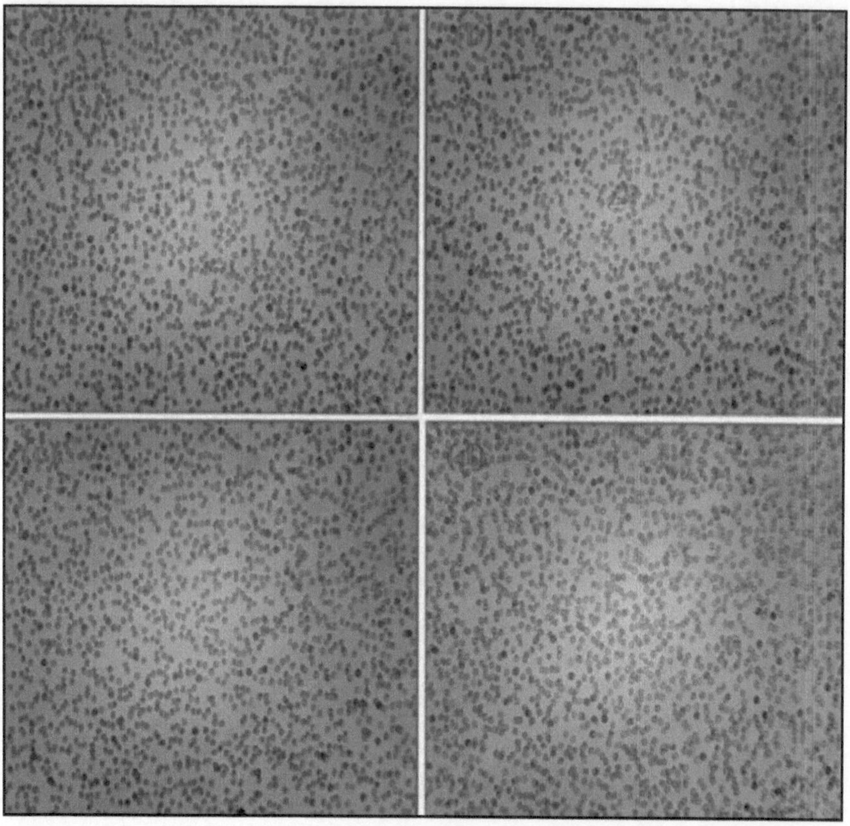

FIGURE 8.5 Photomicrographs of dicationic amphiphile-stabilized AuNP-treated human RBC by light microscopy: (a) control RBC, (b) RBC treated with DCaC-AuNPs, (c) DCaDC-AuNPs, and (d) DCaLC-AuNPs at the highest concentration of 120 μg/mL.

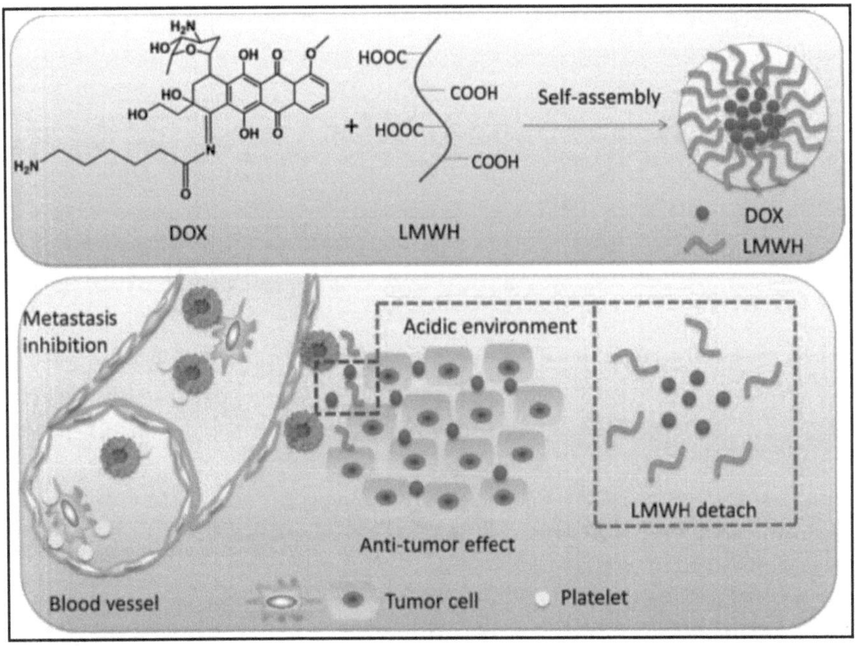

FIGURE 8.6 Schematic design of the multifunctional polymer–drug conjugate.

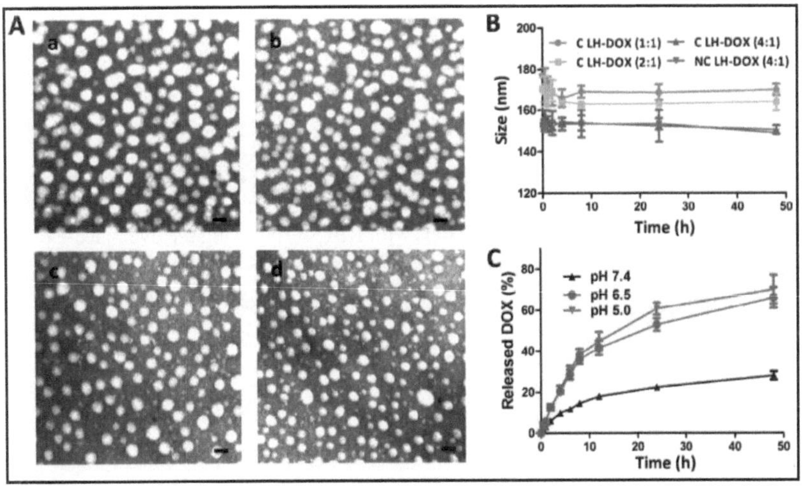

FIGURE 8.7 (A) Representative transmission electron microscopy images of NPs: (a) C (cleavable) LH–DOX 1:1, (b) 2:1, (c) 4:1, (d) NC (noncleavable) LH–DOX. The scale bar represents 150 nm. (B) Size changes of NPs in 50% FBS for 48 h. (C) Drug release profiles of C LH–DOX (4:1) under various pH conditions (pH 5.0, pH 6.5, and pH 7.4) for 48 h. Errors are standard errors of the mean (SEM) for N = independent experiments.

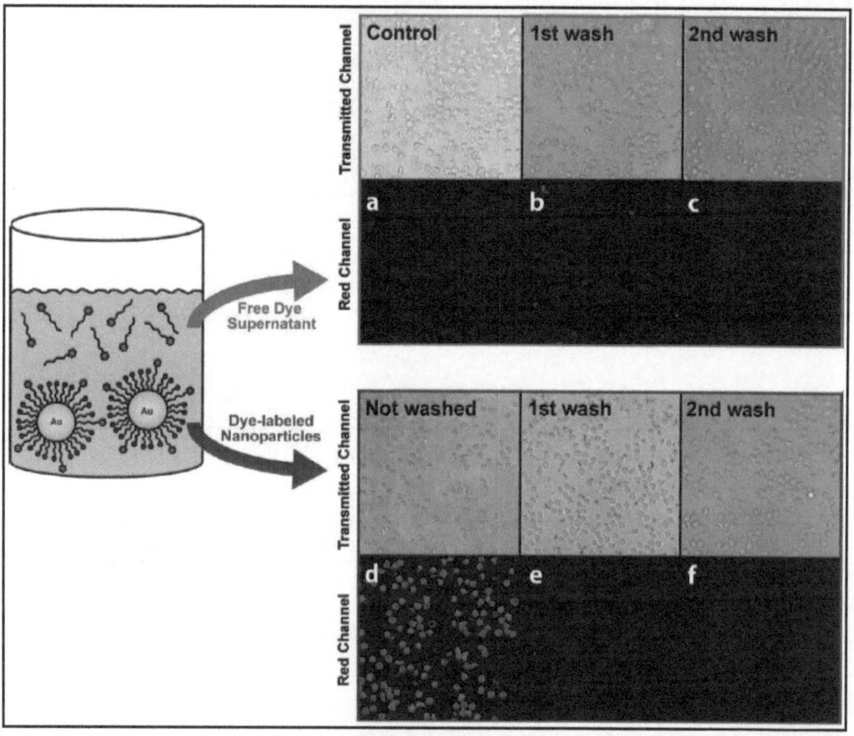

FIGURE 8.9 Dye-labeled NP incubations with RBCs. Following the dye place-exchange reaction, the supernatants (a–c) and dye-labeled NP pellets (d–f) of successive washes were incubated with RBCs. All of the unbound dye is removed in the first wash.

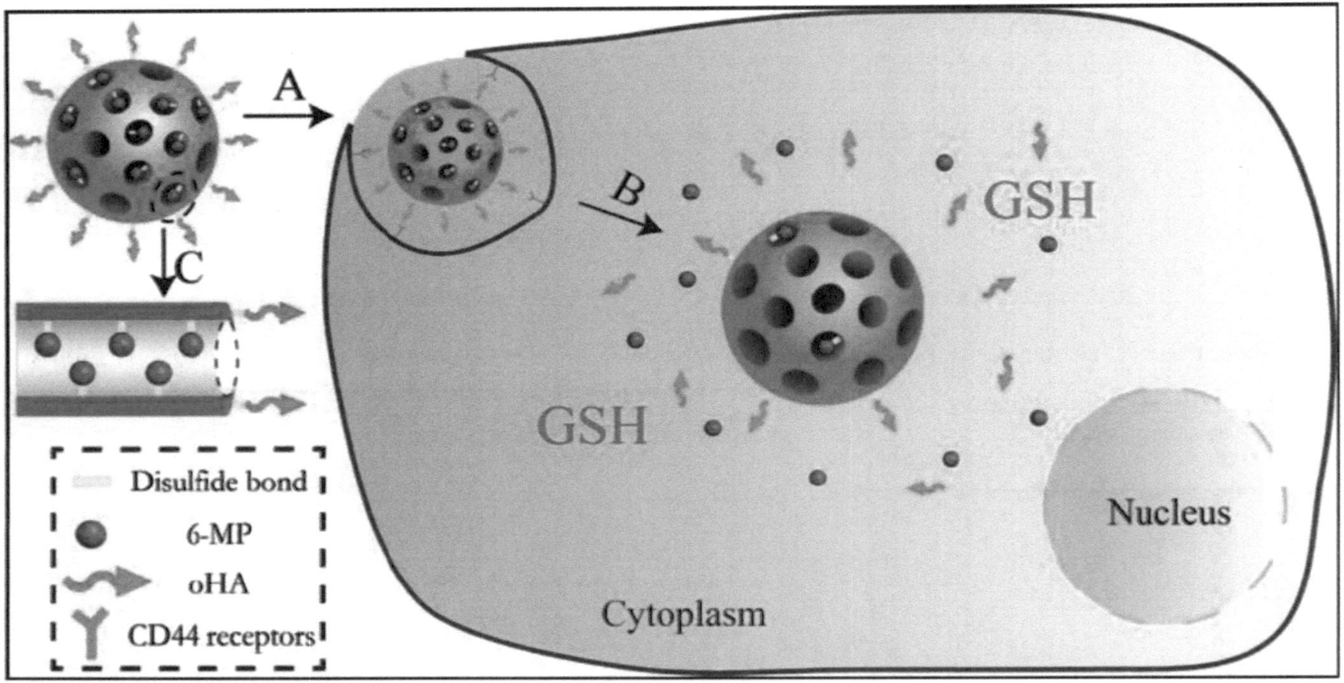

FIGURE 8.10 (A) Cellular uptake through oHA-mediated CD44 interaction, (B) GSH-triggered drug release inside the cell, (C) magnified image of pore structure of CMS-SS-MP/oHA.

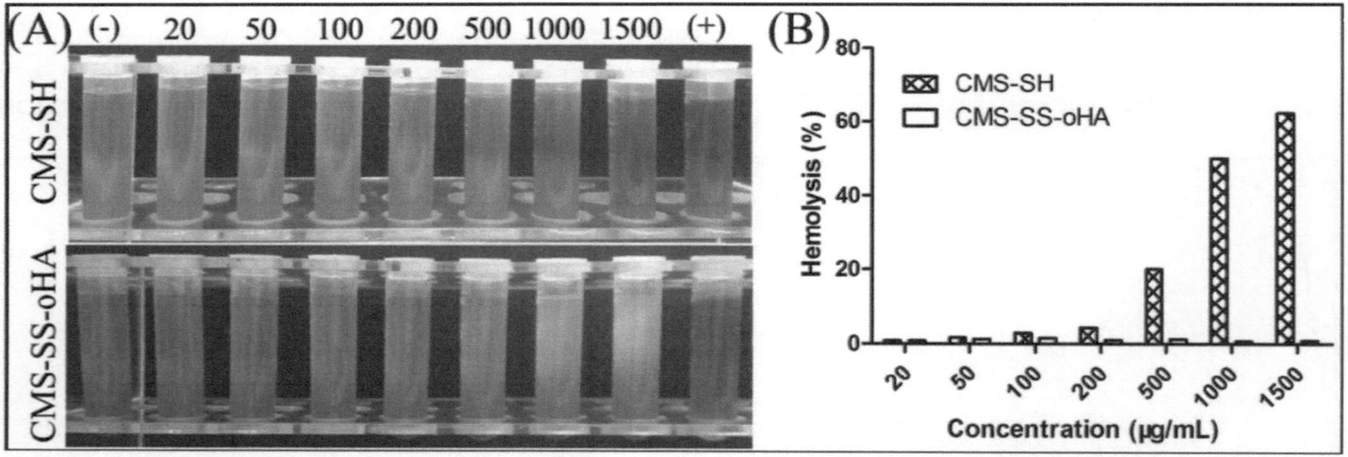

FIGURE 8.11 (A) Hemolytic photographs and (B) hemolysis percentages of drug and drug-loaded oHA at different concentrations (μg mL^{-1}).

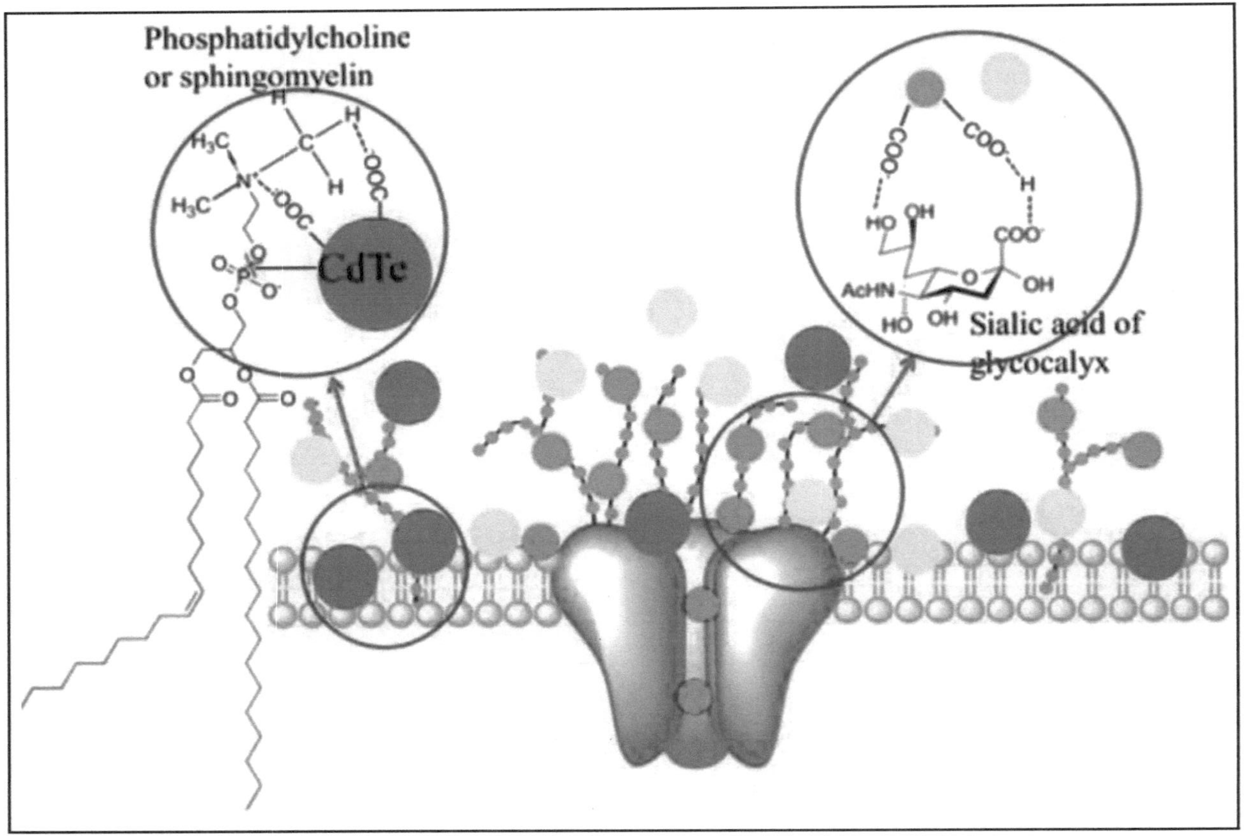

FIGURE 8.19 Possible interaction of MSA-QDs with RBC membrane.

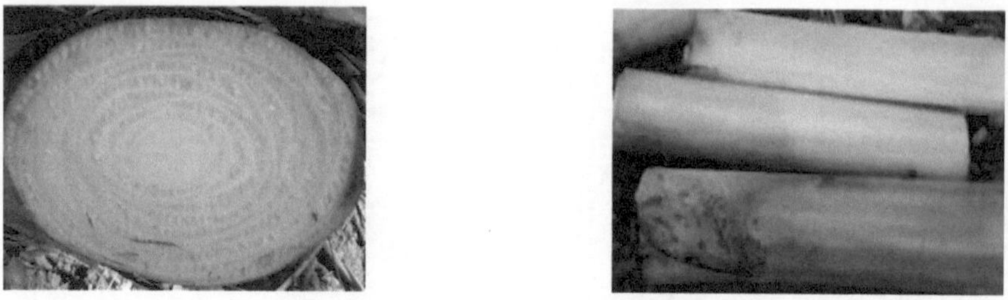

FIGURE 11.1 Banana pseudostem.

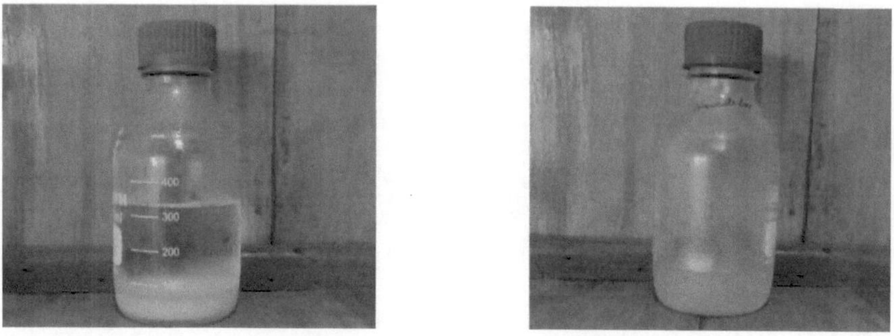

(a) (b)

FIGURE 11.2 (a) Cellulose from pseudostem powder and (b) nanocellulose from pseudostem powder.

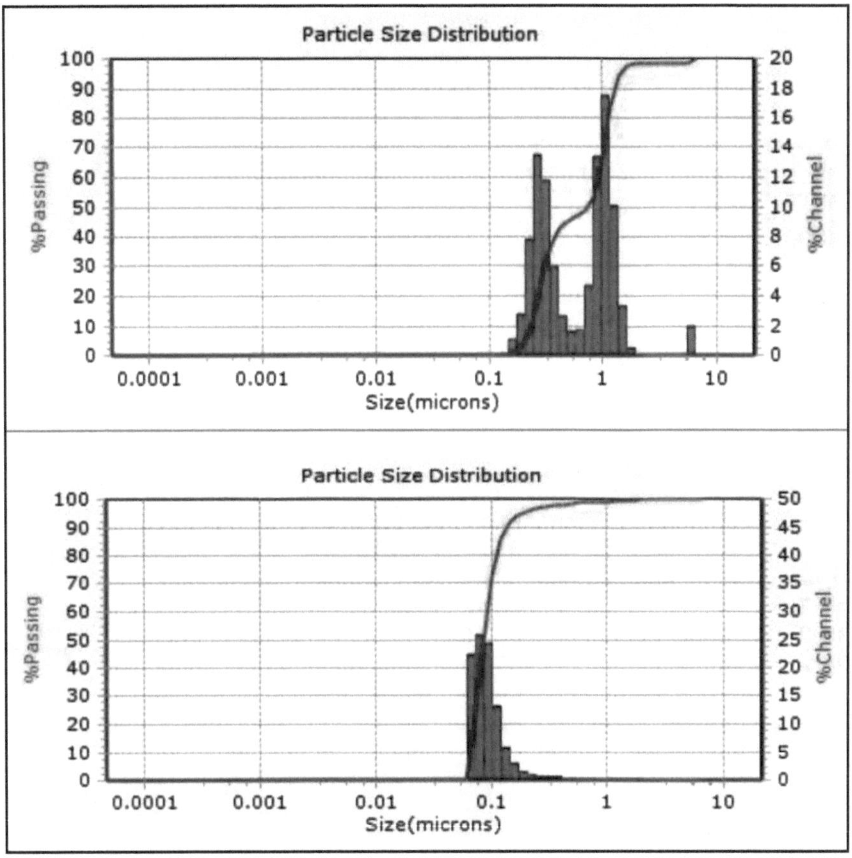

FIGURE 11.3 Particle size of cellulose and nanocellulose.

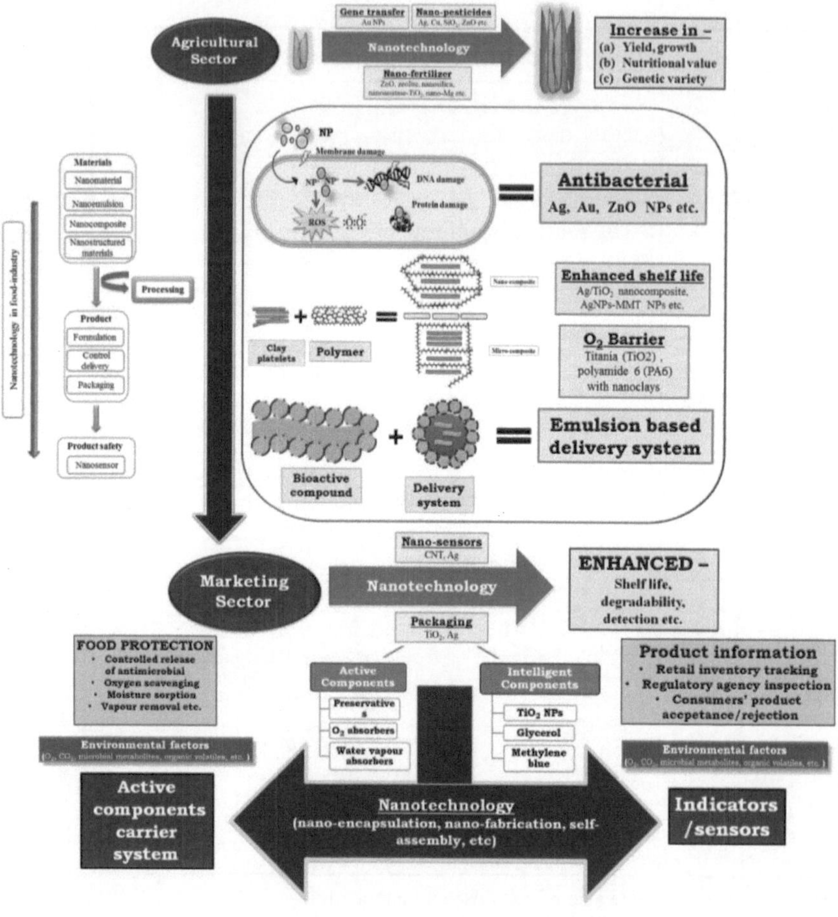

FIGURE 13.1 Schematic representation of role of nanotechnology in various sectors of food industry.

FIGURE 14.2 Base of the reactor.

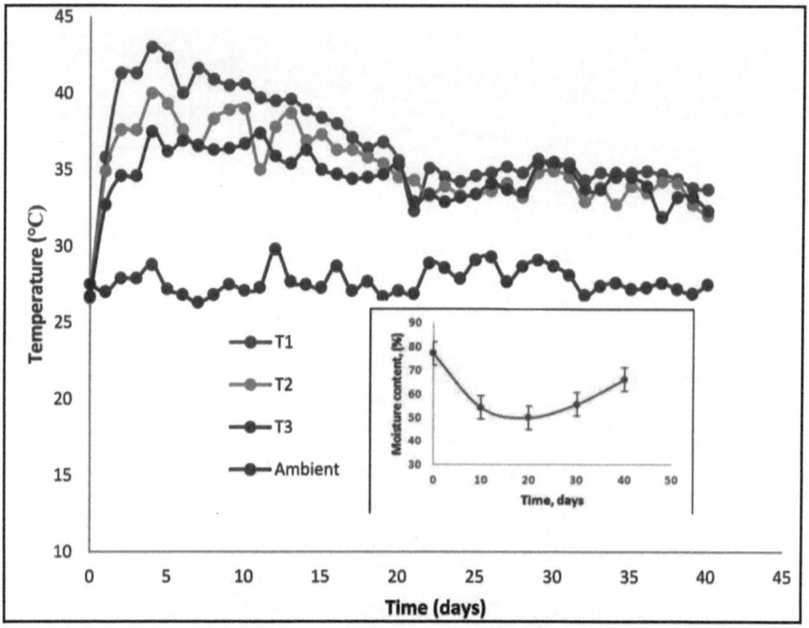

FIGURE 14.3 Temperature trend during the first 40 days of the composting process. Small figure shows the moisture content profile in the compost mixture (standard errors [SE] in vertical bars).

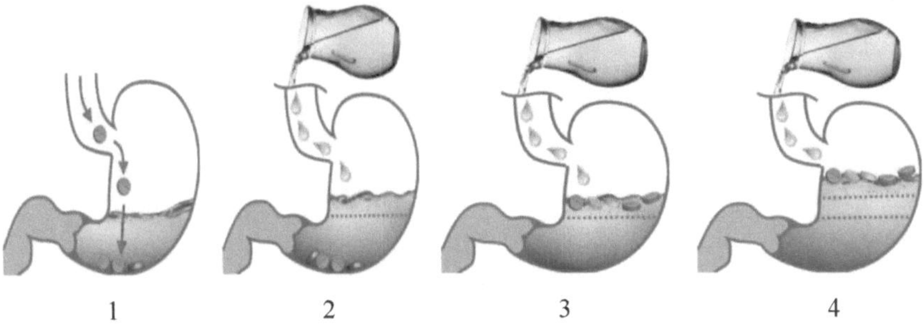

FIGURE 17.1 Influence of gravity and water on placement in the stomach cavity stones and modern tablets (1 and 2) and floating tablets, which is a dry foam (RU Patent 2254121) (3 and 4).

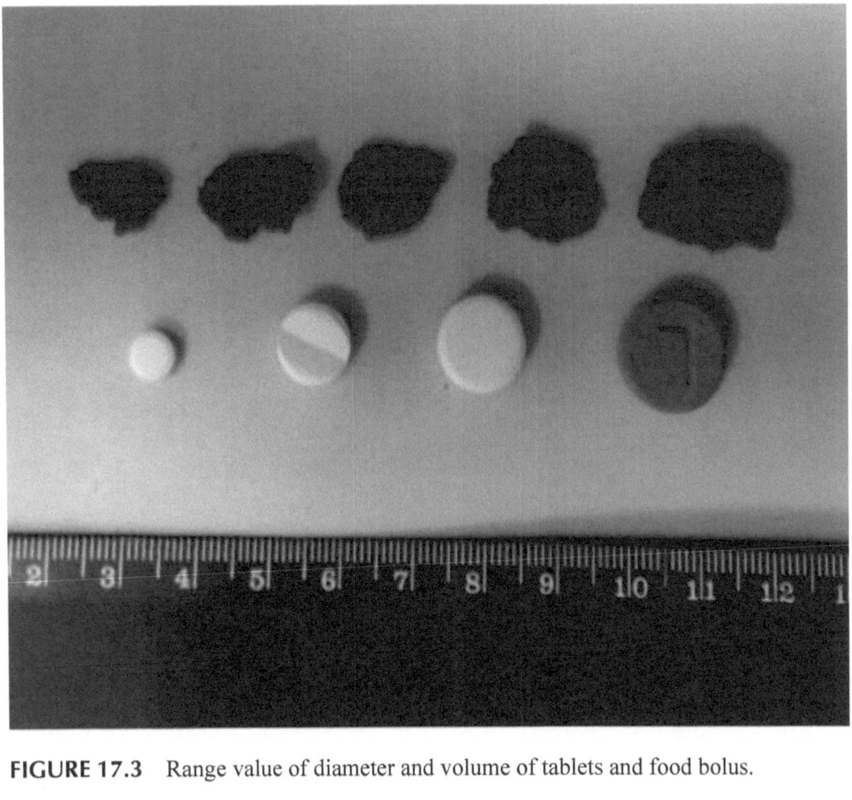

FIGURE 17.3 Range value of diameter and volume of tablets and food bolus.

CHAPTER 11

NANOCELLULOSE FROM BANANA PSEUDOSTEM: ITS CHARACTERIZATION AND APPLICATION

M. GHOSH, Y. A. BEGUM, S. MUCHAHARY, and SANKAR CHANDRA DEKA*

Department of Food Engineering and Technology, Tezpur University, Assam, India

**Corresponding author. E-mail: sankar@tezu.ernet.in*

ABSTRACT

Acid hydrolysis and ultrasound showed itself to be a better way of nanocellulose development. Isolation of cellulose was done from pseudostem powder by giving ultrasound and cellulose was converted to nanocellulose. Nanocellulose is a reinforcing agent applied in development of biodegradable foam/film. Nanocellulose with PVA film (NP) and nanocellulose with PVA, starch, and glycerol (NSG) film are developed at different concentrations of nanocellulose, that is, 5%, 7%, and 10%; and mechanical property, water vapor transmission rate (WVTR), moisture content, thickness, X-ray power diffraction (XRD), thermogravimetric analysis (TGA), and biodegradability tests are done to examine the best combination. Results show that 5% NP and 5% NSG has maximum mechanical property and minimum WVTR, crystalline property increased and thermal degradation temperature also increased. NP and NSG films were compared with PVA control film. However, moisture barrier of the film increased up to 5% but after that it decreased. Mechanical property also increased with the increase of nanocellulose but it decreased the concentration of nanocellulose by more than 5%. Transparency decreases with increase of nanocellulose: it is more than 80% in case of NP but NSG transparency decreases due to addition of starch, nanocellulose, and glycerol by more than 50%. Application of the film in the storage study of black grapes and its effect have been examined on the basis of pH, weight loss, total soluble solid (TSS), TA, color, hardness, and microbial contamination. Result shown that pH increases with the increase in temperature. At 4°C, grapes weight loss up to 12 days is less than 5%, but at 27°C, weight loss is up to 8 days and is less than 5%. No bacteria, mould, and yeast growth is seen in the fruit to up to 12 days of study at 4°C and at 27°C, bacteria growth was (2.6 log cfu/mL) 4 cfu in 10^{-1} dilution. Mould and yeast growth at 27 ◦C was (4 log cfu/mL) 1 at 10^{-1} and (2.6 log cfu/mL, 4 log cfu/mL) 4 at 10^{-1} and 1 at 10^{-3} respectively. TA depends on the packaging type and mainly it increases at low temperature and decreases at high temperature. TSS increases at low temperature but decreases at high temperature. Hardness of grapes decreases during storage study from 12N to 9N and the rate of decrease is very less when at 4°C. Change in color is very less at 4 °C and 27 °C, but at PVA film, change in color is maximum 9.5, 11.34. 1% wt. Nanocellulose with 10%wt PVA, starch, CaCO3, HCl was added for the development of biodegradable foam. Water uptake, SEM, and XRD analysis are done to check the quality of foam. Biodegradable foam water uptake is more and more porous and increases the crystallinity percentage.

11.1 INTRODUCTION

India is the largest producer of plantain and bananas with an annual production of 29.78 MT from an area of 0.83 million ha with 37 MT/ha productivity, thereby contributing 27.8% of the world banana productions.[18] Banana pseudostem is the primary waste after the maturation of banana fruit; it is disposed on the field and drain and is a cause of environmental pollution.

Nanocellulose is referred to as nanostructure cellulosic materials. There has been an increase in the interest in nonmaterial from renewable origins. Nanocellulose can be extracted from the plant by enzymatic, chemical, or physical strategies. Cellulose consists of nanofibrilated cellulose and cellulose nanocrystals. These nanometer-sized single fibers of cellulose are commonly referred to as nanocrystals, whiskers, nanowhiskers, microfibrillated cellulose, microfibril aggregates, or nanofibers.[2,5,10,24] Cellulose nanocrystal is isolated from cellulose fiber after the complete dissolution of the amorphous region by chemical hydrolysis.[22] The size and properties of nanowhiskers depend on the source and hydrolysis conditions of cellulose fibers.[3,4,8] Cellulose nanocrystal is a rigid rod-like structure with widths of a few nanometers and lengths between 100 and 250 nm (from the plant) and 100 nm to several micrometer (from algae and bacteria).[13] Cellulose nanocrystall have high strength, rigidity, and higher modulus—138–150 GPa.[19] It can be isolated from various sources such as wheat straw and orange peel, and used as reinforcements in polymer matrices.[1,9,11,12] Cellulose nanocrystal can be incorporated into a polymeric matrix that results in outstanding properties and has numerous advantages using cellulose nanocrystal as a reinforcement in nanocomposite.

The *Musa* ABB species found locally in northeast is one of the mostly grown species. Specially, the pseudostem of *Musa* ABB species are mostly disposed. No studies have been concerned with the pseudostem of *Musa* ABB of northeast in case of production of biodegradable film and foam. In this work, the aim is to increase the potential of culinary banana pseudostem cellulose as biomaterial with biocompatibility and high mechanical properties and synthetic polymer (PVA) for designing biodegradable foam and film at different concentration of CNC and PVA by freeze-drying process and by hand-casting method. Foam characterization by SEM, X-ray diffraction, water uptake biodegradable foam of cellulose nanocrystals (CNFs), and polyvinyl alcohol and film characterization done by WVTR, thermogravimetric analysis (TGA), moisture absorption, thickness, biodegradability, XRD, and mechanical strength. The study involved application of film storage study of black grapes for 12 days of study at 4°C and 27°C.

Production and characterization of CNFs from banana peel has potential application in commercial field that can add high value to culinary banana.[15]

11.2 MATERIALS AND METHODS

11.2.1 PLANT MATERIAL

Banana plant (*musa* genesis) was collected from Assam, India. Pseudostems are chopped as shown in the Figure 11.1.

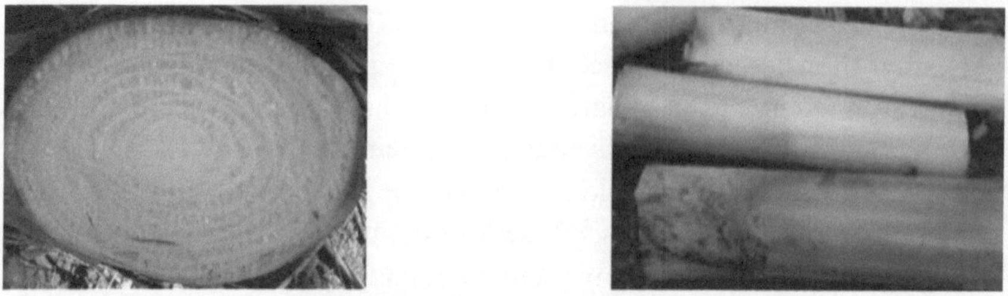

FIGURE 11.1 **(See color insert.)** Banana pseudostem.

11.2.2 PREPARATION OF BANANA PSEUDOSTEM FLOUR

The pseudostems were immediately dipped in solution of 1% potassium metabisulphite for 12 h to inhibit oxidation and enzymatic browning. Subsequently, the pseudostem pieces were dried at 50°C for 24 h by using tray dryer. The dried pieces were first ground with pulverizer, after that they were allowed passing through 0.12 mm mesh screen, and then packed in polyethylene bags.

11.2.3 PROXIMATE ANALYSIS OF BANANA PSEUDOSTEM

The proximate compositions of banana pseudostem were analyzed using the standard AOAC methods.

11.2.4 ESTIMATION OF CELLULOSE CONTENT

The cellulose estimation followed a standard method,[25] first dissolving in acid and then measured in spectrophotometer with Anthrone as coloring agent.

11.2.5 ISOLATION OF CELLULOSE FIBERS AND NANOCELLULOSE

Isolation of cellulose fibers (CF) and nanocellulose was initiated by soaking 20 g of samples in 640 mL of water, then 4 mL of CH_3COOH, and 8 g of $NaClO_2$ were added to the beaker for every hour for duration of 5 h. After first bleaching treatment for 5 h, the insoluble pellets were washed and were made neutral until the yellow color and the odor of chlorine dioxide was removed. First, bleaching process resulted in effective discoloration and confirming the leaching out of phenolic compounds and lignin. Then, the second step was the alkali treatment to convert the holocellulose to cellulose at room temperature, which helped in elimination of residual hemicelluloses. The insoluble pellets obtained from previous step were added with 80 mL of 17.5% NaOH and the mixture was thoroughly stirred with glass rod. After that, for every 5 min intervals, another 40 mL of NaOH solution was added into the mixture for three times. The mixture was allowed to stand for 30 min, by making the total time 45 min and then 240 mL of distilled water was added to the mixture and allowed to stand for 1 h before washing and filtering. Next, 800 mL of 8.3% NaOH solution was added into cellulose for 5 min and rinsed with water. The solution was neutralized by adding 120 mL of 10% acetic acid for 5 min. Finally, the cellulose was filtered, washed, and rinsed with distilled water until the cellulose residue was free from acid.[21]

11.2.6 PREPARATION OF NANOCELLULOSE

The cellulose, as obtained earlier was acid hydrolyzed with 60% (w/v) sulfuric acid (fiber to liquor ratio of 1:20) for 5 h at 50°C under strong agitation. 5-fold excess water was added to the mixture. The resulting mixture was cooled and centrifuged. The fractions were continuously washed by the addition of distilled water and centrifugation. The centrifugation process was stopped after at least five time of washing, and then pH of the suspension was maintained above 5. After chemical pretreatment, about 120 mL of a solution was sonicated for 30 min at power of 1000 w and at frequency of 25 kHz equipped with a cylindrical titanium alloy probe tip 1.5 cm in diameter. The ultrasonic treatment was carried out in an ice bath to avoid overheating, Then, suspension obtained was stored in refrigerator at 4°C.[6]

11.2.7 MEASUREMENT OF THE PARTICLE SIZE DISTRIBUTION OF CNF BY DYNAMIC LIGHT SCATTERING

Dynamic light scattering (DLS) was used to determine the size distribution profile of small particle in suspension. Particle size was measured by laser diffraction using a nano-sized particle analyzer in the range between 0.6 and 6.0 μm, under the following conditions: particle refractive index, 1.59; particle absorption coefficient, 0.01; water refractive index, 1.33; viscosity, 0.8872 cP; and temperature, 25°C.

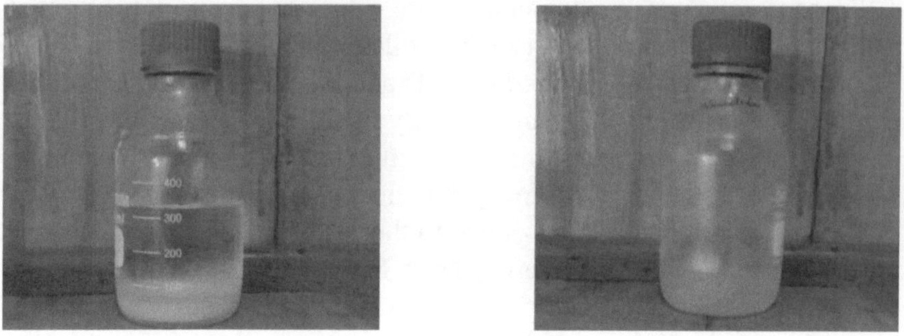

(a) (b)

FIGURE 11.2 (See color insert.) (a) Cellulose from pseudostem powder and (b) nanocellulose from pseudostem powder.

11.2.8 SCANNING ELECTRON MICROSCOPY ANALYSIS

Scanning electron microscopy (SEM) was used to investigate the surface texture or cross-section surfaces of CNC obtained after undergoing different chemical and ultrasonic treatments.

11.2.9 TRANSMISSION ELECTRON MICROSCOPY ANALYSIS

The morphology of CNCs was evaluated using transmission electron microscopy (TEM). The samples were dispersed by placing them in Sonicator for 5 min. Drops of dilute CNCs suspensions were deposited on glow-discharged thin carbon-coated TEM micro-grids. The excess liquid was absorbed using filter paper and allowed to dry at ambient temperature, and then the specimen was completely dried. TEM images were observed at 14,000× (scale bar 2000 nm) at an accelerating voltage of 80 kV. The diameters of CNCs were calculated from TEM images using microscope image analysis system which helped to illustrate size distribution.

11.2.10 THERMOGRAVIMETRIC ANALYSIS

TGA was used to determine the lost weight with the temperature. Thermal degradation characterized the thermal stability of the samples, which were heated from 30°C to 600°C at the heating rate of 10°C/min. All of the measurements were performed under a nitrogen atmosphere.

11.2.11 FOURIER TRANSFORMS INFRARED SPECTROSCOPY

The Fourier transforms infrared spectroscopy (FTIR) spectra of the nanocellulose samples were recorded on a spectrophotometer to characterize the functional group in the range of 500–4500 cm^{-1} with a resolution of 4 cm^{-1}. The samples were ground into powder by a fiber microtome and then blended with KBr followed by pressing the mixture into ultra-thin pellets.

11.3 RESULTS AND DISCUSSIONS

11.3.1 PROXIMATE ANALYSIS OF BANANA PSEUDOSTEM

Culinay banana pseudostem is preconsumption waste and a major form of waste. Biodegradable film is developed for food packages from isolated nanocellulose. It may result in high-value commercial use of the waste. Table 11.1 shows that proximate composition in dry basis, moisture content of pseudostem is about 9.03 ± 0.05, it is also seen that moisture content is high in case of weight basis, the samples taken were immediately analyzed after harvesting.

TABLE 11.1 Proximate Composition.

Sl. no.	Constituent	Concentration (dry basis %)
1	Moisture	9.03 ± 0.05
2	Fat	1.50 ± 0.7
3	Protein	0.93 ± 0.05
4	Crude fiber	23.32 ± 0.23
5	Carbohydrate	54.93 ± 0.15
6	Ash	8.95 ± 0.16
7	Cellulose content	24.23

11.3.2 CHARACTERIZATION OF NANOCELLULOSE

11.3.2.1 PARTICLE SIZE DISTRIBUTION AND ZETA POTENTIAL

Distribution of particle present nanocellulose is determined by the DLS studies. Zeta potential of the particle observed was −9.03 and −21.02 mV for cellulose and nanocellulose. The minimum and maximum particle size obtained was 66.7 and 187.7 nm for nanocellulose. Particle size obtained for cellulose was 261 to 1.362 um. Zeta potential can be measured by tracking the moving rate of negatively or positively charged particles across an electric field and Zeta potential less than −15 mV showed stability.[20,28]

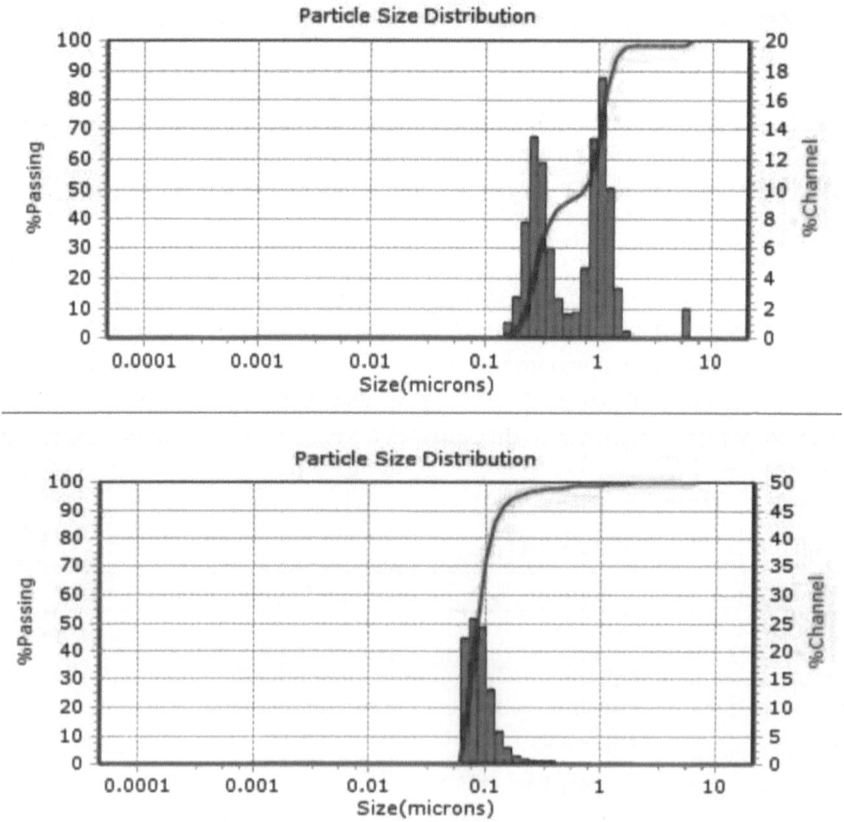

FIGURE 11.3 (See color insert.) Particle size of cellulose and nanocellulose.

11.3.2.2 TRANSMISSION ELECTRON MICROSCOPY

TEM image of cellulose nanofibers from culinary banana pseudostem after 30 min of high intensity ultrasonication treatment at 50 amp confirmed the presence of nanofibers. Chemical treatment removed the entire amorphous compound (lignin, pectin, hemicellulose) and only cellulose nanofiber were present. Cellulose nanofiber from banana pseudostem exhibits well-defined needle-shaped particle. Chemical hydrolysis followed by high-intensity ultrasound is the best method to extract nanofiber, the methods used is very sample. According to study, the length L (nm) of the nanofiber is 290 ± 9.7, diameter, D (nm) of nanofiber is 23 ± 1.2, and aspect ratio (L/D) is 12.60 ± 5.45.

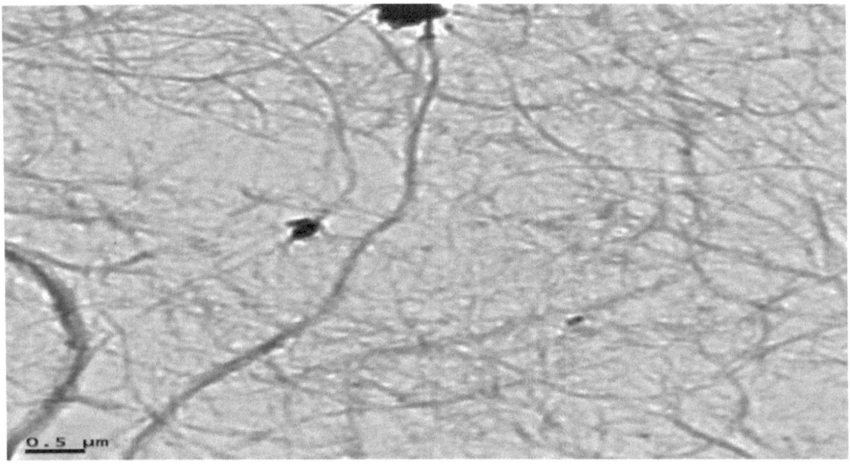

FIGURE 11.4 TEM morphology of nanocellulose.

11.3.2.3 MICROSTRUCTURE EVALUATION BY SCANNING ELECTRON MICROSCOPY

Evaluation of the structure after the chemical and high-intensity ultrasound treatment was carried out. SEM image of the cellulose nanofiber confirm the removal of amorphous component like lignin, pectin, and hemicellulose further, it was confirmed by the TEM image. Surface of untreated banana pseudostem is irregular and hemi-cellulose, lignin, and other non-cellulosic substances, such as pectin and wax, are known to act as natural binders.[7]

The chemical and ultrasonic treatment helps in defibrillation and fragmentation of nanofiber, also possible because of removal of non-cellulosic constituents.[17] Ultrasonic treatment reduces size and tailoring of the CNF.

FIGURE 11.5 SEM micrograph of nanocellulose.

11.3.2.4 THERMOGRAVIMETRIC ANALYSIS

TGA curve of nanocellulose showed that an initial weight loss from 25°C to 150°C, due to moisture evaporation and initial thermal degradation. Result showed that, initial weight loss moisture removal was at 91.04 °C and major thermal degradation started from at 243.27–370.66°C and peak at 200–400°C, attributed to cellulose degradation, it may be due to high thermal stability of the cellulose which can be used as packaging agent, from 500–600°C last trace amount left. It is reported that at 1000 W, thermal degradation of the nanocellulose from culinary banana peel at 295.33°C.[15] It is reported that acid-hydrolyzed nanocellulose from sugarcane; bagasse degradation started at 249°C and the peak rate of degradation is reached at 345°C.[17] Oxidation and break down of carbonized residues at 350–450°C to gaseous products with lower molecular masses.[22]

11.3.2.5 FOURIER TRANSFORM INFRARED SPECTROSCOPY

FTIR spectroscopy has been extensively used in cellulose of obtaining direct information on chemical changes that occur during various chemical treatments. The FTIR spectra of culinary banana pseudostem are elucidated in Figure 11.7 in the region of 3000–3600 cm^{-1} the broadened absorption band distinctive to the –OH stretching was observed.[16,23] The bond observed at 3434.99 cm^{-1} corresponds to intermolecular hydrogen bonding in cellulose II. At 2077.03 cm^{-1}, a small peak was observed which can be attributed to the aliphatic saturated C–H stretching vibration in cellulose and hemicelluloses.[20] Peak found at around 1636.88 cm^{-1}, which indicates the partial reaction of the C=O bonds of hemicelluloses.[26]

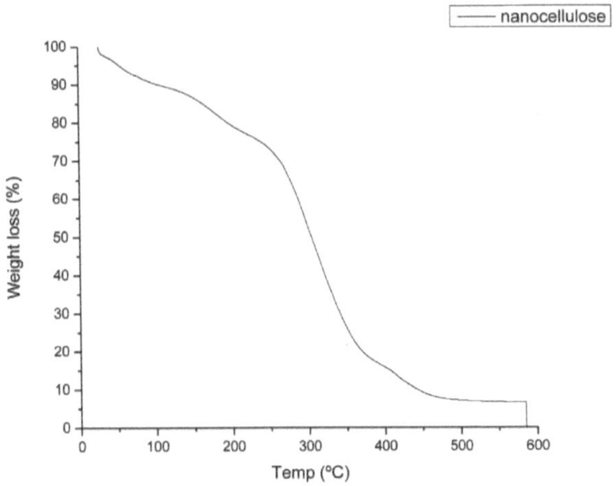

FIGURE 11.6 TGA curve for nanocellulose.

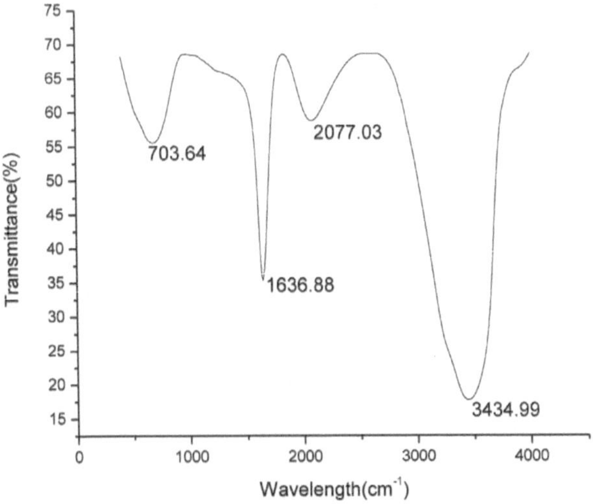

FIGURE 11.7 FTIR spectra of nanocellulose.

11.4 CONCLUSION

The results of the present investigation revealed that nanocellulose developed by acid hydrolysis and ultrasound treatments from pseudostem powder of banana is a potential source for development of biodegradable film or foam.

KEYWORDS

- **pseudostem**
- **ultrasonication**
- **nanocellulose**
- **biodegradable**
- **culinary banana**

REFERENCES

1. Angles, M. N.; Dufresne, A. Plasticized Starch/Tunicin Whiskers Nanocomposites. 1. Structural Analysis. *Macromolecules* **2000,** *33* (22), 8344–8353.
2. Azizi Samir, M. A. S.; Alloin, F.; Dufresne, A. Review of Recent Research into Cellulosic Whiskers, Their Properties and Their Application in Nanocomposite Field. *Biomacromolecules* **2005,** *6* (2), 612–626.
3. Beck-Candanedo, S.; Roman, M.; Gray, D. G. Effect of Reaction Conditions on the Properties and Behavior of Wood Cellulose Nanocrystal Suspensions. *Biomacromolecules* **2005,** *6* (2), 1048–1054.
4. Bondeson, D.; Mathew, A.; Oksman, K. Optimization of the Isolation of Nanocrystals from Microcrystalline Cellulose by Acid Hydrolysis. *Cellulose* **2006,** *13* (2), 171–180.
5. Chen, W.; Yu, H.; Liu, Y. Preparation of Millimeter-long Cellulose I Nanofibers with Diameters of 30–80 nm from Bamboo Fibers. *Carbohydr. Polym.* **2011,** *86* (2), 453–461.
6. Chen, W.; Yu, H.; Liu, Y.; Chen, P.; Zhang, M.; Hai, Y. Individualization of Cellulose Nanofibers from Wood Using High-intensity Ultrasonication Combined with Chemical Pretreatments. *Carbohydr. Polym.* **2011,** *83* (4), 1804–1811.
7. Cordeiro, N.; Belgacem, M. N.; Torres, I. C.; Moura, J. C. V. P. Chemical Composition and Pulping of Banana Pseudo-Stems. *Ind. Crops Prod.* **2004,** *19* (2), 147–154.
8. Dash, R.; Li, Y.; Ragauskas, A. J. Cellulose Nanowhisker Foams by Freeze Casting. *Carbohydr. Polym.* **2012,** *88* (2), 789–792.

9. Dufresne, A.; Cavaillé, J. Y.; Dupeyre, D.; Garcia-Ramirez, M.; Romero, J. Morphology, Phase Continuity and Mechanical Behaviour of Polyamide 6/Chitosan Blends. *Polymer* **1999,** *40* (7), 1657–1666.
10. Eichhorn, S. J.; Dufresne, A.; Aranguren, M.; Marcovich, N. E.; Capadona, J. R.; Rowan, S. J.; Gindl, W. Review: Current International Research into Cellulose Nanofibres and Nanocomposites. *J. Mater. Sci.* **2010,** *45* (1), 1.
11. Favier, V.; Chanzy, H.; Cavaille, J. Y. Polymer Nanocomposites Reinforced by Cellulose Whiskers. *Macromolecules* **1995,** *28* (18), 6365–6367.
12. Helbert, W.; Cavaille, J. Y.; Dufresne, A. Thermoplastic Nanocomposites Filled with Wheat Straw Cellulose Whiskers. Part I: Processing and Mechanical Behavior. *Polym. Comp.* **1996,** *17* (4), 604–611.
13. Hu, Z. Tailoring Cellulose Nanocrystal, Polymer and Surfactant Interactions for Gels, Emulsions, and Foams. Ph.D. Thesis, 2015. *McMaster University, Canada.*
14. Khawas, P.; Deka, S. C. Isolation and Characterization of Cellulose Nanofibers from Culinary Banana Peel using High-Intensity Ultrasonication Combined with Chemical Treatment. *Carbohydr. Poly.* **2016,** *137*, 608–616.
15. Khawas, P.; Das, A. J.; Deka, S. C. Production of Renewable Cellulose Nanopaper from Culinary Banana (Musa ABB) Peel and its Characterization. *Industr. Crops Prod.* **2016,** *86*, 102–112.
16. Li, W.; Zhang, Y.; Li, J.; Zhou, Y.; Li, R.; Zhou, W. Characterization of Cellulose from Banana Pseudo-Stem by Heterogeneous Liquefaction. *Carbohydr. Polym.* **2015,** *132*, 513–519.
17. Mandal, A.; Chakrabarty, D. Isolation of Nanocellulose from Waste Sugarcane Bagasse (SCB) and Its Characterization. *Carbohydr. Polym.* **2011,** *86* (3), 1291–1299.
18. Mohapatra, D.; Mishra, S.; Sutar, N. Banana and Its By-Product Utilization: An Overview. *J. SciInd. Res.* **2010,** *69* (5), 323–329.
19. Ng, H. M.; Sin, L. T.; Tee, T. T.; Bee, S. T.; Hui, D.; Low, C. Y.; Rahmat, A. R. Extraction of Cellulose Nanocrystals from Plant Sources for Application as Reinforcing Agent in Polymers. *Compos. Part B: Eng.* **2015,** *75*, 176–200.
20. Pelissari, F. M.; doAmaralSobral, P. J.; Menegalli, F. C. Isolation and Characterization of Cellulose Nanofibers from Banana Peels. *Cellulose* **2014,** *21* (1), 417–432.
21. Penjumras, P.; Rahman, R. B. A.; Talib, R. A.; Abdan, K. Extraction and Characterization of Cellulose from Durian Rind. *Agricult. Agricult. Sci. Pro.* **2014,** *2*, 237–243.
22. Pereira, A. L. S.; do Nascimento, D. M.; Morais, J. P. S.; Vasconcelos, N. F.; Feitosa, J. P.; Brígida, A. I. S.; Rosa, M. D. F. Improvement of Polyvinyl Alcohol Properties by Adding Nanocrystalline Cellulose Isolated from Banana Pseudostems. *Carbohydr. Polym.* **2014,** *112*, 165–172.
23. Sao, K. P.; Mathew, M. D.; Ray, P. K. Infrared Spectra of Alkali Treated Degummed Ramie. *Text. Res. J.* **1987,** *57* (7), 407–414.
24. Siró, I.; Plackett, D. Microfibrillated Cellulose and New Nanocomposite Materials: A Review. *Cellulose* **2010,** *17* (3), 459–494.
25. Thimmaiah, S. K. *Standard Methods of Biochemical Analysis;* Eds.; Kalyani Publishers: New Delhi, 1999; pp 287–310.
26. Wang, B.; Sain, M.; Oksman, K. Study of Structural Morphology of Hemp Fiber from the Micro to the Nanoscale. *Appl. Comp. Mater.* **2007,** *14* (2), 89.

27. Zhang, P.; Whistler, R. L.; BeMiller, J. N.; Hamaker, B. R. Banana Starch: Production, Physicochemical Properties, and Digestibility—A Review. *Carbohydr. Polym.* **2005,** *59* (4), 443–458.
28. Zhao, N.; Mark, L. H.; Zhu, C.; Park, C. B.; Li, Q.; Glenn, R.; Thompson, T. R. (2014). Foaming Poly (Vinyl Alcohol)/Microfibrillated Cellulose Composites with CO_2 and Water as Co-Blowing Agents. *Industr. Eng. Chem. Res.* **2014,** *53* (30), 11962–11972.

CHAPTER 12

ENCAPSULATION OF POLYPHENOLS INTO MICRO- AND NANOPARTICLES FOR IMPROVED HEALTH EFFECTS

ANURAG MAURYA[1], MONOJ KUMAR DAS[2], ANAND RAMTEKE[2], SANKAR CHANDRA DEKA[3], NEELU SINGH[4], and PAULRAJ RAJAMANI[4,*]

[1]*Department of Botany, Shivaji College, University of Delhi, New Delhi 110027, India*

[2]*Department of Molecular Biology and Biotechnology, Tezpur University, Tezpur, Assam, India*

[3]*Department of Food Engineering and Technology, Tezpur University, Napaam, Tezpur, Assam, India*

[4]*School of Environmental Sciences, Jawaharlal Nehru University, New Delhi 110067, India*

[*]*Corresponding author. E-mail: paulrajrajamani@gmail.com*

ABSTRACT

Polyphenol and other antioxidants show positive health effects in condition, for example, inflammation, cancer development, aging, and some infections, etc. Therefore, their use has increased in the field of food, nutrition, nutraceuticals, cosmetics, and pharmaceuticals in recent past. Polyphenols extracted out of plant matrices are prone to undesirable environmental condition, for example, high temperature, light, and oxygen during processing and storage. This reduces the quality of phenol or antioxidant capacity. To protect polyphenol from undesirable condition, it is encapsulated into microscale or nanoscale particles using biological polymers, for example, proteins and starch, ascoverin material. Besides, encapsulation improves health effect of polyphenol by mediating controlled release, increased bioavailability, targeting for organ or tissue-specific delivery, preventing degradation from digestive enzymes and pH, and increased absorption in intestine. This chapter emphasizes on matrices, various strategies used for encapsulation of polyphenols at microscale and nanoscale, and effect of encapsulation.

12.1 INTRODUCTION

Increasing world population posed new challenges to several aspects of food science, for example, production, processing, transport, and improvising its health benefits with minimum side effects.[18] Application of microscale (10^{-6}) and nanoscale (10^{-9}) technologies for food processing has improved nutritional value of food ingredients, and enhanced their therapeutic efficacy when used as nutraceutical agent. Microscale and nanoscale food particles show superior absorption and bioavailability. This is due to the reduced size of particle. At reduced size, high surface-to-volume ratio of particle exerts altered physical, chemical, and biological properties.[14] Further, food material has been fabricated at microscale and nanoscale to achieve desirable characteristics with respect to its stability, taste, absorption, stability in storage condition, and in body fluid, bioavailability, controlled release, and health benefits. Encapsulation is one of such technique in which active food ingredients is entrapped into a protective shell.[24]

Polysaccharides, protein, lipids, and other complex biological polymers which are generally regarded as safe were used as wall material for encapsulation of bioactive food ingredients, for example, vitamins, antioxidants, polyphenols, probiotic microorganism, and therapeutic nutraceuticals.[39] Encapsulated compound is protected from air (oxygen), processing, and storage condition and digestive juices of gastrointestinal tract. Encapsulation masks bad smell and taste of active ingredient, enhance its stability, and deliver it to desired tissue or cell.[12] Encapsulation also helps in prolonged retention in gastrointestinal tract, better absorption, controlled release, and increased bioavailability of encapsulated material. Polyphenols are medicinally active class secondary metabolite produced in plants. They promote health due to their antioxidant effect and by modulating many molecular signalling pathways in the cell. But these antioxidant compounds are sensitive to oxygen, temperature, and juices of gastrointestinal tract. Therefore, protection of phenolic compounds by encapsulating them into a comparatively inert wall material can enhance the health benefits.[29]

12.2 POLYPHENOLS AS ANTIOXIDANT

Polyphenols are diverse class of secondary metabolites found on plants which are characterized by common benzene ring with several hydroxyl groups. Usually, polyphenols are classified into four subgroups: flavonoids (which include anthocyanidins, catechins, flavones, flavonols, flavanones, and isoflavones), coumarins, stilbenes, and tannins along with other constituents like chalcones and lignans, which exhibit polyphenolic structures.[6]

Flavonoids are derivative of C6-C3-C6 ring structure in which two benzene rings (A and B) are linked together through a three carbon heterocyclic ring (C). Methyl, hydroxyl, glycoside substitution, degree of oxidation in flavonoid (C6-C3-C6) ring, and arrangement of C-3 group determine the kind of individual flavonoid compound. Delocalization of π electrons in phenoxyl radical makes phenol a good electron acceptor or antioxidant. A double bond in ring C helps to delocalize π electrons all across the flavonoid molecule, resulting in a stable aryloxyl radical; this makes flavonoid an excellent antioxidant molecule. Glycoside and hydroxyl substituents make flavonoid ring hydrophilic while methyl and isopentenyl substituent make it lipophilic.[24]

Flavonol, flavone, flavan-3-ol, anthocyanidin, flavanone, and isoflavone are important subgroups of flavonoids. Flavanols are most widely occurring flavonoids in plant kingdom; among them, O-glycoside derivatives are most common dietary flavonol which include quercetin, myricetin, kaempferol, and isorhamnetin. Flavones, which include apigenin and luteolin, are common in celery, parsley, and some herbs. Flavones are 2-phenylchromen-4-one (2-phenyl-1-benzopyran-4-one) skeleton with hydroxyl, methyl, O- and C-alkyl, or glycosyl substituent groups. Green tea antioxidants, catechin, epictaechin, epigallocatechin 3-gallate (EGCG) belong to flavan-3-ol group. C3 group of flavan-3-ol skeleton lack double bond, therefore, has two chiral carbon. Two chiral centers produce four isomers at each level of B-ring hydroxylation. Flavanones are similar to flavan-3-ols structurally; ring C is attached to B ring at C2 atom in the a-configuration in the majority of naturally occurring flavanones. When B ring is attached at C3 rather than C2, compounds are known as isoflavones.[12,27,30]

12.3 METHODS OF ENCAPSULATION

12.3.1 SPRAY DRYING

It is one of the oldest method and most widely adopted method in food industry for encapsulation because of simplicity and cost effectiveness of the process. Encapsulating matrix is dissolved in solvent; most commonly water and active ingredient or core material is homogenized in matrix solution. Homogenization results in solution, dispersion, or suspension of core material in matrix solution. When applied onto spray drier, homogenized mixture is atomized through a nozzle or spinning disc in a hot air chamber. Solvent from small atomized droplets are vaporized and powdered material is deposited at bottom of chamber. Collected particle ranges from 50 to 100 μM in diameter. Usually, shape of particle is spherical. Matrix should be proficiently soluble in water to apply this technique. Spray drying is simple, rapid, and relatively low-cost technique. It produces particles of good stability with high yield. Nonetheless, due to adhesion of particle on spray drier, yield is reduced. Although temperature of drier chamber can be adjusted between 150°C and 300°C, highly sensitive ingredient can be degraded in drying process.[22]

12.3.2 SPRAY COOLING

Similar to spray drying, spray cooling also involves atomization of matrix solution and drying in a chamber. Drying takes place at low temperature preventing thermal decomposition of thermo-labile core material. Dispersion is made in liquefied state of solvent, normally lipids which have high melting point. Dispersion is atomized in cold spray drier chamber where droplets solidify into powder. For example, phytosterol was encapsulated in stearic and hydrogenated vegetable oil through spray cooling. Particles formed in this process were spherical in shape and were ranging from 13.8 to 32.2 μm in diameter.[14]

12.3.3 EXTRUSION

This process involves extrusion of matrix (shell) solution through a nozzle into a gelling solution. Compound of interest premixed with matrix solution is trapped in gelled bead. Most common gelling agent is sodium alginate and hardening solution is calcium chloride. At laboratory scale, extrusion is done through a pipette or syringe, whereas spraying nozzle, vibrating nozzle, spinning disk, double capillary, jet cutter, and atomizing disk are used for large-scale preparation. Size of particle varies according to diameter of nozzle and flow rate of extrusion. Very small size of particle (50 μm) is made through DC current applied during extrusion; this process is termed as electrostatic extrusion. Extrusion method is a mild process for temperature-sensitive active compound. It is performed at low temperature in laboratory. Hydrophilic as well as hydrophobic compounds can be encapsulated by this process. Gelled bead protects bioactive compound from evaporation and oxidation in air, thus providing longer shelf-life. Major disadvantages of the technique is very less number of options for gelling agent, difficulty in producing small size particle, and it is a costlier method at industrial scale.[17,33]

12.3.4 COACERVATION

This technique encapsulates active compound in a hollow sphere at low temperature. In coacervation, a polyelectrolyte separates into an agglomerated colloidal particle from its solution phase. Formation of condensed phase, called coacervate, leaves a dilute solution of polyelectrolyte as continuous phase. Agglomeration occurs on an immiscible active core, hence encapsulates core material. Agglomeration processes is regulated by ionic strength and pH of the solution. However, electrostatic interactions are major driving force for coacervate formation, but hydrophobic interaction and hydrogen bonding are also found accountable in this process. After encapsulation, the active core in coacervate, polyelectrolyte shell can be strengthened by enzymatic or chemical crosslinking. Single polyelectrolyte is used to form a simple coacervate, while a mixture of polyelectrolyte for a complex coacervate. Protein and polysaccharide are commonly used biological polyelectrolyte. For example, a cationic protein and an anionic polysaccharide can be used for complex coacervation. At pH higher than pI, protein bears negative charge and repels negatively charged polysaccharide; therefore, both remain in solution. But at pH lower than pI, the two polymers will interact and will separate from solution phase to form coacervate. Coacervation is widely used technique of encapsulation in food industry because of its high encapsulation efficacy.[27,48]

12.3.5 EMULSION

Emulsions are dispersion of one liquid into another which are normally immiscible, for example, oil and water. Oil is dispersed in aqueous phase and is called oil-in-water system, whereas water droplets dispersed in oil is termed as water-in-oil system. An amphipathic molecule, called emulsifier, is essential to stabilize emulsion. It prevents agglomeration of dispersed phase. Nanoemulsions are colloidal solution of two immiscible liquids having colloidal particles less than 100 nm. High energy homogenization process and large amount of emulsifier is applied for nanoemulsion preparation. High-shear mixer, high-pressure homogenizer, microfluidizer, membrane microchannel homogenizer are the techniques applied for large scale emulsification. A laboratory homogenizer and sonication can be used for small scale emulsification. Emulsifier provides stability in liquid state, while for long term storage, emulsion can be spray dried to powder.[14,30]

A double emulsion is made when emulsion of two liquid is dispersed further into another liquid or continuous phase. Water-in-oil-in-water (W/O/W) double emulsion is the dispersion of small water droplets in large oil droplets, which in turn is dispersed in continuous water phase. W/O/W emulsions are found better as compared with O/W emulsions for encapsulation and protection of hydrophilic polyphenols. Hydrophilic compound trapped in internal water phase cannot diffuse out across oil interface toward outer continuous phase of water. Oil interface also helps to control the release of entrapped polyphenols for prolonged release in the body. W/O/W emulsions have been generated to encapsulate mixture of hydrophilic and lipophilic polyphenols. Major drawback of emulsion system is its sensitivity to environmental stress (heating, freezing, dehydration). These stresses make emulsion unstable, that is, it may undergo flocculation, coalescence, and Ostwald ripening, which impair controlled delivery of encapsulated polyphenol. Due to amphiphilic nature, some polyphenol can slowly diffuse into the oil phase or even outer water phase in W/O/W emulsion. Diffusion of polyphenol alter their release pattern and expected controlled and targeted release.[21,23]

Nanoemulsions are colloidal dispersions of small size droplets (< 100 nm) of one liquid in a different immiscible liquid. These nanodroplets are less prone to gravitational separation and aggregation due to their smaller size and higher interface area, hence are more stable as compared with droplets conventional emulsion.[45]

EGCG, major bioactive polyphenol of green tea, is known to enhance α-secretase activity. This activity was significantly enhanced when EGCG was applied in nanoemulsion-encapsulated neuronal cells in vitro.[43] In in vivo system, nanoemulsion–encapsulation of EGCG significantly increased its systemic absorption and doubled bioavailability as compared with free EGCG.[5] Similarly, nanoemulsion encapsulation of curcumin showed more efficient and rapid in vitro cellular uptake as compared with free compounds. Nanoemulsion encapsulated curcumin had a longer half-life and enhanced bioavailability than free curcumin in vivo.[30]

12.4 SYSTEMS FOR ENCAPSULATION

Bio-based matrix like polysaccharide, protein, and lipid has been applied to prepare various form of carrier, for example, capsules, nanospheres, hydrogel, emulsions, solid–lipid nanoparticles (SLNs), liposomes, micelles for polyphenol encapsulation. Size, shape, and internal structure are important characteristics of these structures which vary according to kind of matrix and method of preparation. By definition, size of carrier should be less than 100 nm for being a nanocarrier. However, colloids as nanosystems within the range of 500 nm are considered in nano range, in practice. The size of 1–1000 μM is recommended for microsystem. Encapsulation is generalized term for loading of active principle in polymer matrix, which include entrapment within polymer fiber, encapsulation in hollow core of particle, or adsorption in porous matrix.[14]

12.4.1 CAPSULES

Capsules are biopolymeric hollow sphere encapsulating bioactive compound in its core. Bioactive compound remains dispersed or solubilized in matrix. It is one of the widely adopted methods of encapsulation of polyphenol. Coacervation is the popular technology to prepare capsules. Encapsulated compound remain protected by shell under storage condition and in biological fluid. This also provides advantage of controlled release in biological fluid. The release depends upon pH, temperature, and ionic strength of external environment and molecular weight of matrix macromolecules.[14,27] Curcumin encapsulated in casein capsule showed enhanced bioactivity.[35]

12.4.2 HYDROGEL

Hydrogel are solid microparticle and nanoparticle of hydrophilic polymeric fibers intertwined in three dimension. Polymeric fibers are associated with each other mainly through noncovalent interaction like Van der Waals interactions, hydrogen bonding, or physical entailment. They are formed mainly through gelation process.[38] They can absorb water and swell up to 30 times of its size. Hydrophilic moieties (i.e., hydroxyl, ethers, carboxyl, amines, and sulphydryl groups) create interaction with water in the wetting process. Hydrophilic active compound remain dispersed or dissolved in absorbed water in gelled particle. The swelling and shrinkage in response to chemical stimuli, for example, pH, solvent composition, and ionic strength and physical forces like temperature, electric or magnetic field, light and pressure is important for controlled release of active compound. For example, a pH-sensitive hydrogel shrinks at low pH in gastric conditions and swells at high pH under intestinal conditions. This hydrogel will protect encapsulated compound in stomach but will release in small intestine where absorption takes place.[41,42]

12.4.3 LIPID-BASED SYSTEMS

Lipids are integral part of plasma membrane which encapsulates cytosolic material. Therefore, lipid can be considered as ideal matrix for encapsulation of hydrophilic as well as hydrophobic compounds. Due to their intrinsic physicochemical properties, diversity and biocompatibility lipidic carriers, emulsions, liposomes, nanoemulsions, SLNs, and nanostructured lipid carriers (NLCs) has been studied for delivery of polyphenols.[13]

12.4.4 EMULSIONS

Emulsion is a stable dispersion of small droplets of one liquid into another immiscible liquid. Compound of interest can be encapsulated in dispersed phase. Emulsions can be classified into water-in-oil (W/O) system where water is dispersed into oil and oil-in-water (O/W) emulsion in which oil is dispersed into water. Oil-in-water has been widely applied for delivery of lipophilic compounds. High energy homogenizer is required for dispersion process and an emulsifier is essential to stabilize the emulsion.[14] Quercetin, ellagic acid, EGCG, curcumin, and resveratrol were entrapped in emulsion prepared by homogenization-solvent removal method. Solvent removal makes the emulsion droplets into solid nanoparticle. Such encapsulation improved intestinal absorption efficiency of polyphenol and exhibited enhanced biological effects of such antioxidant and anti-cancer activity in in vitro and in vivo system. Encapsulation crude polyphenol mixture, for example, tea and bayberry polyphenol showed significant improved stability during their storage while maintaining the antioxidant activity. Furthermore, they showed a delayed and controlled release fin response to external stimulus like pH or enzymes.[9,14]

12.4.5 SOLID–LIPID NANOPARTICLES

SLNs are lipid-based delivery particle which has been widely applied to encapsulate, protect, and release lipophilic active ingredients. SLNs are colloidal emulsion of crystalline lipid in aqueous medium, stabilized by an emulsifier coating. Fatty acids, acyl glycerol, and waxes are used for core while phospholipids, sphingomyelins, bile salts, and sterols are examples of stabilizers in the preparation of SLNs. SLNs encapsulate lipophilic compound in its solid core. Compound entrapped in crystalline core are more protected and show delayed release as compared with when encapsulated in liquid state lipid nanoparticle. However, synthesis of SLNs requires purified crystalline lipids; incorporation of polyphenolic compounds may disrupt crystalline arrangement, introducing instability to the particle during storage. Furthermore, solid core of SLNs limits the loading capacity encapsulation. High energy methods like microfluidization, high-pressure homogenization, or sonication are required for synthesis of SLNs.[26,32]

12.4.6 NANOSTRUCTURED LIPID CARRIERS

To overcome the limitation of loading capacity in SLNs, small proportion (upto 5% by weight) of liquid state lipid was mixed to crystalline one during synthesis. Addition of liquid lipid to the crystalline lipid loosens the crystal lattice, thus providing space for loading. Such structure, called NLCs present higher loading capacity and a sustained release of core compound because they show a low crystallinity index. A comparative study shows that loading capacity for quercetin was greater in NLCs or liquid nanoemulsions (LNE) than SLNs.[2]

12.4.7 LIPOSOMES

Liposomes are hollow vesicle surrounded by lipid bilayer similar to plasma membrane. They can entrap hydrophilic, hydrophobic, and amphiphilic bioactive compounds. Size of liposomes varies from 30 nm to micrometers and can be unilamellar or multilamellar. Each lamellae is a lipid bilayer, two layer of stacked lipids having their nonpolar tail buried in hydrophobic middle part and hydrophilic heads oriented toward exterior aqueous phase. Various studies on liposomal encapsulation of polyphenols such as catechin, curcumin, resveratrol, EGCG, fisetin, quercetin, silymarin showed prolonged antioxidant effect, increased bioavailability, enhanced biological effect, increased solubility as compared with free compound in in vitro and in vivo system.[31] Schematic representation of encapsula-tion of bioactive compound and there release on altered environmental conditionsare shown in Figure 12.1

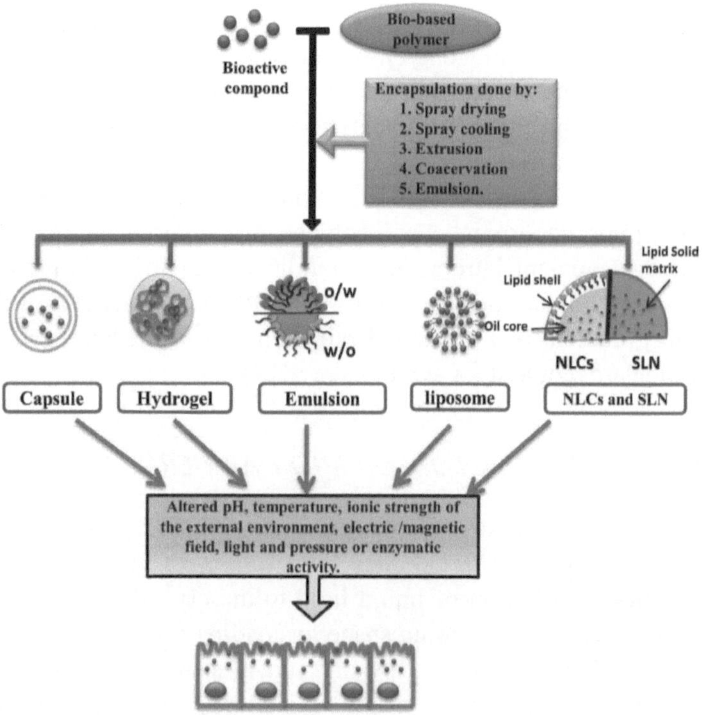

FIGURE 12.1 Schematic representation of encapsulation of bioactive compound and there release on altered environmental conditions.

12.5 ENCAPSULATION OF POLYPHENOLIC COMPOUNDS

12.5.1 ANTHOCYANIN

Most of anthocyanin encapsulation researches were performed with plant extract which is a mixture of many anthocyanins. Bilberry anthocyanin was encapsulated by a combination of emulsion and heat gelation method in whey protein and sunflower oil. In brief, a solution of whey protein and bilberry extract at low pH was poured in sunflower oil. Mixture was heated at 80°C with continuous stirring for heat induced gelation of protein.[10] Microcapsules formed by emulsion/gelation was collected as pellet after centrifugation of the mixture. Size of microcapsules ranged from 0.5 to 2.5 mm. Size was reduced to 70 μm by addition of emulsifier. Hydrogel encapsulation of anthocyanin rich extract exhibited controlled release in the simulated intestinal fluids (SIF) condition.[11] Another study in which encapsulation of bilberry extract was tested in simulated gastric fluids (SGF) show that W/O/W double emulsion protects anthocyanin at low pH in simulated gastric juice. Outer layer of water is crucial for protection at low pH.[19] Outer aqueous phase containing whey protein protect better than polysaccharide mixed phase.[20] Pure anthocyanin, cyanidin-3-O-glucoside (C3G) was encapsulated in soy protein ferritin. Mixture of C3G was with ferritin at pH 2.0. pH was increased to 7 to induce gelation. Ferritin provided thermal stability to encapsulated anthocyanin at high temperature up to 50°C.[47]

12.5.2 QUERCETIN

Quercetin is another polyphenol, encapsulated and tested for stability on SIFs. Following liquid–liquid dispersion strategy, quercetin solution in dimethyl sulfoxide (DMSO) was mixed in aqueous bovine serum albumin (BSA) solution; DMSO was evaporated through freeze-drying leaving dry nanoparticulate (70 nm) powder. Antioxidant activity of nanoencapsulated quercetin powder remain stable under SIF conditions.[15,16] Another strategy, called antisolvent precipitation, was adopted to encapsulate quercetin into dry zein-casein capsule. Quercetin and zein were dissolved in an ethanol/ water solution. This solution was mixed with sodium caseinate aqueous solution. Vacuum drying of this mixture resulted particles ranging from

130 to 161 nm. This encapsulation was proved to protect quercetin from alkaline pH and UV radiation.[36] A different study showed that entrapment of quercetin in pullulan fiber enhanced its antioxidant effect.[1]

12.5.3 CATECHINS

Gelation is a popular method for encapsulation of catechin. In one study, catechins were encapsulated in β-lactoglobulin (β-Lg) matrices to be cold-set gelation process. The nanoparticle (size 50 nm) formed by this process retarded oxidation and degradation of catechin and showed delayed release under SGF condition.[40] In a report, caseinophosphopeptide and chitosan solution was subjected to ionic gelation. After gelation, coacervate nanoparticles were formed encapsulating catechin.[25] In another study, complex coacervate core micelles (C3Ms) were prepared with gelatin–dextran conjugates encapsulation of tea polyphenols, which showed high encapsulation efficiency.[48] Moreover, catechin was emulsion encapsulated in oil-in-water system. Aqueous phase consists of a mixture of catechins, BSA or ovalbumin, and emulsifier Tween-20; oil phase was added drop by drop to aqueous phase.[4]

12.5.4 CURCUMIN

Curcumin is lipophilic compound and shows limited solubility in water. Therefore, its absorption from intestine endothelium and its bioavailability is very less. Encapsulation increased solubility and bioavailability of curcumin in simulated gastrointestinal fluid. In one study, curcumin was encapsulated in W/O emulsion and under simulated gastrointestinal fluid, it showed delayed release. Encapsulated nanopowder (40–250 nm) prepared through spray drying of curcumin/sodium caseinate emulsions showed enhanced antioxidant activity (Rao and Khanum, 2016).

12.5.5 PHENOLIC ACIDS

Phenolic acids are good antioxidant but they are prone to oxidation in air. Encapsulation protects phenolic acid from oxidation along with providing other desirable effect. Ferulic acid was entrapped in zein fiber utilizing coaxial electrospinning method with the addition of unspinnable acetic acid. This encapsulation showed controlled release of ferulic acid.[46] Gallic acid was electrospinned with zein fiber for encapsulation. The nanofiber (size ranged from 327 to 387 nm) retained antioxidant activity of gallic acid.[34] BSA O/W emulsion was implied to encapsulate caffeic acid which provided stability to the antioxidant activity during storage.[3] Ellagic acid was loaded in threonine base peptide microtube. This formulation was effective against *Escherichia coli* and *Staphylococcus aureus*.[7]

12.5.6 PROANTHOCYANIDINS

High temperature and oxygen is detrimental for antioxidant activity of proanthocyanidins. Bitterness and astringency are another reason which makes it necessary to encapsulate proanthocyanidins.[8] Sorghum condensed tannins (SCT) were encapsulated in coacervate prepared by mixing ethanolic solution of kafirin and SCT followed by addition of water.[44] Cranberry proanthocyanidins (CPs) were encapsulated in zein nanoparticle (392–447 nm).[48] Pomegranate ellagitannins having anticancer property were entrapped in gelatin nanoparticles (200 nm) prepared by gelation method.[28]

12.6 CONCLUSION

Polyphenols are class of naturally occurring molecules with benzene/phenyl ring in the structure. They are good antioxidant molecule and show biological activity against inflammation, cancer development, cardiovascular disease, and antimicrobial properties, etc. Major subclasses of polyphenols are flavonoids, anthocyanins, proanthocyanidins, phenolic acids, and tannins. Once extracted from parent tissue, polyphenol are prone to deterioration by oxidizing air, temperature, light, and storage condition. Besides, their bioavailability is limited in humans because of several factors like poor solubility in water, degradation in gastrointestinal tract, impaired absorption in intestinal wall. To overcome these issues, polyphenols are encapsulated in bio-based polymers including proteins, polysaccharides, and lipids. Following various method of encapsulation such as extrusion, homogenization, emulsion, spray drying, gelation, coacervation microscale and nanoscale particles are formed. Each method has some advantages and disadvantages but encapsulated polyphenol by any of the method showed improved desirable effect. Nanoscale particle shows better stability under storage, processing, and simulated gastrointestinal condition as compared with microscale particles. Future research in this area is required to circumvent the disadvantages of encapsulation techniques, for particle with improved required properties, and for stable nanoscale particle.

KEYWORDS

- **bioavailability**
- **encapsulation**
- **microscale**
- **nanoscale**
- **polyphenols**
- **drug delivery**
- **antioxidant**

REFERENCES

1. Aceituno-Medina, M.; Mendoza, S.; Rodriguez, B. A.; Lagaron, J. M.; Lopez-Rubio, A. Improved Antioxidant Capacity of Quercetin and Ferulic Acid During In Vitro Digestion Through Encapsulation Within Food-grade Electrospun Fibers. *J. Funct. Foods* **2015,** *12*, 332–341.
2. Aditya, N. P.; Shim, M.; Lee, I.; Lee, Y.; Im, M. H.; Ko, S. Curcumin and Genistein Coloaded Nanostructured Lipid Carriers: In Vitro Digestion and Antiprostate Cancer Activity. *J. Agric. Food Chem.* **2013,** *61* (8), 1878–1883.
3. Almajano, M. P.; Gordon, M. H. Synergistic Effect of BSA on Antioxidant Activities in Model Food Emulsions. *J. Am. Oil Chem. Soc.* **2004,** *81* (3), 275–280.
4. Almajano, M. P.; Delgado, M. E.; Gordon, M. H. Albumin Causes a Synergistic Increase in the Antioxidant Activity of Green Tea Catechins in Oil-In-Water Emulsions. *Food Chem.* **2007,** *102* (4), 1375–1382.
5. Anand, P.; Nair, H. B.; Sung, B.; Kunnumakkara, A. B.; Yadav, V. R.; Tekmal, R. R., et al. Design of Curcumin-loaded PLGA Nanoparticles Formulation with Enhanced Cellular Uptake, and Increased Bioactivity In Vitro and Superior Bioavailability In Vivo. *Biochem. Pharmacol.* **2010,** *79*, 330e338.
6. Anantharaju, P. G.; Gowda, P. C.; Vimalambike, M. G.; Madhunapantula, S. V. An Overview on the Role of Dietary Phenolics for the Treatment of Cancers. *Nutr. J.* **2016,** *15* (1), 99.

7. Barnaby, S. N.; Fath, K. R.; Tsiola, A.; Banerjee, I. A. Fabrication of Ellagic acid incorporated Self-assembled Peptide Microtubes and their Applications. *Colloids Surf. B Biointerfaces* **2012,** *95*, 154–161.
8. Berendsen, R.; Guell, C.; Ferrando, M. A Procyanidin-rich Extract Encapsulated in Water-In-Oil-In-Water Emulsions Produced by Premix Membrane Emulsification. *Food Hydrocoll.* **2015,** *43*, 636–648.
9. Berton-Carabin, C. C.; Coupland, J. N.; Elias, R. J. Effect of the Lipophilicity of Model Ingredients on their Location and Reactivity in Emulsions and Solid Lipid Nanoparticles. *Colloids Surf A Physicochem. Eng. Asp.* **2013,** *431*, 9–17.
10. Betz, M.; Kulozik, U. *Microencapsulation of Bioactive Bilberry Anthocyanins by Means of Whey Protein Gels.* 11th International Congress on Engineering and Food (Icef11), 1, 2047–2056.
11. Betz, M.; Steiner, B.; Schantz, M.; Oidtmann, J.; Mader, K.; Richling, E.; Kulozik, U. Antioxidant Capacity of Bilberry Extract Microencapsulated in Whey Protein Hydrogels. *Food Res. Int.* **2012,** *47* (1), 51–57.
12. Bora, A. F. M.; Ma, S.; Li, X.; Liu, L. Application of Microencapsulation for the Safe Delivery of Green Tea Polyphenols in Food Systems: Review and Recent Advances. *Food Res. Int.* **2018,** *105*, 241–249.
13. Conte, R.; Calarco, A.; Napoletano, A.; Valentino, A.; Margarucci, S.; Di Cristo, F.; Peluso, G. Polyphenols Nanoencapsulation for Therapeutic Applications. *J. Biomol. Res. Ther.* **2016,** *5* (2).
14. de Souza Simões, L.; Madalena, D. A.; Pinheiro, A. C.; Teixeira, J. A.; Vicente, A. A.; Ramos, Ó. L. Micro-and Nano Bio-based Delivery Systems for Food Applications: In Vitro Behavior. *Adv. Coll. Interface Sci.* **2017,** *243*, 23–45.
15. Fang, R.; Hao, R. F.; Wu, X.; Li, Q.; Leng, X. J.; Jing, H. Bovine Serum Albumin Nanoparticle Promotes the Stability of Quercetin in Simulated Intestinal Fluid. *J. Agric. Food Chemistry* **2001,** *59* (11), 6292–6298.
16. Fang, R.; Jing, H.; Chai, Z.; Zhao, G.; Stoll, S.; Ren, F.; Leng, X. Design and Characterization of Protein-quercetin Bioactive Nanoparticles. *J. Nanobiotechnol.* **2011,** 9.
17. Fang, Z.; Bhandari, B. Encapsulation of Polyphenols–A Review. *Trends Food Sci. Technol.* **2010,** *21* (10), 510–523.
18. Floros, J. D.; Newsome, R.; Fisher, W.; Barbosa-Cánovas, G. V.; Chen, H.; Dunne, C. P.; Knabel, S. J. Feeding the World Today and Tomorrow: The Importance of Food Science and Technology. *Compr. Rev. Food Sci. Food Saf.* **2010,** *9* (5), 572–599.
19. Frank, K.; Kohler, K.; Schuchmann, H. P. Formulation of Labile Hydrophilic Ingredients in Multiple Emulsions: Influence of the Formulation's Composition on the Emulsion's Stability and on the Stability of Entrapped Bioactives. *J. Dispers. Sci. Technol.* **2011,** *32* (12), 1753–1758.
20. Frank, K.; Walz, E.; Graf, V.; Greiner, R.; Kohler, K.; Schuchmann, H. P. Stability of Anthocyanin-rich W/O/W-emulsions Designed for Intestinal Release in Gastrointestinal Environment. *J. Food Sci.* **2012,** *77* (12), N50–N57.
21. Garti, N. Double Emulsions-scope, Limitations and New Achievements. Colloids Surf A Physicochem. Eng. Asp. **1997,** *123*, 233e246.
22. Gharsallaoui, A.; Roudaut, G.; Chambin, O.; Voilley, A.; Saurel, R. Applications of spray-drying in microencapsulation of food ingredients: An overview. *Food Res. Int.* **2007,** *40* (9), 1107–1121.

23. Hemar, Y., Cheng, L. J., Oliver, C. M., Sanguansri, L., & Augustin, M. Encapsulation of Resveratrol Using Water-In-Oil-In-Water Double Emulsions. *Food Biophys.* **2010,** *5* (2), 120–127.
24. Hu, B.; Liu, X.; Zhang, C.; Zeng, X. Food Macromolecule Based Nanodelivery Systems for Enhancing the Bioavailability of Polyphenols. *J. Food Drug Anal.* **2017,** *25* (1), 3–15.
25. Hu, B.; Ting, Y. W.; Zeng, X. X.; Huang, Q. R. Bioactive Peptides/Chitosan Nanoparticles Enhance Cellular Antioxidant Activity of (-)-Epigallocatechin-3-gallate. *J. Agric. Food Chem.* **2013,** *61* (4), 875–881.
26. Jenning, V.; Gohla, S. H. Encapsulation of Retinoids in Solid Lipid Nanoparticles (SLN). *J. Microencapsul.* **2001,** *18* (2), 149–158.
27. Jia, Z.; Dumont, M. J.; Orsat, V. Encapsulation of Phenolic Compounds Present in Plants Using Protein Matrices. *Food Biosci.* **2016,** *15*, 87–104.
28. Li, Z.; Percival, S. S.; Bonard, S.; Gu, L. W. Fabrication of Nanoparticles Using Partially Purified Pomegranate Ellagitannins and Gelatin and their Apoptotic Effects. *Mol. Nutr. Food Res.* **2011,** *55* (7), 1096–1103.
29. Li-Chan, E. C. Bioactive Peptides and Protein Hydrolysates: Research Trends and Challenges for Application as Nutraceuticals and Functional Food Ingredients. *Curr. Opin. Food Sci.* **2015,** *1*, 28–37.
30. Lu, W.; Kelly, A. L.; Miao, S. Emulsion-based Encapsulation and Delivery Systems for Polyphenols. *Trends Food Sci. Technol.* **2016,** *47*, 1–9.
31. Mignet, N.; Seguin, J.; Chabot, G. G. Bioavailability of Polyphenol Liposomes: A Challenge Ahead. *Pharmaceutics* **2013,** *5* (3), 457–471.
32. Mukherjee, S.; Ray, S.; Thakur, R. S. Solid Lipid Nanoparticles: A Modern Formulation Approach in Drug Delivery System. *Ind. J. Pharm. Sci.* **2009,** *71* (4), 349.
33. Nedovic, V.; Kalusevic, A.; Manojlovic, V.; Levic, S.; Bugarski, B. An Overview of Encapsulation Technologies for Food Applications. *Procedia Food Sci.* **2011,** *1*, 1806–1815.
34. Neo, Y. P.; Ray, S.; Jin, J.; Gizdavic-Nikolaidis, M.; Nieuwoudt, M. K.; Liu, D. Y.; Quek, S. Y. Encapsulation of Food Grade Antioxidant in Natural Biopolymer by Electrospinning Technique: A Physicochemical Study Based on Zein-gallic Acid System. *Food Chem.* **2013,** *136* (2), 1013–1021
35. Pan, K.; Zhong, Q.; Baek, S. J. Enhanced Dispersibility and Bioactivity of Curcumin by Encapsulation in Casein Nanocapsules. *J. Agric. Food Chem.* **2013,** *61* (25), 6036–6043.
36. Patel, A. R.; Heussen, P. C. M.; Hazekamp, J.; Drost, E.; Velikov, K. P. Quercetin Loaded Biopolymeric Colloidal Particles Prepared by Simultaneous Precipitation of Quercetin with Hydrophobic Protein in Aqueous Medium. *Food Chem.* **2012,** *133* (2), 423–429.
37. Prüsse, U.; Bilancetti, L.; Bučko, M.; Bugarski, B.; Bukowski, J.; Gemeiner, P.; Nedovic, V. Comparison of Different Technologies for Alginate Beads Production. *Chem. Papers* **2008,** *62* (4), 364–374.
38. Ramos, O. L.; Pereira, R. N.; Martins, A.; Rodrigues, R.; Fuciños, C.; Teixeira, J. A.; Vicente, A. A. Design of Whey Protein Nanostructures for Incorporation and Release of Nutraceutical Compounds in Food. *Crit. Rev. Food Sci. Nutr.* **2017,** *57* (7), 1377–1393.
39. Rao, P. J.; Khanum, H. A Green Chemistry Approach for Nanoencapsulation of Bioactive Compound: Curcumin. *LWT-Food Sci. Technol.* **2016,** *65*, 695–702.

40. Santiago, L. G.; Castro, G. R. Novel Technologies for the Encapsulation of Bioactive Food Compounds. *Curr. Opin. Food Sci.* **2016,** *7*, 78–85.
41. Shpigelman, A.; Israeli, G.; Livney, Y. D. Thermally-induced Protein–Polyphenol Co-assemblies: Beta Lactoglobulin-based Nanocomplexes as Protective Nanovehicles for EGCG. *Food Hydrocoll.* **2010,** *24* (8), 735–743.
42. Singh, N. K.; Lee, D. S. In Situ Gelling Ph-and Temperature-Sensitive Biodegradable Block Copolymer Hydrogels for Drug Delivery. *J. Controlled Rel.* **2014,** *193*, 214–227.
43. Singh, S. K.; Dhyani, A.; Juyal, D. Hydrogel: Preparation, Characterization and Applications. *Pharma Innov.* **2017,** *6* (6, Part A), 25.
44. Smith, A.; Giunta, B.; Bickford, P. C.; Fountain, M.; Tan, J.; Shytle, R. D. Nanolipidic Particles Improve the Bioavailability and Alpha-secretase Inducing Ability of Epigallocatechin-3-gallate (EGCG) for the Treatment of Alzheimer's Disease. *Int. J. Pharm.* **2010,** *389*, 207e212.
45. Taylor, J.; Taylor, J. R. N.; Belton, P. S.; Minnaar, A. Kafirin Microparticle Encapsulation of Catechin and Sorghum Condensed Tannins. *J. Agric. Food Chem.* **2009,** *57* (16), 7523–7528.
46. Wang, X.; Wang, Y. W.; Huang, Q. Enhancing Stability and Oral Bioavailability of Polyphenols Using Nanoemulsions.
47. Yang, J. M.; Zha, L. S.; Yu, D. G.; Liu, J. Y. Coaxial Electrospinning with Acetic Acid for Preparing Ferulic Acid/Zein Composite Fibers with Improved Drug Release Profiles. *Colloids Surf. B Biointerfaces* **2013,** *102*, 737–743.
48. Zhang, T.; Lv, C. Y.; Chen, L. L.; Bai, G. L.; Zhao, G. H.; Xu, C. S. Encapsulation of Anthocyanin Molecules Within A Ferritin Nanocage Increases their Stability and Cell Uptake Efficiency. *Food Res. Int.* **2014,** *62*, 183–192.
49. Zhou, H. H.; Sun, X. Y.; Zhang, L. L.; Zhang, P.; Li, J.; Liu, Y. N. Fabrication of Biopolymeric Complex Coacervation Core Micelles for Efficient Tea Polyphenol Delivery via a Green Process. *Langmuir* **2012,** *28* (41), 14553–14561.

CHAPTER 13

NANOTECHNOLOGY-BASED CHALLENGES AND SCOPE IN THE FOOD INDUSTRY: FROM PRODUCTION TO PACKAGING

NEELU SINGH[1], MONOJ KUMAR DAS[2], ANAND RAMTEKE[2,*], PAULRAJ RAJAMANI[2], SANKAR CHANDRA DEKA[3], AFTAB ANSARI[4], DAMBARUDHAR MAHANTA[4,] and ANURAG MAUYRA[5]

[1]*School of Environmental Sciences, Jawaharlal Nehru University, New Delhi 110067, India*

[2]*Department of Molecular Biology and Biotechnology, Cancer Genetics and Chemoprevention Research Group, Tezpur University, Napaam, Tezpur 784028, Assam, India*

[3]*Department of Food Engineering and Technology, Tezpur University, Napaam, Tezpur 784028, Assam, India*

[4]*Department of Physical Sciences, Nanoscience and Soft Matter Laboratory, Tezpur University, Napaam, Tezpur 784028, Assam, India*

[5]*Department of Botany, Shivaji College, University of Delhi, New Delhi 110027, India)*

[*]*Corresponding author. E-mail: anand@tezu.ernet.in*

ABSTRACT

In recent years, we have seen rapid advancement in the field of nanotechnology, releasing nanoproducts which are commonly used in our day-to-day life ranging from cosmetics to electronics, etc. Application of nanotechnology is proven beneficial for the food industry and found various application at various stages from agriculture field (used as nanofertilizer, nanopesticides, etc.) to the market (in food packaging, as food additive, enhancing texture, and aroma of food) as it ensures the food security from field to the market which has become possible due to the usage of nanoparticles in various formulations like nanolaminates, nanoemulsions, nanoliposomes which helps in controlled release of nutrients in the vicinity, which is easily absorbed in git, thus increasing their bioavailability. Usage of nanomaterial in nanosensor facilitates the consumer to know the current status of the food item, wither any spoilage or manhandling takes place during transportation or at storage condition. Altogether nanomaterial not only increases the yield but also offers pest resistance further act as antimicrobial checking the growth of any microbes, increasing the shelf life, flavor, aroma of the food item, whereas any change in the moisture content or mistreatment with the food item can be easily detected by the application of nanosensor. Thus, in order to ensure the food security for the growing population through retaining their quality and quantity both, however, irrational usage of nanomaterial raises severe health concerns as several studies reports the harmful effect of nanomaterial which cannot be overlooked; therefore, there is an urgent need to formulate some regulatory body to inspect the nanogoods prior release to the market.

13.1 INTRODUCTION

Increasing population, seasonal variation, pollution in air, soil and water, indiscriminate usage of pesticides and fertilizers are certain factors which directs us toward a future where food shortage with deprived nutritional value is most certain. To meet up with the continuing consumers' demand of a consistently growing population, food industry has to evolve with the uprising technology and methods/disciplines such as biotechnology, information technology (IT), cognitive sciences, and recently by nanotechnology, which revolutionizes the food sector from basal agriculture field to the food packaging. Nanotechnology deals with study and application of particles, whose size ranges from 1 to 100 nm at least in one dimension. Past few decades witnessed the booming usage of these nanomaterials in various fields like cosmetics to electronics, medicine to defence, and textile to agriculture, including the food industry. The exclusive properties displayed by these nanoscale materials are highly advanced over their counter bulk part, therefore replacing their place in market. The unique noble physicochemical properties, distinctive from bulk counterpart, are by the virtue of their small size (i.e., greater surface-to-volume ratio) which leads to higher reactivity. Furthermore, forerunning mainstream technological fields working in coherence with nanotechnology allows major breakthrough in various sector of the food beginning from the fields like production to the processing, transportation, storage, safety, traceability, and food security.[14]

Among the various industries, food industry bags a huge amount in the global market, being a basic requirement in addition to clothing and shelter and with rise in standard of living, and awareness/consciousness of health/nutrition, demands for high quality and nutritious value of the food have risen. Thus, offering a fascinating area of research to the food industry to safeguard the food produced from the agriculture field to the processing, packaging and their delivery to the marketplace from microbial decay, instead also enhance the food taste, texture, thereby including quality by improving the nutritional status coupled with the generation of a good quality product. In agricultural sector, various nanoparticles are being used such as macronutrients are coated with nanoparticles, viz., zinc oxide to fulfil their nutritional requirement by enhanced absorption, nanoclay for retention of water, and liquid agrochemical which was further maintained by their control release, filters coated with titanium dioxide can be used for efficient removal of toxicant from contaminated water, etc. Thus, nanomaterials become an innate choice which claims not only to improve the texture and quality of food with consistent taste but also increase their shelf life and productivity. Various formulations such as emulsions, biopolymer matrices, and associated colloids in nanotechnology ensure not only the delivery of bioactive compound but also maintain their level precisely such as ferritin nanocages, whereas nanocapsules used to disguise not only the odor or taste but also regulate the controlled interaction of bioactive compound with the food and further protect from moisture, heat, chemical, biological decay during processing, cargo space, and consumption. Micelles (5–100 nm) can be used to deliver antioxidants and vitamins to food and beverages without altering the basic constituents. Nanoemulsions are being produced at high pressure and found be to highly efficient in delivery of compound like proteins, polysaccharide, and phospholipids which are trapped in the core and later delivered; these emulsions are highly resistant to environment degradation as compared with conventional oil in water emulsions. Liposomes are lipid vesicle, encapsulating vitamins, enzymes to increase the ripening of dairy product like cheese, thus augmenting their nutritional value. Use of nanoclays like montmorillonite in addition with biopolymer shields oxygen and water vapor well in packed food, thus extends the shelf life of the food products. Similarly, nanolaminates are used as coating material in edibles to improve texture, nutrients, and flavors, antimicrobials and antioxidants, also nanobased supplements, such as Chinese nano-tea claimed to increase the uptake of selenium. Also, nanosensors are being used to determine any contaminants or mycotoxins, pathogen, or gas if present in food, engage to degrade the quality of groceries such as liposome vesicle to detect peanut allergic proteins in chocolate, zinc oxide in E-nose for as gas sensors. Likewise, nano-tongue has been devised to detect adulterant even if present in parts per trillion, further reflected by color change in labels if contaminated. Gold-coated quartz crystals, a DNA-based sensor, are used to detect bacteria like *E. coli.* Devices like nanoelectromechanical system (NEMS) technology are already in use to detect trans-fat content, thus can control the storage condition and serves as active sell by device which is portable as well as also offer low cost on site detection with smart communication for monitoring any adulteration in wrapping and storage ambience. Nanocantilever devices are also being used in detection of toxins, chemical contaminants, and antibiotics in food item, pathogen detection due to their ability to vibrate with intensity which varies with pathogen load in water or in food.

Nanotechnology in food industry offers a soaring business which was started with US$2.6 billion in 2006, expected to reach US$20.4 billion in 2015. But widespread rejection of genetically modified foods, the food industry is unpredictable owing on the foods modified at nanoscale. In fact, there are no globally accepted rules and regulations for nanotechnology, thus manufacturers' have to go through the existing guidelines to introduce new nanoproducts in the market which can be time consuming and difficult too.

Although it is quite thrilling to have innovative nanofood products being launched in the market on a regular basis, various serious questions can be raised regarding the safety issue which cannot be otherwise neglected. Reports reveal that accidental release of toxic nanoproducts at work place causes potential harm to the occupational worker. Furthermore, continued exposure for long time, even in small concentration of nanogoods induces considerable damage. Works on various animal models show aggregation of nanoparticles in brain, thereby posing serious concern. Therefore, a need of an eagle eye is a requisite to watch over and make the industries accountable regarding the usage of nanomaterials within safe limits so that the advantages are rightfully enjoyed and drawbacks, hence, should be openly discussed.

13.2 NANOMATERIALS: AGRICULTURAL SECTOR TO MARKETING

Nanomaterials have found various application in food industry starting from agricultural field to the market, serving as a nutritional supplement as nanofertilizers, as pesticides known as nanopesticides, in food processing, storage, packaging, bioavailability of nutrients, as nano-ensors, etc.

Increasing environmental instability, shortage of cultivable lands and water reduced soil fertility due to inappropriate usage of pesticides, and growing population are some of the common problems for countries with agri-based economies.[44] Therefore, there is an urgent need to grow foods which is not only drought and pesticide resistant but also nutrient rich and thus can fulfil the nutritional requirement of the growing population. Nanotechnology proves to solve many of such agri-based issues, thus becoming an important tool in modern agriculture to boost production through various means such as nanoderivatives of fertilizers and also as pesticides in crop improvement. Nanosensors for disease identification in crop, genetic manipulation for crop improvement, postharvest management are some of other ways in which nanotechnology can be incorporated in food production efficiently.[82] Details Schematic representation of role of nanotechnology invarious sectors of food industry are shown in Figure 13.1 and Table 13.1 shows various Application of Nanomaterial Ranging, Its Uses as Pesticides, Fertilizers, and in Food Packaging.

13.2.1 NANOPESTICIDES

Nanopesticides is not a single entity as it consist of various surfactants, metal nanoparticles (inorganic), and polymers (organic); therefore, rightly said as nanoformulations which falls in the nanometer range, overcoming the various limiting factors such as water solubility, availability further increases the controlled release of active ingredient such as nanoformulations of Vali-nano-CC (nano-sized calcium carbonate loaded with validamycin pesticide) shown germicidal effect against *Rhizoctonia solani*,[79] nanoscale TiO_2 doped with silver (TiO_2/Ag) or with zinc (TiO_2/Zn; AgriTitan) is effective against *Xanthomonas perforans* causes bacterial spot disease of tomato,[71] (MgO NPs) is effective against *Ralstonia solanacearum* both in vitro and in vivo.[10]

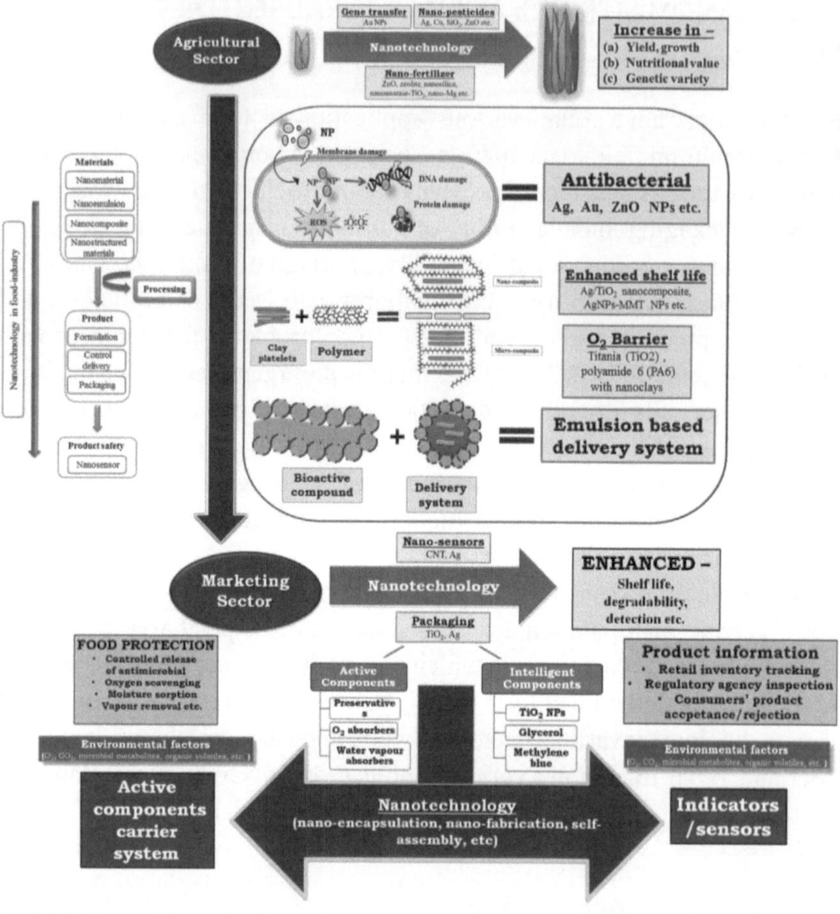

FIGURE 13.1 **(See color insert.)** Schematic representation of role of nanotechnology in various sectors of food industry.

13.2.2 NANOFERTILIZERS

Nanofertilizers, consisting one or more types of nutrients to fulfil the nutritional requirement of plants support their growth.[17] Nano fertilizers can be broadly categorized as follows:

i) **Macronutrient nanofertilizers:** These are the nutrients which are required by the plants in large quantity such as urea-modified hydroxyapatite nanoparticles synthesized as a source of N, and

TABLE 13.1 Various Application of Nanomaterial Ranging, Its Uses as Pesticides, Fertilizers, and in Food Packaging.

1(a) Nanomaterials as/in pesticides	Type	Activity	References
Deltamethrin	Solid–lipid nanoparticles	Decreased direct and indirect photodegradation	62, 63
Thiamethoxam-coated liposomes	Electrospun nanofibers	Efficient at 50% of the recommended dosage	98
TiO_2	Inorganic nanoparticle	Better efficacy compared to standard treatment	69, 70
Al_2O_3	Inorganic nanoparticle	Similar or greater insecticide activity than most effective commercially available Diatomaceous earth formulation	87
Imidacloprid and avermectin	Nanocatalyst-conjugated in microcapsules	Faster decomposition in soil and/or plant	35
Nanoenapcsulated glyphosate or sulfonylurea herbicide	Nanocapsules	Targeted delivery	Perez et al. (2008)
TiO_2-M262 polymer metaflumizone	Nanocapsules with catalyst ai conjugate	protection against premature degradation	42
Nanopermethrin	Nanoemulsion	enhanced uptake/efficacy	3
Nanodispersed triclosan	Nanodispersions	enhanced toxicity to target organism (lower dose)	105
ZnO	Nanoparticle	Dramatic increase in root growth of peanut plant after nano-ZnO treatment	77
Zeolite	Nanoparticle	Nitrogen, phosphorus, boron, zinc, potassium, sulphur, adsorbent	65, 88
Kaoline	Nano-subnanocomposites	Strong adsorption and thickness to macronutrients and organic C	50
Nanoporous silica	Nanoparticle	Controlled release	38
Nanosilica	Nanoparticle	Enhance the plant's resistance to stress	DeRosa et al. (2010)

TABLE 13.1 *(Continued)*

1(a) Nanomaterials as/in pesticides	Type	Activity	References
Zn-Al layered double-hydroxide	Nanocomposites	Controlled release of chemical compounds	41
Nano-anatase TiO_2	Nanoparticle	Improved crop yield through the photoreduction of nitrogen gas	100
ZnO	Coated nanoparticle	Increased biomass production	61
ZnO + CuO + B2O	Nanocomposites	Increased production	Dimpka et al. (2017)
Nano-coated urea	Nanoparticle	Reduction in N_2O	47
Nano-CaO	Nanoparticle	Increased Ca accumulation;	22
Nano-Mg	Nanoparticle	Promotion of photosynthesis	23

1(c) Nanomaterials in food packaging			
Nanomaterial	**In food product**	**Activity**	**References**
LDPE nanocomposite containing Ag and ZnO NP	Orange juice	Lactobacillus plantarum	27
Titania (TiO_2)		O_2 scavenging	99
TiO_2	Lactuca astiva	*E. coli* inhibition	15
AgNP in absorbant pads	Poultry meat	*Staphylococcus aureus*, *E. coli*	30
Ag-MMT NPs	Fresh cut Carrot	*Pseudomonas* spp.	19
Nanocomposite of Cellulose/Ag	Fresh cut melon fruit and beef meat	*Pseudomonas* spp.	52
Carbon nanotube (CNT) with allyl isothiocyanate (AIT)	cooked chicken meat	*Salmonella choleraesuis*	24

TABLE 13.1 *(Continued)*

Nanomaterial	In food product	Activity	References
Silver-ethylene–vinyl alcohol (EVOH)	Apple peel, chicken breast, pork, peels	*L. monocytogenes*, *Salmonella* spp.	Martínez-Abad et al., 2012
(LDPE)/ Ag and ZnO	Chicken breasts	Lactobacillus, Enterobacteriaceae, and mesophiles	68
Olyamide 6 (PA6) with nanoclays	Beef steaks	Improved UV block capability, oxygen barrier property	75
Au-Np	Canned tuna	Freshness of the food example	20
Carbon nanotubes (CNT)	Meat	Pathogen detection	102
$CACO_3$ NPs	Fish sample	Freshness in sample	83
Ag-MMT NPs	Fresh cut Carrot	Freshness in sample	19
Ag-MMT NPs	Fruit salad	Fruit freshness	18
AgNPs-MMT NPs	Fior di latte cheese	Shelf life	32
PVC) film with ZnO NPs	Fresh cut 'Fuji' apple	Fruit decay rate lowered	49
Ag-LDPE (Low density polyethylene)	Barberry	Freshness of sample	88
polyamide 6 (PA6) with nanoclays	Beef steaks	Improved UV block capability, oxygen barrier property	75
Ag-NP/MAP	Fiordilatte cheese	Prolong shelf life and antimicrobial activity	59
Ag/TiO_2 nanocomposite	Bread	Enhanced shelf life	60

showed their prolonged and controlled release of nitrogen.[46] Synthetic nanohydroxyapatite nanoparticles serves as source of phosphorus found to increase the yield of *Glycine max*.[51]

ii) **Micronutrient nanofertilizer:** These are the nutrients which are required by the plants in minute concentration but are essential for their growth. Nanoformulations of micronutrients further enhance their growth such as spray of Mn nanoparticles on mung bean (*Vigna radiata*) increases the biomass by 38%,[76] ZnO NP found to increase shoot and root length, biomass, chlorophyll content in various plants such as on *Vigna radiate* and *Cicer arietinum*.[55]

iii) **Nanoparticulate fertilizer:** These nanoparticles have shown growth promoting activity in various plants. TiO_2 shows growth promoting activity, thereby increases yield in spinach,[33] single-walled carbon nanotubes (CNTs) enhances root elongation in onion and cucumber,[11] etc.

13.3 OTHER SYNTHETIC NANOSTRUCTURES

Synthetic nanostructures which are used in food not only enhance bioavailability of bioactive components facilitating their control release but also protect them during processing and transportation, increasing their shelf life with improved texture and taste. Nanoemulsions, microemulsions, liposomes, and polymeric NPs are the commonly used nanostructured system used frequently in food.[12]

13.3.1 NANOEMULSIONS

Nanoemulsion is a mixture of two or more liquids which are not easily miscible, these dispersed droplets varies in diameter ranging from 50 to 100 nm.[37] These nanoemulsions can encase the functional ingredients within these droplets, preventing their chemical degradation and active constituent will only be released under specific environment.[80] Due to their small size, they are advantageous over conventional methods as they show high surface area to volume ratio and therefore are thermally stable. Moreover, they easily interact with gastrointestinal tract (GIT) and lipases digest the small droplet of nanoemulsions easily.[104] Nanoemulsions are used in the various food products as salad dressing beverages, flavored oils, sweeteners which are released under various environmental triggers such as pH, heat, ultrasonic waves, etc.[37] Unlike regular emulsions, many of nanoemulsions are transparent, thus can be easily added in the drinks,[78] also used flavoring agent further. Yu and Huang showed nanoemulsions can be used for improving the digestibility of the food such as nanoemulsion of curcumin in oil phase facilitates easy digestion as compared with taken directly.[103]

13.3.2 NANOLAMINATES

Nanolaminates consist of thin grade film (1–100 nm/layer) of two or more layers which is physically or chemically bonded in various dimensions. Therefore, various applications in edible films in packaging of wide variety of foods such as fruits, vegetables, candies, meats, baked goods, etc. have been found.[80] These films protect edibles from moisture, gases, and lipids, enhance texture, flavors, taste, nutrients, colors, and show antimicrobial activity.

13.3.3 BIOPOLYMERIC NANOPARTICLES

Food grade biopolymers such as polysaccharides/proteins can be used to produce nanosize particles, later can be used for the encapsulation of functional ingredients released only under certain environment specificity. Polylactic acid (PLA) is most commonly used biopolymer by various manufacturers for the encapsulation of drug, proteins, and vaccines but has shown certain shortcoming due to quickly removed from the blood stream.[72] Nanoliposomes are colloidal arrangement that exhibit amphiphilicity, which primarily constitutes of phospholipids in aqueous phase and are highly biocompatible and biodegradable, therefore being used in various fields including food and pharmaceutical sectors.

13.4 NANOMATERIALS IN FOOD PACKAGING

Food packaging is an important industry that responses not only to changing demand of consumers but also to the ecofriendly nature of the product, incorporating suitable developments and advancements in material research. With the objectives for effective nontransient distribution and preservation of products, the packaging industry has to perform unfavorable adverse conditions with the likes of mechanical shock and vibrations during distribution process as well as contaminants as dust, microorganisms that can hamper quality of the product. Moreover, on an economic level, packaging must provide proper advertising and information features related to the product. In order to resolve these issues, two categorical methods can be put forward, namely, active (one that meets up the preservation and safeguard aspects) can be implemented and intelligent (one that meets up the commercial, i.e., marketing and feedback aspects).[49]

13.4.1 BARRIERS IN FOOD PACKAGING: CURRENT SCENARIO

Despite acceptance by a large section of society, the major challenges a food packaging industry face are cost-effective/ecofriendly production as well as inadequate mechanical steadiness. The use of metal and glass are materials that answer the challenges but only to limited extent. Polymers and plastics, hence, offer to be a popular solution, given its lighter weight, cost-effective production, and its unique frame adaptability with greater flexibility. As a result, industries utilize more than 20% of plastics for food packaging.[104] Being petroleum-based product, the most important challenge plastic utilization faces is to safeguard human health and biodegradability.

Other barriers in packaging include the degradation of product quality over time due to variable moisture content in the environment as well as exposure to atmosphere (containing different mixture of gases). To provide a longer shelf-life service of specific products requires inclusion of commodities with facets of thermoplastics, introducing new challenges in cost-effectiveness and increased toil in recycling the same. Poor shelf life imparts low sales and results in generation of waste. Also inadequate shelf life leads to foodborne diseases, thereby giving a negative impact on the brand name of the product. The proliferation of microorganism due to contamination and temperature abuse, decrease of nutritional values due to oxidation, loss of organoleptic, and/or nutritional qualities due to interaction with deleterious extrinsic factors (light, oxygen, and water) are some of the examples of food quality and safety regulations that still need to be overcome (Mihindukulasuriya and. Lim, 2014).

Exploitation of nanotechnology has widened the scope of addressing several barriers and challenges through extensive research at large.

13.4.2 DEVELOPMENT OF NANO-COMPOSITES FOR FOOD PACKAGING

Utilization of petroleum-based products such as plastics resolves worries related to long-term adaptability of a packaging product and improving shelf life of the food product, but probes complications as regards biodegradability and food safety. Use of biopolymers could maintain food quality by providing better shelf life as well as accounts for adequate biodegradability; however, the mechanical strength of the packaging material needs to be compromised.

In contrast, nanocomposites have emerged as potential candidates that might overcome shortcomings of both environmental friendly polymers and petroleum-based polymers, integrating the proficient assets of both. In general, nanocomposites refer to composite system comprised a single or a mixture of polymers (at least one organic/inorganic filler), which has a dimension lower than 100 nm. Several nanoscale materials in biopolymer composite structure are already in use, namely, $SiO_{2,}$ $TiO_{2,}$ $KMnO_4$, etc., improving its mechanical properties. As reported in the literature, mechanical properties such as stiffness, tensile strength, toughness, shear strength, delamination resistance, fatigue, barrier properties, thermal stability, and optical properties of polymers[106–109] are substantially improved due to incorporation of nanomaterials into their matrices.

Proper distribution of the nanoscale materials in biopolymer matrices and the permeability are the factors that dictate the improved effective mechanical nature of the nanocomposite as well as results in potential to overcome barriers such as degradation due to moisture, etc. Extensive study has been done as regards packaging over clay/polymer nanocomposite for instance, Xie et al. have reported a seven-fold improvement in oxygen barrier properties of a low density polyethylene (LDPE) due to incorporation of organic montmorillonite (OMMT) as compared with conventional polyethene.[123] Polysaccharides and proteins are the most popularly used bio-based polymers which are attractive due to their enhanced biodegradability aspects.

13.4.3 NANOTECHNOLOGY IN ACTIVE PACKAGING

The food safety and quality can be ensured via proper monitoring of exposure to elements in nature that probe threat to preservation conditions. This can be achieved with suitable application of nanotechnology assimilating the active components into nanocomposites. These nanocomposites prove to be a carrier system that interacts with factors such as O_2, CO_2, microbial metabolites, organic volatiles, etc. ensuring improved shelf life. Such active component carrier systems are arranged by methods such as nanoencapsulation, nanofabrication, nanocomposites, self-assembly, etc., of active components such as preservatives, O_2 absorbers and moisture, water vapor absorbers, etc. Usually, inorganic metal nanosized particles have proved to be a better nanocarrier for active components. These nanoscale materials are extensively studied for putting into purpose their antimicrobial properties due to their reduced dimension and increased reactivity, resulting in low proliferation rate of pathogenic microorganisms which leads to increased protection toward degeneration and decay. Silver nanoparticles (AgNPs) are known to show antimicrobial activity, with high penetrability into outer and inner cell membranes, thereby disrupting its metabolic activity and hampering its DNA replication process as well.[110] AgNPs are effective against both Gram-positive and negative bacteria, where generation of reactive oxygen species (ROS) can be controlled in an effective manner, providing suitable antimicrobial environment showing immense potential in application as a nanocomposite as a food composite as well as a packaging material.[104,105]

Studies suggest that both the dimension of the nanoparticles and the matrix for a nanocomposite are important factors having an influence on antimicrobial and antibacterial properties. In particular, AgNPs with a diameter of ~41 nm had greater antimicrobial response than 100 nm in hydroxypropyl methylcellulose films against *E. coli* and *S. aureus*.[111] Although far-reaching studies on AgNPs have been performed already, there still exists scope as regards the enhanced efficacy and precise regulation against bacteria.

As a user-friendly matrix for a nanocomposite, PVA has proven to be suitable for candidate/host for nanoparticles with the likes of Ag and TiO_2. Polyolefin is another polymer ordinarily used in food packaging (as a sealant against moisture barrier), but in a broad vision fails to provide adequate oxygen barrier for products which are susceptible to oxygen.

High density polyethylene (HDPE) films modified by incorporation of iron (containing kaolinite) provide better oxygen scavenging packaging films.[112] The oxygen scavenging ability was attributed to oxygen trapping and increased tortuous/twisted diffusion path.

It is also the shape of the nanomaterial that plays a significant role in dictating the antimicrobial property along with the dimension of the nanomaterial. Jing et al. have reported comparative account both nanoparticle and nanotube titania against *E. coli*.[124] TiO_2 nanoparticles used as a coating material for packaging substrates (glass, acetate films) are proved to be effective in creating deoxygenated closed environment (Li et al., 2004). In this regard, photocatalytic nature of TiO_2 has been extensively exploited by researchers considering different targets. Kubacka et al. in their study provided manifestations that cells exposed to TiO_2 photocatalysis exhibit rapid cell inactivation at the regulatory and signalling levels along with strong decrease of the coenzyme-independent respiratory chains. The ability to adapt and transport Fe and P was also found to be lowered.[4]

Moreover, use of nanotechnology in active packaging also involves enhancement of the water and moisture-repellent properties, a shortfall of non-petroleum/paper-based packaging materials. TiO_2 and Ag nanoparticle incorporated papers/films provide maintaining a nanostructured hydrophobic surface. Teisala et al. used TiO_2 for inducing hydrophobicity by liquid flame spray (Teisala et al., 2012). In another study, polystyrene (PS) nanocomposites containing TiO_2 and Ag nanoparticles were prepared and applied as a coating onto a paper substrate prepared from rice straw.[114] Their results indicated that these nanocomposites are largely owing to nonabsorbant promising antibacterial aspects.

13.4.4 NANOTECHNOLOGY IN INTELLIGENT PACKAGING

The communication aspect of packaging is also as important as ensuring the quality of food, not only from points of view of commercial benefit and the industrial viewpoint but as well as to the consumers' need. Nanotechnological methods such as nanoencapsulation, nanofabrication, self-assembly, etc., help in building indicators/sensors, with the incorporation of intelligent components (TiO_2 nanoparticles, glycerol, and methylene blue), resulting in identity tags radiofrequency (ITR), time–temperature indicators (TTI), oxygen and carbon dioxide sensors, freshness indicators, etc.[115, 116] These indicators generate signals (usually visual or electrical) while correlating the physiological state of food product as well as product information (as regards the retail and regulation agency).

TiO_2 nanoparticles also serve as UV-activated indicator/intelligent tag once fabricated in the forms of coating and film.[17,118] UV-activated indicators can be activated in-package after sealing only as long as the package is transparent to UV radiation, and are consequently used for prevention of premature activation of oxygen measurement. Reports show that encapsulation of the active components within electrospun poly(ethylene oxide) fibers (submicron in diameter) result in higher sensitivity of the membrane.[119]

Indicators that signal bacterial inclusion involves utilization of nontoxic down-conversion phosphors–quantum dots, which show efficient fluorescence, are stable against photobleaching, have long decay lifetime, and also possess higher sensitivity. Antibodies conjugated to nanomaterials–quantum dots and UCNPs are being used today in labelling, imaging, and therapy areas.[56,96] Yang and Li (2006) investigated the use of QDs for simultaneous detection of *E. coli* O157:H7 and *Salmonella typhimurium*. Reportedly, with their approach, there was no cross-talk between the two emission peaks, allowing simultaneous detection of the two pathogens. The quantum dot conjugate tactic, thereby has several benefits, including those of considerably higher fluorescence intensity, reduced photobleaching, provision for multiplexed detection, and possibility of absorption over a wide spectral range.

Moreover, up-conversion fluorescent nanoparticles (UCNPs) can be excited by near-infrared, which emits higher energy visible radiation, allowing them to overcome some of the autofluorescence and photodamage problems associated with conventional quantum dot and organic fluorescence compounds. The UCNPs can be used as highly sensitive sensors for protein, enzyme, DNA, bacteria, pH, ammonia, carbon dioxide, and other analytes.[120] Considering their ability to detect the presence of microorganisms and other biological compounds with great specificity and sensitivity, it is expected that these concepts will demand further research and development at large. More specifically, for the development of smart and advanced intelligent labels for the detection of pathogens and their toxins in food would receive priority. As food safety standards continue to rise, the need for such biological indicators is believed to shoot up.

Time–temperature indicators (TTIs) are attractive in evaluating the rate of degradation given that food quality is comprehensively dependent on the temperature abuse. Ag triangular nanoplates are an example of a nanomaterial type which proves to be efficient colorimetric indicators for time–temperature history.[121] Ag nanoplates have sharp corners and displayed strong in-plane dipole resonance mode in visible radiation spectrum. In due course of time, the corners of the nanoplate become rounder, resulting in a gradual blue shift due to change in the resonance peak position, thereby altering its color from cyan to blue. The rate of this color transformation is temperature dependent. The advantages of this development involve lower cost and easy production, exhibiting better visual feedback as well. The main challenges of applying TTIs to reflect the quality of food is that the kinetics of the TTI's response must match with those of the key degradation reactions in food. AgNPs coated on paper or plastic film also serve as freshness indicators (Smolander, Hurme, Koivisto, and Kivinen, 2004). Fresh meats undergoing spoilage produce sulphide volatiles upon reacting with which the thin coating turns into distinctive dark color.

13.4.5 BENEFITS OF NANOTECHNOLOGY IN FOOD PACKAGING INDUSTRY

13.4.5.1 ENHANCED BIOAVAILABILITY OF NUTRIENTS

Nutraceuticals are bioactive proteins used to impregnate in the food, imparting health benefits to the customer in addition to the food itself.[13] Reduction in the size of the bioactive compounds further increases the solubility, delivery, and accessibility of the compounds across the GIT into the blood.[16,84] Therefore, the bioaccessibility and improved absorption can be achieved by altering the size and solubility of bioactives by modulating the surface-to-volume ratio which occurs during digestion. Some bioactive compounds to improve the bioavailability as nutritional supplements such as vitamins, coenzyme Q10 (CoQ10),[93] calcium ,[45] iron ,[74] curcumin[54] have already been extremely tested in nanodelivery system. There are various nanodelivery means such as biopolymeric nanoparticles,[39] nanofibers,[89,97] nanolaminates,[40] nanoemulsions,[58] lipid-based nanoencapsulators/nanocarriers,[29] which act as means of transport in delivering the cargo such as bioactive compounds. CoQ10 shows poor water solubility, and consequently, bioavailability increases by lipid-free nano-CoQ10 system customized with different surfactants formulated to improve their bioaccessibility through oral administration.[106] Further this governs their interactions with biologic setting and biodistribution. Surface-modified nanodelivery schemes have been developed for which chemical grafting of hydrophilic molecules can be done, knowing thatpolyethylene glycol is the generally known hydrophilic molecule.[92]

13.4.5.2 OXYGEN SCAVENGING AND NANOTECHNOLOGY

Interaction of oxygen with food results in induction of degradative reactions like browning, rancidity, depletion of vitamins, growth of aerobic microorganisms, loss of essential flavor compounds from the food. Furthermore, certain compounds are released by the fresh fruit such as ethylene which enhances the postharvest ripening process and thus reduces the shelf life. Active packaging came up with the concept to scavenge the ethylene and oxygen; therefore, high-density polyethylene (HDPE) films modified with iron containing kaolinite scavenge oxygen significantly. Furthermore, packaging substrate coated with TiO_2 is known to be effective in producing deoxygenated closed environment. Also, TiO_2 layered polypropylene films are effective in removal of ethylene vapor in packed horticultural goods utilizing the photocatalytic property of TiO_2.[56] Ethylene gas produced during ripening fruit can be easily absorbed by AgNPs which are therefore suitable for extending the shelf life if vegetables and fruits.[21] Of course, the amount of gas release differs from fruit to fruit and in this regard, both qualitative and quantitative assessment is necessary.

13.4.5.3 NUTRIENT SUPPLEMENT BY NANOSCALED MATERIALS

To enhance the taste, nutrients, absorption, and increased bioavailabilty, potential use of nanoselenium in green tea products has shown numerous health benefits. Furthermore, nanoencapsulation of various ingredients (such as antioxidants, probiotics, vitamins, preservatives, omega fatty acids, carotenoids, peptides, proteins, lipids, and carbohydrates) are frequently used to be incorporated in nanodelivery system. Zein, a major protein found in corn, has the potential to form a tube like structure which can be used as a vehicle for flavor compounds which has been explored to use as nutritional supplements.[86] Partial hydrolysis of milk in presence of specific protease leads to the self-assembly of α-lactalbumin into nanotubes, as in case of corn protein where zein is found which may facilitate the encapsulation of nutrients.[34] Such advanced method of packaging with pH triggered controlled release enhances satiability and extends shelf life and provides uninterrupted delivery of several active ingredients)[72].

13.4.5.4 NANOMATERIALS WITH ANTIMICROBIAL PROPERTIES

Microbial contagion is the leading cause for degrading the food quality and is of major concern while food processing, transportation, and storage. In this line, nanomaterial came up with a very positive and promising role in shielding the food items during processing, relocation, and storage with increased shelf life. Antimicrobial mode of action of nanomaterials is based on three models: (1) induction of oxidative stress by generating ROS, (2) metal ion release, (3) nonoxidative mechanisms, all of which can occur simultaneously.[94] ROS-mediated cell membrane damage is the principle route serving as antibacterial activity by various nanomaterials such as TiO_2, resulting in formation of numerous pits leads to cytoplasmic leakage and finally death of the microbes,[122] ZnO causes reduced enzymatic activity and thus acts as a bactericidal,[66] others like iron NPs cause cell aggregation which causes inactivation due to compression.[90] Furthermore, gradual release of metal ions from the metal oxides are later adsorbed on the bacterial surface, having close contact with various proteins and nucleic acids, thereby disrupting the cellular machinery resulting in inhibition of growth of the microorganisms.[94] However, some NPs like MgO-mediated inhibition in bacteria is neither by ROS mediated nor by metal ion interaction but they manage to alter the metabolic processes significantly inhibiting their growth.[48]

13.4.5.5 NANO-SENSORS FOR DETECTION OF CONTAMINANTS

Deviation from the assigned storage, transportation environmental conditions could lead to early deterioration of the food, which may harm the consumer and therefore when nanosensors incorporated during food packaging can easily detect the marker of the pathogen metabolism, which further alerts the consumer from the toxin released by the pathogen. This will virtually eliminate the need of an expiry date from the food product and thus help to know the accurate state of the food. However, manufacturer can easily predict the manhandling of the product from packaging to the market and ensure the suitable state of the product.[21] Therefore, with the help of this smart sensor, any deterioration in food quality can be easily detected such as by color change or any gas produced during spoilage.[57] There are various types of nanosensors which are being employed by the food industries, for instance, array biosensor, electronic noses, nanocantilevers, nanotest strips, and nanoparticle in solution.[67]

In order to slow down the biochemical processes like oxidation, degradation, etc., further detection of organic contaminant in the food with incorporation of nanomaterials which offer effective resistance for the diffusion of gases like carbon dioxide, oxygen is required. AuNPs are used in the detection of melamine content in the raw milk, exhibiting concentration-dependent change in color from red to blue.[1] Nanosized TiO_2 or SnO_2 with redox active-dye (methylene blue) are used as photo-activated indicator ink for in-package oxygen, which changes color even on minute exposure to oxygen.[26] Detection of moisture can be done by analysing the swelling process in the polymer matrix causing interparticle separation, which results in sensor strips to either absorb or reflect color and consequently easily detected as regard the moisture content in the food.[53] Furthermore, foodborne pathogen can cause serious illness which may eventually lead to death. The optical and magnetic properties of nanomaterial therefore prove to be quite substantial in devising nano-based microbe detection sensors, as well as in isolation of pathogens. Two-photon Rayleigh scattering (TPRS) optical property of AuNPs conjugated with *E. coli* antibody has already been used for specific and selective detection of *E. coli*.[85] Furthermore, immune-genetic separation (IMS) is detected using magnetic nanoparticles, for instance, iron oxide nanoparticle-surface functionalized with antibodies specific for *L. monocytogenes,* can effectively segregate the bacteria from artificially tainted milk and later be detected using RT-PCR.[101] Therefore, usage of nanomaterial improves not only shelf life and detect the environment of the packed food but also helps in detection and isolation of microorganisms.

13.4.5.6 ENHANCEMENT OF APPEARANCE AND TEXTURE

Nanoemulsions serve as color additives, such as β-carotene which is used to colorize aqueous-based foods using alginic acid. Calcium ions further allows to be used as a natural fat soluble colorant in a very unique way.[5]

TiO_2, SiO_2, and Al_2O_3 are also used in food colorization as dispersing aids, with a toxic limit ~ 2%. Further to enhance the texture of the food item, thicken the paste and maintain the flow characteristics in powdered foods, nanoparticles like SiO_2 are used as anticaking agent, as well as used for carrier of flavor or fragrance.[37]

13.5 RISK ASSOCIATED WITH NANO-INCLUSIONS

Despite of the advantages of nanotechnology in various food industrial sectors, the potential risk to the population should not be overlooked. Various studies come up with the harm induced by nanoparticles both in vitro and in vivo, such as high content of cerium has been detected in the fruits of tomato expose to cerium oxide nanoparticles at the concentration of 10 mg/L as compared with controls. This study reflects how nanoparticles migrate from the zone of application from the root to the shoot and finally accumulated in the fruits seems to be the potential sink for the nanoparticles.[95]

Therefore, these nanomaterials came in close vicinity of the food during processing of edibles directly/indirectly so there is an urgent requirement of a regulatory mechanism with a strong hold and an objectiveto keep an eagle's eye on any risk conjectured with nano-foods and application of nanotechnology in the food industry sector.[82] Therefore, risk assessment should be done with an aim to identify and quantify the associated risk with nanomaterials such as migration test which analyze the migration of nanoparticles from the packing stuff to the food material and later to the human, thus predicting the safe limit of exposure of nanomaterial on the basis of various animal trials, also their circulation and retention in the system. Moreover, health risk of the nanomaterials in food also depends on various other factors like the type of nanomaterial used in packaging and its associated toxicity, rate of migration from packing material to the food, and also on the rate of consumption of particular food.[21]

13.6 CONCLUSION

With the ingress of nanotechnology in the field of food-packaging industry, food products are exposed directly/indirectly to nanomaterials, thereby resulting in effective and efficient food production, processing, and delivery methods. Various studies have already been made and the incorporation of nanoscience in food industry is a continuous and evolving process. Although nanotechnology has proved to better the pre-existing facilities related to food technology, the negative impacts of nanomaterial should not be disregarded. Most of the citizens are unaware about the helpfulness of nanotechnology in food and hence products' quality is being exaggerated in the name of nanotechnology,[13] which needs to be avoided to the fullest extent for consumers' safety with regards the harm posed by toxic nanomaterials. A continuous supply of nano-derived product in the market without proper authentication may pose severe health concerns. Therefore, for precautionary measure and substantial standardize, the labelling and safety assessment of nanoscale materials are used in a variety of food applications. Further to develop regulatory standards, various characteristics should be taken in account such as measurement of size, shape, and measurement of the physicochemical analysis, processing, as well as safety and risk simultaneously.[13] Therefore, to harness the benefit offered by these novel nanomaterials, all the aspects should be studied in detail.

KEYWORDS

- **biotechnology**
- **food industry**
- **nanogoods**
- **nanotechnology**
- **packaging**

REFERENCES

1. Ai, K.; Liu, Y.; Lu, L. Hydrogen-bonding Recognition-induced Color Change of Gold Nanoparticles for Visual Detection of Melamine in Raw Milk and Infant Formula. *J. Am. Chem. Soc.* **2009,** *131* (27), 9496–9497.
2. Alfadul, S. M.; Elneshwy, A. A. Use of Nanotechnology in Food Processing, Packaging and Safety–Review. *African J. Food Agric. Nutr. Dev.* **2010,** *10* (6).
3. Anjali, C. H.; Khan, S. S.; Margulis-Goshen, K.; Magdassi, S.; Mukherjee, A.; Chandrasekaran, N. Formulation of Water-dispersible Nanopermethrin for Larvicidal

Applications. *Ecotoxicol. Environ. Saf.* **2010,** *73*, 1932–1936; DOI:10.1016/j.ecoenv.2010.08.039.
4. Kubacka, A., et al. Understanding the Antimicrobial Mechanism of Tio2-based Nanocomposite Films in a Pathogenic Bacterium. *Sci. Rep.* **2014,** *4*, 4134. DOI:10.1038/srep04134.
5. Astete, C. E.; Sabliov, C. M.; Watanabe, F.; Biris, A. Ca2+ Cross-linked Alginic Acid Nanoparticles for Solubilization of Lipophilic Natural Colorants. *J. Agric. Food Chem.* **2009,** *57* (16), 7505–7512.
6. Becaro, A. A.; Puti, F. C.; Correa, D. S.; Paris, E. C.; Marconcini, J. M.; Ferreira, M. D. Polyethylene Films Containing Silver Nanoparticles for Applications in Food Packaging: Characterization of Physico-chemical and Anti-microbial Properties. *J. Nanosci. Nanotechnol.* **2015,** *15* (3), 2148–2156.
7. Beigmohammadi, F.; Peighambardoust, S. H.; Hesari, J.; Azadmard-Damirchi, S.; Peighambardoust, S. J.; Khosrowshahi, N. K. Antibacterial Properties of LDPE Nanocomposite Films in Packaging of UF Cheese. *LWT-Food Sci. Technol.* **2016,** *65*, 106–111.
8. Bodaghi, H.; Mostofi, Y.; Oromiehie, A.; Ghanbarzadeh, B.; Hagh, Z. G. Synthesis of Clay–Tio_2 Nanocomposite Thin Films with Barrier and Photocatalytic Properties for Food Packaging Application. *J. Appl. Polym. Sci.* **2015,** *132* (14).
9. Cai, L.; Chen, J.; Liu, Z.; Wang, H.; Yang, H.; Ding, W. Magnesium Oxide Nanoparticles: Effective Agricultural Antibacterial Agent Against Ralstonia Solanacearum. *Front. Microbiol.* **2018,** *9*, 790.
10. Canas, J. E.; Long, M.; Nations, S.; Vadan, R.; Dai, L.; Luo, M.; Olszyk, D. Effects of Functionalized and Nonfunctionalized Single-walled Carbon Nanotubes on Root Elongation of Select Crop Species. *Environ. Toxicol. Chem.* **2008,** *27* (9), 1922–1931.
11. Chang, Y. C.; Chen, D. H. Adsorption Kinetics and Thermodynamics of Acid Dyes on a Carboxymethylated Chitosan-conjugated Magnetic Nano-adsorbent. *Macromol. Biosci.* **2005,** *5* (3), 254–261.
12. Chau, C. F.; Wu, S. H.; Yen, G. C. The Development of Regulations for Food Nanotechnology. *Trends Food Sci. Technol.* **2007,** *18* (5), 269–280.
13. Chaudhry, Q.; Scotter, M.; Blackburn, J.; Ross, B.; Boxall, A.; Castle, L.; Watkins, R. Applications and Implications of Nanotechnologies for the Food Sector. *Food Addi. Contam.* **2008,** *25* (3), 241–258.
14. Chawengkijwanich, C.; Hayata, Y. Development of TiO_2 Powder-coated Food Packaging Film and Its Ability to Inactivate *Escherichia coli* In Vitro and in Actual Tests. *Int. J. Food Microbiol.* **2008,** *123* (3), 288–292.
15. Chen, H.; Weiss, J.; Shahidi, F. Nanotechnology in Nutraceuticals and Functional Foods. *Food Technol.* **2006.**
16. Chhipa, H. Nanofertilizers and Nanopesticides for Agriculture. *Environ. Chem. Lett.* **2017,** *15* (1), 15–22.
17. Costa, C.; Conte, A.; Buonocore, G. G.; Del Nobile, M. A. Antimicrobial Silver-montmorillonite Nanoparticles to Prolong the Shelf Life of Fresh Fruit Salad. *Int. J. Food Microbiol.* **2011,** *148* (3), 164–167.

18. Costa, C.; Conte, A.; Buonocore, G. G.; Lavorgna, M.; Del Nobile, M. A. Calcium-alginate Coating Loaded with Silver-montmorillonite Nanoparticles to Prolong the Shelf-Life of Fresh-cut Carrots. *Food Res. Int.* **2012,** *48* (1), 164–169.
19. Cubukçu, M.; Timur, S.; Anik, Ü. Examination of Performance of Glassy Carbon Paste Electrode Modified with Gold Nanoparticle and Xanthine Oxidase For Xanthine and Hypoxanthine Detection. *Talanta* **2007,** *74* (3), 434–439.
20. Cushen, M.; Kerry, J.; Morris, M.; Cruz-Romero, M.; Cummins, E. Nanotechnologies in the Food Industry–Recent Developments, Risks and Regulation. *Trends Food Sci. Technol.* **2012,** *24* (1), 30–46.
21. Deepa, M.; Krishna, T. G.; Prasad, T. N. V. K. V. First Evidence on Phloem Transport of Nanoscale Calcium Oxide in Groundnut Using Solution Culture Technique. *Appl. Nanosci.* **2015,** *5*, 545−551. 10.1007/s13204-014-0348-8.
22. Delfani, M.; Firouzabadi, M. B.; Farrokhi, N.; Makarian, H. Some Physiological Responses of Black-Eyed Pea to Iron and Magnesium Nanofertilizers. *Commun. Soil Sci. Plant Anal.* **2014,** *45*, 530−540.
23. Dias, M. V.; Nilda de Fátima, F. S.; Borges, S. V.; de Sousa, M. M.; Nunes, C. A.; de Oliveira, I. R. N.; & Medeiros, E. A. A. Use of Allyl Isothiocyanate and Carbon Nanotubes in an Antimicrobial Film to Package Shredded, Cooked Chicken Meat. *Food Chem.* **2013,** *141* (3), 3160–3166.
24. Dimkpa, C.; Bindraban, P.; Fugice, J.; Agyin-Birikorang, S.; Singh, U.; Hellums, D. Composite Micronutrient Nanoparticles and Salts Decrease Drought Stress in Soybean. *Agron. Sustainable Dev.* **2017,** *37*, 5. 10.1007/s13593-016-0412-8.
25. Duncan, T. V. Applications of Nanotechnology in Food Packaging and Food Safety: Barrier Materials, Antimicrobials and Sensors. *J. Coll. Inter. Sci.* **2011,** *363* (1), 1–24.
26. Emamifar, A.; Kadivar, M.; Shahedi, M.; Soleimanian-Zad, S. Effect of Nanocomposite Packaging Containing Ag and Zno on Inactivation of *Lactobacillus plantarum* in Orange Juice. *Food Control* **2011,** *22* (3–4), 408–413.
27. Esmailzadeh, H.; Sangpour, P.; Shahraz, F.; Hejazi, J.; Khaksar, R. Effect of Nanocomposite Packaging Containing Zno on Growth of *Bacillus subtilis* and Enterobacter Aerogenes. *Mater. Sci. Eng. C* **2016,** *58*, 1058–1063.
28. Fathi, M.; Mozafari, M. R.; Mohebbi, M. Nanoencapsulation of Food Ingredients Using Lipid Based Delivery Systems. *Trends Food Sci. Technol.* **2012,** *23* (1), 13–27.
29. Fernández, A.; Soriano, E.; López-Carballo, G.; Picouet, P.; Lloret, E.; Gavara, R.; Hernández-Muñoz, P. Preservation of Aseptic Conditions in Absorbent Pads by Using Silver Nanotechnology. *Food Res. Int.* **2009,** *42* (8), 1105–1112.
30. Foster, H. A.; Ditta, I. B.; Varghese, S.; Steele, A. Photocatalytic Disinfection Using Titanium Dioxide: Spectrum and Mechanism of Antimicrobial Activity. *Appl. Microbiol. Biotechnol.* **2011,** *90* (6), 1847–1868.
31. Gammariello, D.; Conte, A.; Buonocore, G. G.; Del Nobile, M. A. Bio-based Nanocomposite Coating to Preserve Quality of Fior Di Latte Cheese. *J. Dairy Sci.* **2011,** *94* (11), 5298–5304.
32. Gao, F.; Hong, F.; Liu, C.; Zheng, L.; Su, M.; Wu, X.; Yang, P. Mechanism of Nano-anatase Tio 2 on Promoting Photosynthetic Carbon Reaction of Spinach. *Biol. Trace Elem. Res.* **2006,** *111* (1–3), 239–253.

33. Graveland-Bikker, J. F.; De Kruif, C. G. Unique Milk Protein Based Nanotubes: Food and Nanotechnology Meet. *Trends Food Sci. Technol.* **2006,** *17* (5), 196–203.
34. Guan, H.; Chi, D.; Yu, J.; Li, H. Dynamics of Residues from a Novel Nano-imidacloprid Formulation in Soybean Fields. *Crop Prot.* **2010,** *29*, 942−946. 10.1016/j.cropro.2010.04.022 10.1002/ps.1732.
35. Hamad, A. F.; Jong-Hun, H. A. N.; Kim, B. C.; Rather, I. A. The Intertwine of Nanotechnology with the Food Industry: the Intertwine of Nanotechnology with the Food Industry. *Saudi J. Biol. Sci.* **2017**.
36. He, X.; Hwang, H. M. Nanotechnology in Food Science: Functionality, Applicability, and Safety Assessment. *J. Food Drug Anal.* **2016,** *24* (4), 671–681.
37. Hossain, K. Z.; Monreal, M.; Sayari, K. Adsorption of Urease on Pe-Mcm-41 and Its Catalytic Effect on Hydrolysis of Urea. *Colloid Surf B* **2008,** *62*, 42–50. 10.1016/j.colsurfb.2007.09.016.
38. Hu, B.; Huang, Q. R. Biopolymer Based Nano-delivery Systems for Enhancing Bioavailability of Nutraceuticals. *Chinese J. Polym. Sci.* **2013,** *31* (9), 1190–1203.
39. Hu, M.; Li, Y.; Decker, E. A.; Xiao, H.; McClements, D. J. Impact of Layer Structure on Physical Stability and Lipase Digestibility of Lipid Droplets Coated by Biopolymer Nanolaminated Coatings. *Food Biophys.* **2011,** *6* (1), 37–48.
40. Hussein, M. Z., et al. Nanotechnology in Fertilizers. *J. Control. Release* **2002,** *82*, 417–427. 10.1016/S0168-3659(02)00172-4.
41. Ishaque, M.; Schnable, G.; Anspaugh, D. Agrochemical Formulations Comprising a Pesticide, an Organic UV-Photoprotective Filter and Coated Metal-oxide Nanoparticles. WO/2009/153231, 2009.
42. Jianhui, Y.; Kelong, H.; Yuelong, W.; Suquin, L. Study on Antipollution Nanopreparation of Dimethomorph and Its Performance. *Chin. Sci. Bull.* **2005,** *50*, 108−112.
43. Joseph, T.; Morrison, M. Nanotechnology in Agriculture and Food. Ananoforum Report 2006.
44. Kim, M. K.; Lee, J. A.; Jo, M. R.; Kim, M. K.; Kim, H. M.; Oh, J. M.; Choi, S. J. Cytotoxicity, Uptake Behaviors, and Oral Absorption of Food Grade Calcium Carbonate Nanomaterials. *Nanomaterials* **2015,** *5* (4), 1938–1954.
45. Kottegoda, N.; Munaweera, I.; Madusanka, N.; Karunaratne, V. A Green Slow-release Fertilizer Composition Based on Urea-modified Hydroxyapatite Nanoparticles Encapsulated Wood. *Curr. Sci.* **2011,** 73–78.
46. Kundu, S.; Adhikari, T.; Mohanty, S. R.; Rajendiran, S.; Coumar, M. V.; Saha, J. K.; Patra, A. K. Reduction in Nitrous Oxide (N_2O) Emission from Nano Zinc Oxide and Nano Rockphosphate Coated Urea. *Agrochimica* **2016,** *60*, 2.
47. Leung, Y. H.; Ng, A.; Xu, X.; Shen, Z.; Gethings, L. A.; Wong, M. T.; Lee, P. K. Mechanisms of Antibacterial Activity of Mgo: Non-ROS Mediated Toxicity of Mgo Nanoparticles Towards Escherichia Coli. *Small* **2014,** *10* (6), 1171–1183.
48. Li, X.; Li, W.; Jiang, Y.; Ding, Y.; Yun, J.; Tang, Y.; Zhang, P. Effect of Nano-Zno-coated Active Packaging on Quality of Fresh-cut 'Fuji'apple. *Int. J. Food Sci. Technol.* **2011,** *46* (9), 1947–1955.
49. Liu, X.; Zhang, M. Characteristics of Nano-subnanocomposites and Response of Soil and Plant Nutrition to Them. Ph.D. Dissertation, Chinese Academy of Agricultural Sciences, 2005.

50. Liu, R.; Lal, R. Synthetic Apatite Nanoparticles as a Phosphorus Fertilizer for Soybean (Glycine Max). *Scientific Reports* **2014,** *4*, 5686.
51. Lloret, E.; Picouet, P.; Fernández, A. Matrix Effects on the Antimicrobial Capacity of Silver Based Nanocomposite Absorbing Materials. *LWT-Food Sci. Technol.* **2012,** *49* (2), 333–338.
52. Luechinger, N. A.; Loher, S.; Athanassiou, E. K.; Grass, R. N.; Stark, W. J. Highly Sensitive Optical Detection of Humidity on Polymer/Metal Nanoparticle Hybrid Films. *Langmuir* **2007,** *23* (6), 3473–3477.
53. Magro, M.; Campos, R.; Baratella, D.; Lima, G.; Holà, K.; Divoky, C.; Zbořil, R. A Magnetically Drivable Nanovehicle for Curcumin with Antioxidant Capacity and MRI Relaxation Properties. *Chem. Eur. J.* **2014,** *20* (37), 11913–11920.
54. Mahajan, P.; Dhoke, S. K.; Khanna, A. S. Effect of Nano-Zno Particle Suspension on Growth of Mung (Vigna Radiata) and Gram (Cicer Arietinum) Seedlings Using Plant Agar Method. *J. Nanotechnol.* **2011**.
55. Maneerat, C.; Hayata, Y. Gas-phase Photocatalytic Oxidation of Ethylene with Tio_2-coated Packaging Film for Horticultural Products. *Trans. ASABE* **2008,** *51* (1), 163–168.
56. Mannino, S.; Scampicchio, M. Nanotechnology and Food Quality Control. *Vet. Res. Comm.* **2007,** *31* (1), 149–151.
57. Mason, T. G.; Wilking, J. N.; Meleson, K.; Chang, C. B.; Graves, S. M. Nanoemulsions: Formation, Structure, and Physical Properties. *J. Phys. Condens. Matter* **2006,** *18* (41), R635.
58. Mastromatteo, M.; Conte, A.; Lucera, A.; Saccotelli, M. A.; Buonocore, G. G.; Zambrini, A. V.; Del Nobile, M. A. Packaging Solutions to Prolong the Shelf Life of Fiordilatte Cheese: Bio-based Nanocomposite Coating and Modified Atmosphere Packaging. *LWT-Food Sci. Technol.* **2015,** *60* (1), 230–237.
59. Mihaly Cozmuta, A.; Peter, A.; Mihaly Cozmuta, L.; Nicula, C.; Crisan, L.; Baia, L.; Turila, A. Active Packaging System Based on Ag/Tio_2 Nanocomposite Used for Extending the Shelf Life of Bread. Chemical and Microbiological Investigations. *Packaging Technol. Sci.* **2015,** *28* (4), 271–284.
60. Mukherjee, A., et al. Differential Toxicity of Bare and Hybrid Zno Nanoparticles in Green Pea (Pisum Sativum L.): A Life Cycle Study. *Front. Plant Sci.* **2016,** *6*, 1242. 10.3389/fpls.2015.01242.
61. Nguyen, H. M.; Hwang, I. C.; Park, J. W.; Park, H. J. Photoprotection for Deltamethrin Using Chitosan-Coated Beeswax Solid Lipid Nanoparticles. *Pest Manag. Sci.* **2012,** *68* (7). 1062–8. 10.1002/ps.3268.
62. Nguyen, H. M.; Hwang, I. C.; Park, J. W.; Park, H. J. Enhanced Payload And Photo-Protection For Pesticides Using Nanostructured Lipid Carriers With Corn Oil As Liquid Lipid. *J. Microencapsul.* **2012,** *29* (6), 596–604. 10.3109/02652048.2012.668960.
63. Observatory Nanotechnologies for Nutrient and Biocide Delivery in Agricultural Production. Working Paper April 2010.
64. Selva Preetha, P.; Balakrishnan, N. A Review of Nano Fertilizers and Their Use and Functions in Soil. *Int. J. Curr. Microbiol. App. Sci.* **2017,** *6* (12), 3117–3133. 10.20546/ijcmas.2017.612.364.
65. Padmavathy, N.; Vijayaraghavan, R. Interaction of ZnO Nanoparticles with Microbes—A Physio and Biochemical Assay. *J. Biomed. Nanotechnol.* **2011,** *7* (6), 813–822.
66. Pal, M. *J. Food Microbiol. Safety Hygiene* **2017**.

67. Panea, B.; Ripoll, G.; González, J.; Fernández-Cuello, Á.; Albertí, P. Effect of Nanocomposite Packaging Containing Different Proportions of Zno and Ag on Chicken Breast Meat Quality. *J. Food Eng.* **2014,** *123*, 104–112.
68. Paret, M. L.; Palmateer, A. J.; Knox, G. W. Evaluation of a Light-activated Nanoparticle Formulation of Titanium Dioxide with Zinc for Management of Bacterial Leaf Spot on Rosa 'Noare'. *Hortsci.* **2013,** *48* (2), 189–192.
69. Paret, M. L.; Vallad, G. E.; Averett, D. R.; Jones, J. B.; Olson, S. M. Photocatalysis: Effect of Lightactivated Nanoscale Formulations of Tio_2 on Xanthomonas Perforans and Control of Bacterial Spot of Tomato. *Phytopathology* **2013,** *103* (3), 228–236. 10.1094/phyto-08-12-0183-r.
70. Pathakoti, K.; Manubolu, M.; Hwang, H. M. Nanostructures: Current Uses and Future Applications in Food Science. *J. Food Drug Anal.* **2017,** *25* (2), 245–253.
71. Pereira, D. I.; Bruggraber, S. F.; Faria, N.; Poots, L. K.; Tagmount, M. A.; Aslam, M. F.; Powell, J. J. Nanoparticulate Iron (III) Oxo-hydroxide Delivers Safe Iron That is Well Absorbed and Utilised in Humans. *Nanomed. Nanotechnol. Biol. Med.* **2014,** *10* (8), 1877–1886.
72. Picouet, P. A.; Fernandez, A.; Realini, C. E.; Lloret, E. Influence of PA6 Nanocomposite Films on the Stability of Vacuum-aged Beef Loins During Storage in Modified Atmospheres. *Meat Sci.* **2014,** *96* (1), 574–580.
73. Pradhan, S.; Patra, P.; Das, S.; Chandra, S.; Mitra, S.; Dey, K. K.; Goswami, A. Photochemical Modulation of Biosafe Manganese Nanoparticles on Vigna Radiata: A Detailed Molecular, Biochemical, and Biophysical Study. *Environ. Sci. Technol.* **2013,** *47* (22), 13122–13131.
74. Prasad, T. N. V. K. V; Sudhakar, P.; Sreenivasulu, Y.; Latha, P.; Munaswamya, V.; Raja Reddy, K.; Sreeprasad, T. S.; Sajanlal, P. R.; Pradeep, T. Effect of Nanoscale Zinc Oxide Particles on the Germination, Growth and Yield of Peanut. *J. Plant Nutr.* **2012,** *35*, 905–927. 10.1080/01904167.2012.663443.
75. Prasad, R.; Bhattacharyya, A.; Nguyen, Q. D. Nanotechnology in Sustainable Agriculture: Recent Developments, Challenges, and Perspectives. *Front. Microbiol.* **2017,** *8*, 1014.
76. Qian, K.; Shi, T.; Tang, T.; Zhang, S.; Liu, X.; Cao, Y. Preparation and Characterization of Nano-sized Calcium Carbonate as Controlled Release Pesticide Carrier for Validamycin Against Rhizoctonia Solani. *Microchim. Acta* **2011,** *173* (1–2), 51–57.
77. Ravichandran, R. Nanotechnology Applications in Food and Food Processing: Innovative Green Approaches, Opportunities and Uncertainties for Global Market. *Int. J. Green Nanotechnol. Phys. Chem.* **2010,** *1* (2), P72–P96.
78. Ritzoulis, C.; Scoutaris, N.; Papademetriou, K.; Stavroulias, S.; Panayiotou, C. Milk Protein-based Emulsion Gels for Bone Tissue Engineering. *Food Hydrocoll.* **2005,** *19* (3), 575–581.
79. Sekhon, B. S. Nanotechnology in Agri-food Production: An Overview. *Nanotechnol. Sci. Appl.* **2014,** *7*, 31.
80. Shan, D.; Wang, Y.; Xue, H.; Cosnier, S. Sensitive and Selective Xanthine Amperometric Sensors Based on Calcium Carbonate Nanoparticles. *Sens. Actuators B Chem.* **2009,** *136* (2), 510–515.

81. Shegokar, R.; Müller, R. H. Nanocrystals: Industrially Feasible Multifunctional Formulation Technology for Poorly Soluble Actives. *Int. J. Pharm.* **2010,** *399* (1–2), 129–139.
82. Singh, A. K.; Senapati, D.; Wang, S.; Griffin, J.; Neely, A.; Naylor, K. M.; Ray, P. C. Gold Nanorod Based Selective Identification of *Escherichia coli* Bacteria Using Two-photon Rayleigh Scattering Spectroscopy. *ACS Nano.* **2009,** *3* (7), 1906–1912.
83. Sozer, N.; Kokini, J. L. Nanotechnology and its Applications in the Food Sector. *Trends Biotechnol.* **2009,** *27* (2), 82–89.
84. Stadler, T.; Buteler, M.; Weaver, D. K.; Sofie S. Comparative Toxicity of Nanostructured Alumina and a Commercial Inert Dust for Sitophilus Oryzae (L.) and Rhyzopertha Dominica (F.) at Varying Ambient Humidity Levels. *J. Stored. Prod. Res.* **2012,** *48*, 81–89. 10.1016/j.jspr.2011.09.004.
85. Subramanian, K. S.; Sharmila Rahale, C. Nano-fertilizers—Synthesis, Characterization and Application. In: *Nanotechnology in Soil Science and Plant Nutrition*; Adikari, T., Subba, R., Eds; New India Publishing Agency: New Delhi, India, 2013.
86. Sui, X.; Yuan, J.; Yuan, W.; Zhou, M. Preparation of Cellulose Nanofibers/Nanoparticles via Electrospray. *Chem. Lett.* **2007,** *37* (1), 114–115.
87. Sultana, P.; Das, S.; Bhattacharya, A.; Basu, R.; Nandy, P. Development of Iron Oxide and Titania Treated Fly Ash Based Ceramic and Its Bioactivity. *Mater. Sci. Eng. C* **2012,** *32* (6), 1358–1365.
88. Motlagh, V. N.; Hamed Mosavian, M. T.; Mortazavi, S. A. Effect of Polyethylene Packaging Modified with Silver Particles on the Microbial, Sensory and Appearance of Dried Barberry. *Packaging Technol. Sci.* **2013,** *26* (1), 39–49.
89. van Vlerken, L. E.; Vyas, T. K.; Amiji, M. M. Poly (ethylene glycol)-modified Nanocarriers for Tumor-targeted and Intracellular Delivery. *Pharm. Res.* **2007,** *24* (8), 1405–1414.
90. Wajda, R.; Zirkel, J.; Schaffer, T. Increase of Bioavailability of Coenzyme Q10 and Vitamin E. *J. Med. Food* **2007,** *10* (4), 731–734.
91. Wang, L.; Hu, C.; Shao, L. The Antimicrobial Activity of Nanoparticles: Present Situation and Prospects for the Future. *Int. J. Nanomed.* **2017,** *12*, 1227.
92. Wang, Q.; Ma, X.; Zhang, W.; Pei, H.; Chen, Y. The Impact of Cerium Oxide Nanoparticles on Tomato (Solanum Lycopersicum L.) and Its Implications for Food Safety. *Metallomics* **2012,** *4* (10), 1105–1112.
93. Wolfgang, J.; Parak, et al. Biological Applications of Colloidal Nanocrystals. *Nanotechnology* **2003,** 14 R15. 10.1088/0957-4484/14/7/201.
94. Wongsasulak, S.; Patapeejumruswong, M.; Weiss, J.; Supaphol, P.; Yoovidhya, T. Electrospinning of Food-grade Nanofibers from Cellulose Acetate and Egg Albumen Blends. *J. Food Eng. 98* (3), 370–376.
95. Xiang, C.; Taylor, A. G.; Hinestroza, J. P.; Frey, M. W. Controlled Release of Nonionic Compounds from Poly(Lactic Acid)/Cellulose Nanocrystal Nanocomposite Fibers. *J. Appl. Polym. Sci.* **2013,** *127* (1), 79–86. 10.1002/app.36943.
96. Xiao-e, L.; Green, A. N.; Haque, S. A.; Mills, A.; Durrant, J. R. Light-driven Oxygen Scavenging by Titania/Polymer Nanocomposite Films. *J. Photochem. Photobiol. A Chem.* **2004,** *162* (2–3), 253–259.
97. Yang, F., et al. The Improvement of Spinach Growth by Nano-anatase TiO_2 Treatment Is Related to Nitrogen Photoreduction Biol. *Trace Elem. Res.* **2007,** *119*, 77–88. 10.1007/s12011-007-0046-4.

98. Yang, H.; Qu, L.; Wimbrow, A. N.; Jiang, X.; Sun, Y. Rapid Detection of Listeria Monocytogenes by Nanoparticle-based Immunomagnetic Separation and Real-time PCR. *Int. J. Food Microbiol.* **2007,** *118* (2), 132–138.
99. Yang, M.; Kostov, Y.; Rasooly, A. Carbon Nanotubes Based Optical Immunodetection of Staphylococcal Enterotoxin B (SEB) in Food. *Int. J. Food Microbiol.* **2008,** *127* (1–2), 78–83.
100. Yu, H.; Huang, Q. Improving the Oral Bioavailability of Curcumin Using Novel Organogel-based Nanoemulsions. *J. Agric. Food Chem.* **2012,** *60* (21), 5373–5379.
101. Zarif, L. Nanocochleate Cylinders for Oral & Parenteral Delivery of Drugs. *J. Liposome Res.* **2003,** *13*, 109–110.
102. Zhang, H. F, et al. Formation and Enhanced Biocidal Activity of Water-dispersible Organic Nanoparticles. *Nat. Nanotechnol.* **2008,** *3*, 506–511; 10.1038/nnano.2008.188.
103. Zhou, H.; Liu, G.; Zhang, J.; Sun, N.; Duan, M.; Yan, Z.; Xia, Q. Novel Lipid-free Nanoformulation for Improving Oral Bioavailability of Coenzyme Q10. BioMed. Res. Int. **2014**.
104. Rhim, J. W.; Park, H. M.; Ha, C. S. Bio-nanocomposites for Food Packaging Applications. *Prog. Polym. Sci.* **2013,** *38* (10–11), 1629–1652.
105. Busolo, M. A.; Fernandez, P.; Ocio, M. J.; Lagaron, J. M. Novel Silver-based Nanoclay as an Antimicrobial in Polylactic Acid Food Packaging Coatings. *Food Addit. Contam.* **2010,** *27* (11), 1617–1626.
106. Frounchi, M.; Dadbin, S.; Salehpour, Z.; Noferesti, M. Gas Barrier Properties of PP/EPDM Blend Nanocomposites. **2006,** *J. Membrane Sci. 282*(1-2), 142–148.
107. Hemati, F.; Garmabi, H. Compatibilised LDPE/LLDPE/Nanoclay Nanocomposites: I. Structural, Mechanical, and Thermal Properties. *Can. J. Chem. Eng.* **2011,** *89* (1), 187–196.
108. Hyun, K.; Chong, W.; Koo, M.; Chung, I. J. Physical Properties of Polyethylene/Silicate Nanocomposite Blown Films. **2003,** *J. Appl. Polym. Sci. 89* (8), 2131–2136.
109. Zhong, Y.; Janes, D.; Zheng, Y.; Hetzer, M.; De Kee, D. Mechanical and Oxygen Barrier Properties of Organoclay-polyethylene Nanocomposite Films. *Polym. Eng. Sci.* **2007,** *47* (7), 1101–1107.
110. Lok, C. N.; Ho, C. M.; Chen, R.; He, Q. Y.; Yu, W. Y.; Sun, H.; Che, C. M. Proteomic Analysis of the Mode of Antibacterial Action of Silver Nanoparticles. *J. Proteome Res.* **2006,** *5* (4), 916–924.
111. De Moura, M. R.; Mattoso, L. H.; Zucolotto, V. Development of Cellulose-based Bactericidal Nanocomposites Containing Silver Nanoparticles and Their Use as Active Food Packaging. *J. Food Eng.* **2012,** *109* (3), 520–524.
112. Busolo, M. A.; Lagaron, J. M. Oxygen Scavenging Polyolefin Nanocomposite Films Containing an Iron Modified Kaolinite of Interest in Active Food Packaging Applications. *Innov. Food Sci. Emerg. Technol.* **2012,** *16*, 211–217.
113. Teisala, H.; Tuominen, M.; Stepien, M.; Haapanen, J.; Mákelá, J. M.; Saarinen, J. J.; Kuusipalo, J. Wettability Conversion on the Liquid Flame Spray Generated Superhydrophobic TiO2 Nanoparticle Coating on Paper and Board by Photocatalytic Decomposition of Spontaneously Accumulated Carbonaceous Overlayer. *Cellulose* **2013,** *20* (1), 391–408.

114. Youssef, A. M.; Kamel, S.; El-Samahy, M. A. Morphological and Antibacterial Properties of Modified Paper by PS Nanocomposites for Packaging Applications. *Carbohydrate Polym.* **2013,** *98* (1), 1166–1172.
115. Kerry, J. P.; O'grady, M. N.; Hogan, S. A. Past, Current and Potential Utilisation of Active and Intelligent Packaging Systems for Meat and Muscle-based Products: a Review. *Meat Science* **2006,** *74* (1), 113–130.
116. von Bültzingslöwen, C.; McEvoy, A. K.; McDonagh, C.; MacCraith, B. D.; Klimant, I.; Krause, C.; Wolfbeis, O. S. Sol-Gel Based Optical Carbon Dioxide Sensor Employing Dual Luminophore Referencing for Application in Food Packaging Technology. *Analyst* **2002,** *127* (11), 1478–1483.
117. Lee, S. K.; Sheridan, M.; Mills, A. Novel UV-activated Colorimetric Oxygen Indicator. *Chem. Mater.* **2005,** *17* (10), 2744–2751.
118. Mills, A.; Hazafy, D. Nanocrystalline SnO2-based, UVB-activated, Colourimetric Oxygen Indicator. Sens. Actuators B: Chem. **2009,** *136* (2), 344–349.
119. Mihindukulasuriya, S. D.; Lim, L. T. Oxygen Detection Using UV-activated Electrospun Poly (Ethylene Oxide) Fibers Encapsulated with TiO2 Nanoparticles. *J. Mater. Sci.* **2013,** *48* (16), 5489–5498.
120. Gnach, A.; Bednarkiewicz, A. Lanthanide-doped Up-converting Nanoparticles: Merits and Challenges. *Nano Today* **2012,** *7* (6), 532–563.
121. Zeng, J.; Roberts, S.; Xia, Y. Nanocrystal-based Time -Temperature Indicators. *Chem. -A Europ. J.* **2010,** *16* (42), 12559–12563.
122. Foster, H. A.; Ditta, I. B.; Varghese, S.; Steele, A. Photocatalytic Disinfection Using Titanium Dioxide: Spectrum and Mechanism of Antimicrobial Activity. *Appl. Microbiol. Biotechnol.* **2011,** *90* (6), 1847–1868.
123. Xie, L.; Lv, X. Y.; Han, Z. J.; Ci, J. H.; Fang, C. Q.; Ren, P. G. Preparation and Performance of High-barrier Low Density Polyethylene/Organic Montmorillonite Nanocomposite. *Polym. Plast. Technol. Eng.* **2012,** *51* (12), 1251–1257.
124. Jing, Z.; Guo, D.; Wang, W.; Zhang, S.; Qi, W.; Ling, B. Comparative Study of Titania Nanoparticles and Nanotubes as Antibacterial Agents. *Solid State Sci.* **2011,** *13* (9), 1797–1803.

CHAPTER 14

PASSIVE AERATED COMPOSTING OF LEAVES AND PREDIGESTED OFFICE PAPERS

A. Y. ZAHRIM*, S. SARIAH, R. MARIANI, I. AZREEN, Y. ZULKIFLEE, and A. S. FAZLIN

Chemical Engineering Programme, Faculty of Engineering, University Malaysia Sabah, Jalan UMS, 88400 Kota Kinabalu, Sabah, Malaysia

**Corresponding author. E-mail: zahrim@ums.edu.my*

ABSTRACT

Composting studies were conducted in lab scale for 40 days. The substrates of the compost are anaerobically treated palm oil mill effluent (AnPOME), leaves, and paper. The composting process was conducted in cuboid reactor with effective volume of 0.4 m^3. During composting process, the highest temperature of 43°C was achieved on day 4. The physiochemical analyses were conducted to determine the stability and maturity of the compost. Phytotoxicity experiment was also conducted by using cabbage seed (*Brassica oleracea*) to analyze the maturity of the compost. As a result, the parameters which are moisture content, TOC, and C/N were significantly decreasing. The overall mass reduction was 66% and the organic matter degradation was 72.13%. The germination index of the compost was increasing at the end of the period. It shows that the compost was phytotoxic free. Overall, nutrients in the compost product of this study were in the ideal range but some of the parameters are not in the range to achieve the maturity of the compost.

14.1 INTRODUCTION

Palm oil industry is the fourth largest revenue-generating sector in Malaysia.[16] In 2016, oil palm planted area reached 5.74 million hectares, with 1.7% increment reported as compared to the previous year. Sabah is the largest oil palm planted State, with 1.55 million hectares or 27% of the total oil palm planted area, followed by Sarawak with 1.51 million hectares or 26% and Peninsular Malaysia (with 11 states) accounted for 2.68 million hectares or 47% of the total planted area.[25]

The palm oil industries have been significantly contributed to the economic growth and escalate the standard of living among the South East Asian countries especially in Malaysia. Increasing of the crude palm oil (CPO) production from 13.2% to 17.32 Mt had positively influenced the exports price in the major market by 7.3% to RM64.58 billion from 60.17 billion in 2015.[25] In other perspective, palm oil mill processes industry also yield a significant pollution load. Fortunately, a large portion of the pollution load can be discharged by the generation of palm oil mill effluent (POME) alongside.[39] It was reported that one tonne of processed fresh fruit bunches (FFB) will generate different types of wastes including the empty fruit bunch (EFB) (23%), mesocarp fiber (12%), shell (5%), and POME (60%).[2]

POME is one of the final by-products of the palm oil production in a mill. It is described as a liquid effluent discharge from the palm oil mill which appears as a thick brownish liquid at temperature ranging between 80 and 90°C and a pH between 4 and 5.[54] It comprises a mixture of water, oil, total solids (TS), and total suspended solids (TSS) which are about 28% of FFB.[2] Besides that, it has a high content of suspended and dissolved organic matter (OM), and has a chemical oxygen demand (COD) of about 15,000–100,000 mg/L, averaged at 51,000 mg/L.[35] About 36% of total POME is from the combination of the wastewater discharged from sterilizer condensate, 60% of total POME is from the clarification wastewater and the remaining 4% of POME is from the hydrocyclone wastewater.[55]

POME is the most difficult and expensive waste to manage by the mill operators. This is due to large volumes in tonnes that generated at a time. The easiest and cheapest method for disposal, raw POME, or partially treated POME is being settled into nearby river or land. Nevertheless, excessive amount of untreated POME deplete the water's oxygen and suffocate aquatic life.[35] This situation creates an environmental issue for the palm oil mill industry due to its overwhelming polluting characteristics. Many small and big rivers have been devastated by such disposal method as people living downstream are usually affected. POME give adverse impact to the environment because it has a high chemical and biochemical oxygen demand and mineral content such as nitrogen and phosphorous which can cause severe pollution to the environment.[3] The dark brown color of POME also gives a bad muddy water perception especially among the villagers in a rural area.

Composting of POME is an attractive approach in converting waste into organic fertilizers. Composting can be defined as an aerobic microbiological process that converts the organic substances of wastes into stabilized humus and less complex compounds.[50] Co-composting is different from composting as co-composting is the simultaneous composting of two or more types of waste material.[8,29] From co-composting, the potential added benefit of enhancing end compost quality.[28] In addition, composting or co-composting of wastes will provide conversion of wastes into a valuable product that will serve as a soil conditioner or fertilizer.[1] There were many contributing factors such as carbon to nitrogen ratio, moisture content, pH, and aeration rate that affected the composting process.[50]

Some of the challenging task for solid waste disposal are disposing the palm leaves and the paper product especially in metropolitan areas in most developing countries.[24] It will give adverse impact to the environment if this waste is not treated well. Several composting studies related to leaves and paper have been reported was best processed using recycling method.[9,31,44,50] Composting method using anaerobically digested POME (AnPOME) is also well-known in a number of studies.[6,16,48]

In this study, the AnPOME is utilized as a hydrolyzing agent for leaves–paper mixture. The performance of the process was evaluated based on temperature, moisture content, OM losses, total organic carbon (TOC), mass reduction, pH level, electrical conductivity, zeta potential, and phytotoxicity.

14.2 MATERIALS AND METHODS

14.2.1 MATERIALS

Shredded paper was collected around the offices at University Malaysia Sabah. Meanwhile, POME was collected from the anaerobic pond number two in Beaufort Palm Oil Mill. The mill is located at Beaufort, Kota Kinabalu, Sabah. The fresh leaves were collected from Taman Indah Permai, Kota Kinabalu residential area. The initial weight of the compost was 249.9 kg. Table 14.1 shows the weight of each material used for composting process.

TABLE 14.1 Weight of Compost Material.

Compost material	Weight (kg)
Paper	40.0
Leaves	25.0
POME	181.4
Compost starter	3.5
TOTAL	249.9 kg

14.2.2 COMPOSTING

Prior to composting process, the paper was digested with POME for 3 days. Next, the digested paper was mixed fresh mango leaves, followed by adding 3.5 kg of compost starter to enhance biodegradation process.[50] Then, mixture of composting materials was poured into a reactor (Fig. 14.1).

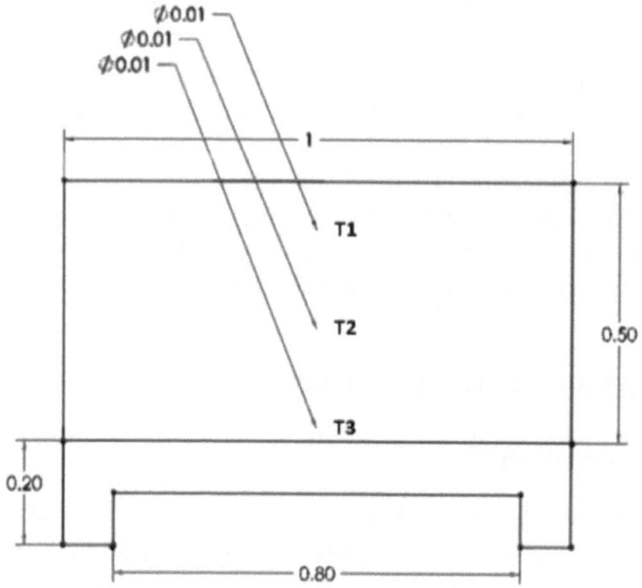

FIGURE 14.1 Front view of bioreactor (unit for dimension is meter).

A laboratory scale of composting reactor (effective volume = 0.4 m^3) was built using wood. The base was built using cement to provide stability to the reactor. The reactor specification is 1 m × 1 m × 0.5 m (length × width × height). There were holes at the base of the reactor for leachate discharging as shown in Figure 14.2.

FIGURE 14.2 **(See color insert.)** Base of the reactor.

The composting study was conducted at Environmental Lab, Block B, Faculty of Engineering, University Malaysia Sabah. The composting process was conducted for a period of 40 days. The sampling was taken every 10 days for physicochemical analysis.

14.2.3 PHYSIOCHEMICAL ANALYSIS

Temperature of the compost was measured daily over the period of 40 days, by using digital thermometer (Prima Long Thermometer). Temperature reading was taken at three different positions: upper (T1), core (T2), and bottom (T3) (Fig. 14.1). The ambient temperature was also recorded.

The sampling was conducted every 10 days of the composting. About 100 g of samples were collected from the center of the composting reactor. The analysis of the fresh samples was performed immediately after taking them out of the reactor. The leaves and mixture of paper and POME were weighed separately. Leaves were cut into small pieces and mixed together with the mixture of paper and POME.

Moisture content analysis was performed by drying 10 g of samples in an oven at 105°C for 24 h.[47] The dried sample was then weighed and the moisture content, % was calculated by using eq 14.1:

$$\text{Moisture content},\% = \frac{W_{\text{crucible+sample (before drying)}} - W_{\text{crucible+sample(after drying)}}}{W_{\text{sample}}} \times 100. \quad (14.1)$$

OM and ash concentrations were determined by dry combustion at 550°C for 4 h in furnace (Thermolyne, 46100 model).[47] The OM was determined as volatile solid. The percentage of ash of the sample is calculated by using the following eq 14.2:

$$\text{Ash, }\% = \frac{W_{\text{crucible+sample(after burning)}} - \text{W}_{\text{crucible}}}{W_{\text{sample}}} \times 100\%. \quad (14.2)$$

The percentage of TOC and percentage of OM were determined by eqs 14.3 and 14.4[52,56]:

OM_{loss} (%) = 100 – 100[X_1 (100 – X_n)] / [X_n(100 – X_1)]

$$\text{TOC, }\% = \frac{100 - \text{Ash},\%}{1.8} = \frac{\%\text{Organic matter}}{1.8} \quad (14.3)$$

$$OM\text{loss }\% = 100 - 100\frac{\left[X_1(100 - X_n)\right]}{\left[\text{X}_n(100 - X_1)\right]}, \quad (14.4)$$

where X_1 and X_n were the initial ash content and the ash content on each corresponding day.

10 g of sample was mixed with 100 mL of distilled water and then filtered to obtain water-soluble extract. The water-soluble extract was used for the pH, conductivity, zeta potential analysis, and phytotoxicity test.[41,50] A multiparameter meters (Hanna Instrument, Model:HI 9811-5) was used to determine the pH and conductivity of the solution.

14.2.4 ZETA POTENTIAL

The zeta potential of the compost was determined on days 0, 10, 20, 30, 40, and 50, by analyzing water-extract solution. The solution was tested by using a Malvern Zetasizer Nano Series model ZS machine to determine the zeta potential of the compost.

14.2.5 PHYTOTOXICITY ASSAY

Seed germination technique based on Cabbage seeds (*Brassica oleracea*) was applied to determine the phytotoxicity of the compost. The cabbage seeds were soaked in water for 72 h to let the seeds sprouting. Only the sprouted seeds were used in the experiment.

Water-soluble extract (15 mL) and 15 cabbage seeds were placed on petri dish that previously covered with filter paper. Control sample was also run simultaneously by using 5 mL of distilled water instead of compost extract. The petri dishes were placed in the dark space at room temperature for 72 h. The number of germinated seeds was calculated and the growth of roots was monitored. The percentage of relative seed germination (RSG), relative root growth (RRG), and germination index (GI) were calculated using eqs 14.5, 14.6, and 14.7:

$$RSG\,(\%) = \frac{\text{number of seeds germinated in sample extract}}{\text{number of seeds germinated in control}} \times 100 \qquad (14.5)$$

$$RRG\,(\%) = \frac{\text{root length in sample extract}}{\text{root length in control}} \times 100 \qquad (14.6)$$

$$RRG\,(\%) = \frac{\text{root length in sample extract}}{\text{root length in control}} \times 100 \qquad (14.7)$$

14.2.6 NUTRIENT ANALYSIS OF THE COMPOST

The nutrient content of the compost was analyzed by Sime Darby Research Sdn. Bhd located at Balung Tawau, Sabah, Malaysia.

14.2.7 STATISTICAL ANALYSIS

The average value and standard deviation of the data were calculated using Microsoft Excel. The standard error was computed and errors bars were determined for the data.

14.3 RESULTS AND DISCUSSION

14.3.1 TEMPERATURE PROFILE

The temperature variation of the compost (T1, T2, and T3) in comparison with ambient temperature over the period of 40 days composting are shown in Figure 14.3. The data show typical temperature profiles of the compost as reported previously in a number of other composting studies.[26,33,36,50] The temperature profiles can be represented as the stages of composting, microbial activity and is used to describe the conditions suitable for the proliferation of different microbial groups, that is, meso- and thermophilis.[21]

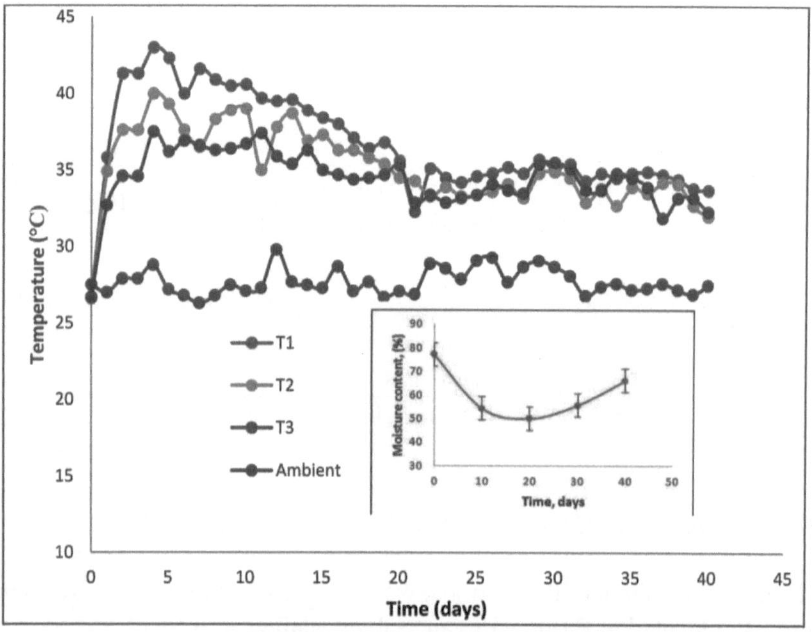

FIGURE 14.3 (See color insert.) Temperature trend during the first 40 days of the composting process. Small figure shows the moisture content profile in the compost mixture (standard errors [SE] in vertical bars).

In this study, the compost temperatures were varied between 26.6°C and 43.0°C, while the ambient temperature ranged from 26.7°C to 29.8°C. Basically, there are four phases that exist in the dynamics of composting, namely temperature evolution, mesophilic, thermophilic, cooling, and maturation.[33] An increase in temperature from 26.6°C to 33.7°C during the first day indicated that the system was in the first stage. A further increase in temperature was recorded on the following day which resulted in the liberation of heat on the sides of the composter indicating the presence of biological activity. The compost temperature (T1) reached the highest temperature of 43.0°C on day 4. This possibly is due to less ventilation in the upper part of the reactor and thus more heat might be trapped that led to higher temperature. The energy produced in the system is due to microbial decomposition of OM is the main cause of increase in temperature.[46] On the other hand, the maximum compost temperature (T3) at the bottom part of the reactor did not exceed 40°C. This part mainly undergoes mesophilic composting, which maintain a composting temperature less than 40°C. Since this compost (T3) reached thermophilic temperature, it can effectively destroy pathogens and weed seed, and stable humus-like substance can be formed by converting the biodegradable solid organic into it.[53] After thermophilic phase, temperatures declined slowly to mesophilic temperature and indicated that the microbial activity has become weaker.

14.3.2 MOISTURE CONTENT

Moisture content is necessary to provide a medium for the transport of dissolved nutrients required for the metabolic and physiological activities of microorganisms.[13] Physical and chemical properties of the waste material change with moisture content which acts as a transporting medium of nutrients for microbial activity.[23] The variation in moisture content values is shown in Figure 14.3. The trend of the moisture content observed in this work is similar to the findings reported by Zahrim et al., Rawoteea et al., and Villegas and Huiliñir.[33,43,50] According to Lin,[20] the minimal requirement for obtaining adequate microbial activities is 50%. The initial moisture content of the paper–leaves AnPOME was 77.22% and the value decreases gradually to 50% on the 20th day due to evaporation caused by thermophilic fermentation. One indication that OM was decomposed, the moisture content will continuously decrease during composting.[17] This is the observation obtained during the first 20 days of the composting process. From day 20 to day 40, the moisture content of the compost increased by 16% due to the lower heat generation. Consequently, the condensed vapor that attached to the back of the reactor's lid fell back into the mixture.[42]

From the results, final compost moisture content obtained was 66.15%. Since the compost moisture content did not fell below 50%, water was not added into the reactor in order to maintain 50–60% moisture content. It should be noted that adequate environment must be maintained to enhance microbial fermentation.

14.3.3 ORGANIC MATTER DEGRADATION DURING THE COMPOSTING PROCESS

The degradation of OM was calculated through the contents of ash (Table 14.2) to evaluate the co-composting performance. This analysis is a valuable indication for a successful composting process.[52] OM mineralization was investigated by determining the losses of OM with the time course during co-composting. It was found that the losses of OM were increased rapidly with time and this OM loss was detected at the end of composting process (72.13%). The OM loss in this study was comparable to the results reported by Zhang et al.,[52] which had a maximum OM loss of 61.66%. Other than that, Petric and Mustafić[30] reported values of 37–50% in composting of wheat straw with poultry manure. Moreover, the OM loss in composting of olive-mill pomace and poultry manure with tomato harvest stalks also reported to be in the range of 21.7–46.1%.[37]

TABLE 14.2 Ash Content in Composts During Compost Processes (Dry Weight).

Composting time (days)	0	10	20	30	40
Average ash content (%)	15.28	27.60	18.16	28.0	38.85

The rate of decomposition in the composting process can also be determined through measurement of TOC. The TOC is used as the energy sources for microorganisms during composting process and the degradation of TOC could be used to illustrate the level of compost maturity.[52] From Figure 14.4, it can be seen that the TOC (%) decreases slightly throughout the composting process. The decrease in TOC is related to the microbial respiration, or amount of carbon dioxide released which depends on the degree of mineralization through microbial activities.[17,18] Thus, a larger decrease in TOC would correspond to a larger degree of decomposition by the microorganisms. In this study, a significant TOC loss was observed during the first 10 days of composting process. This is possibly due to the increased degradation of easily degradable fractions, whereas the steady decrease from day 20 onwards may be influenced by hardly degradable fractions.[30]

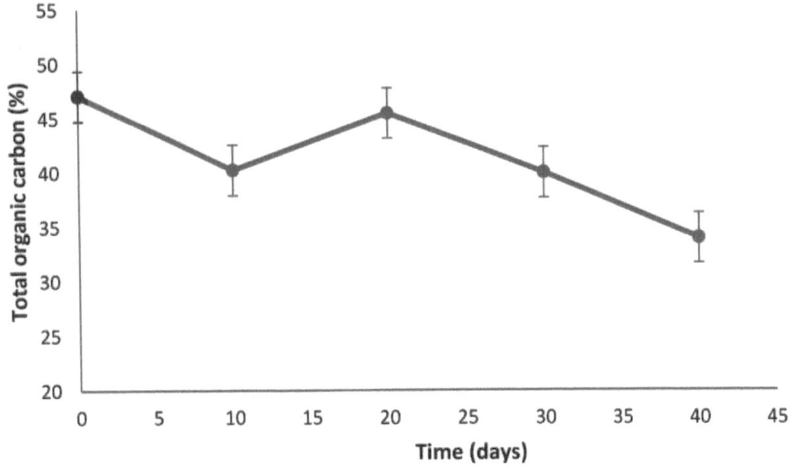

FIGURE 14.4 Profile of total organic carbon during composting process (SE in vertical bars).

14.3.4 MASS REDUCTION

Reactor composting was efficient in treating the total amount of 249.9 kg of paper–leaves AnPOME in 40 days, resulted in a 85 kg of compost. The mass reduction (66%) after 40 days is very significant and the value is in the range of other studies (Table 14.3). The reason for these losses could be due to evaporation of water and microbial activity (C and N gas emissions).[36]

14.3.5 pH AND ELECTRIC CONDUCTIVITY

Figure 14.5 shows the pH profile and conductivity of the sample during the composting process. Based on results, the value of pH was decreasing until 6.9 on day 30. The declining in pH value at the early stage was due to establishment of anaerobic condition causes the formation of organic acids.[7] At the end of the composting process, the pH value was increased

TABLE 14.3 Mass Reduction and OM Losses in Several Composting Process.

Substrate	Composting time (days)	Composting system	Highest temperature (°C)	Mass reduction (%)	OM losses (%)	Reference
Paper (16%), leaves (10%), anPOME (73%), and compost starter (1%)	40	Passive aerated (0.4 m^3)	43	66	72.13	this study
Paper (31%), grass-clipping (46%), and anPOME (23%)	40	Passive aerated (0.05 m^3)	31	18	n.d	[50]
Vegetable (70%), meat (10%), and paper (20%)	140	Passive aerated (0.2 m^3)	58	7.1	65.13	[5]
Olive leaves (8%) and olive humid husks (92%)	90	Passive aerated (15 m^3)	22	n.d	6.4	[4]
Food scrap (88%) and dry leaves (12%)	154	Passive aerated (0.2 m^3)	45.6	73.27	n.d	[15]
Leaves	52	Forced aerated (0.002 m^3)	60	n.d	46	[14]
Poultry manure (75%) with chestnuts leaves and burrs (25%)	103	Passive aerated (0.1088 m^3)	60	n.d	6.27	[11]
Dairy cattle manure (71%) + wallboard paper (29%)	28	Passive aerated (0.431 m^3)	70-75	11.45	n.d	[34]

to alkaline range. This was due to the consumption of protons during the decomposition of volatile fatty acids, the generation of carbon dioxide, and the mineralization of organic nitrogen.[7]

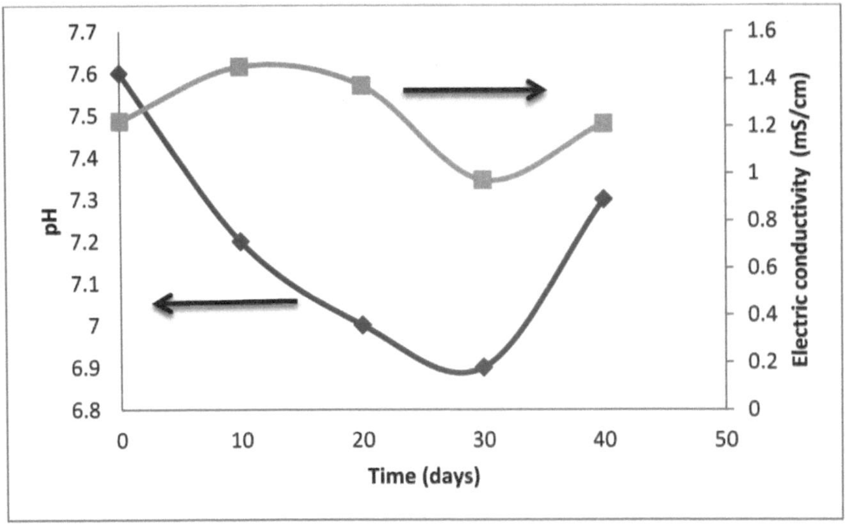

FIGURE 14.5 pH profile and electric conductivity during the composting process.

Electric conductivity reflects the salinity and suitability of the compost product to be used in agricultural industry. The electric conductivity of this study was at the range of 1.45–0.97 mS/cm. The decreasing trend of electric conductivity was due to the increasing concentration of nutrients, such as nitrate and nitrite.[7] However, the electric conductivity was at an acceptable level in terms of safe applications for plant growth which the limit was 2.5 mS/cm.[12]

14.3.6 ZETA POTENTIAL

The zeta potential is the potential difference between the dispersion medium and the stationary layer of fluid attached to the dispersed particles.[50] It indicates the surface characteristics and electric potential variation of the residue surface during decomposition period[19,45] Figure 14.6 shows the zeta potential profile over the composting period. In this study, negative values of surface charge (zeta potential) were obtained in the whole composting process. As the composting progress, the zeta potential value of the compost decreases gradually from −8.855 mV to −20.4 mV in the first 40 days. The lowest zeta potential value of −25.6 mV was observed on the last day of composting. The zeta potential of compost would become decrease since POME is negatively charged from the beginning and tends to be more negative in the following treatments, that is, in aerobic ponds, due to the degradation of OM.[48]

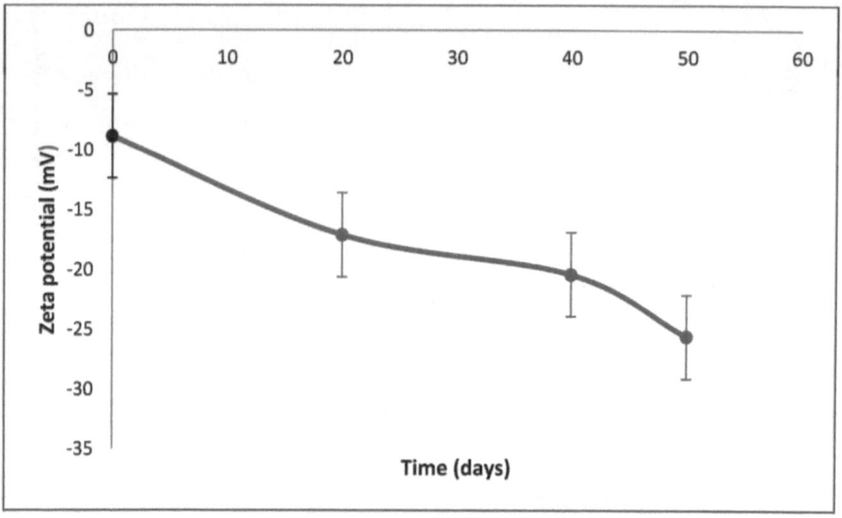

FIGURE 14.6 Zeta potential profile during composting processes (SE in vertical bars).

14.3.7 PHYTOTOXICITY

Figure 14.7 shows increasing trend of GI profile during the first 20 days of composting period. The reason for that is the declining of the electric conductivity. Excessive salinity affects the phytotoxicity of the compost directly but depending on the types of seeds used in the germination test. Different types of plant having different salinity tolerance. Salinity effects can be negligible when the electric conductivity readings are below 2 mS/cm.[40]

However, the GI was reduced to 94.7% at day 30. The factor that could be the causes of this behavior was the pH value. The pH value at day 30 was 6.9 which were slightly acidic. The acidic condition shows that the accumulation of organic acids as the results of biodegradation of soluble carbon. The organic acids are toxic compounds and have the adverse effect on the seed germination.[10,40]

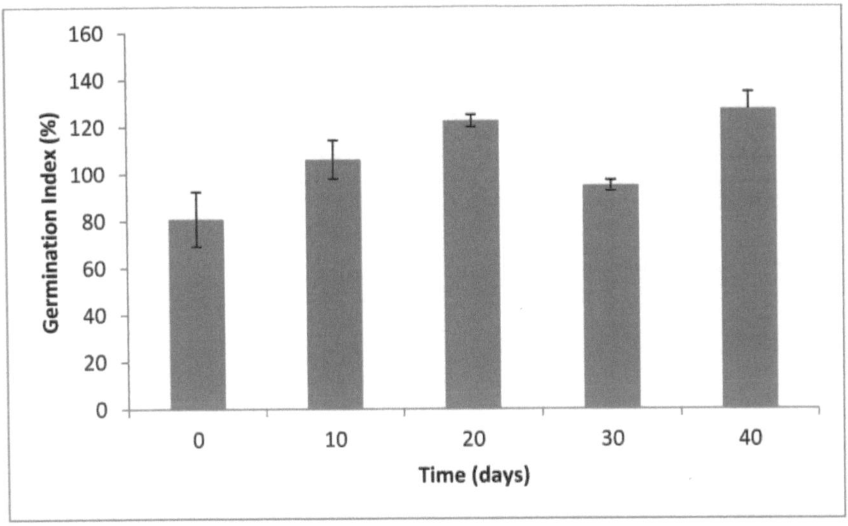

FIGURE 14.7 Profile of germination index during composting (SE in vertical bar).

After all, the highest GI of 127.26 was achieved at day 40. Throughout the composting period, the GI is above 80% which indicates the phytotoxin free and having completed maturity.[32,41]

There are many reasons effected the phytotoxicity of the compost including concentration of volatile organic acids, concentration of NH_4–N, oxygen depletion or presence of heavy metal, and the molecular weight of organic compound.[12,32] Most of these factors influence seed germination simultaneously and it is difficult to assess which parameter determines the greatest influence.[40] Besides, the GI should be interpreted with caution because the value was affected by the type of seed used and the application extraction rate.[32]

14.3.8 COMPOSTING NUTRIENT

Table 14.4 shows the initial and final nutrients content of compost produced in this work, in comparison with previous research. It can be seen that the moisture content and pH of the compost are within the range of other findings which is 50–70% for moisture content and 6–9 for pH. The decreasing value of C/N shows that the biodegradation of the OM occurred during the composting process. Nutrients in compost can improve soil condition and consequently help to reduce the dependency of inorganic fertilizers. Generally, the nutrient content including nitrogen, potassium, and sodium was increasing at the end of the composting period.

TABLE 14.4 Initial and Final Nutrient Content of the Compost Product.

Composition	Initial	Final	Recommended value for final ideal substrate
Moisture content (%)	77	66	40–55[51]
pH	7.6	7.3	5.5–6.5[47]
Conductivity (mS/cm)	1.22	1.21	< 3.0[38]
C/N	44	29	15–20[7]
TOC (%)	47	33	–
N (%)	1.06	1.12	–
P (%)	0.07	0.16	> 0.5 (as P_2O_5)[47]
K (%)	0.21	0.72	> 1.5 (as K_2O)[47]
Mg (%)	0.1	0.24	>70 mg L^{-1}[22]
Ca (%)	ND	5.08	> 200 mg L^{-1}[22]
B (%)	ND	0.112	0.05–0.5 mg L^{-1}[22]

ND—not determined.

14.4 CONCLUSION

In this study, a composting process was successfully conducted over 40 days using a mixture of paper, leaves, and An-Pome as the substrate. Up to 66% of mass reduction of the compost was obtained. During composting process, the highest compost temperature achieved was 43°C on day 4 of composting. The moisture contents decrease from 77% to 66%, and the pH value of the compost produced is 7.3. Both of the TOC and C/N were decreasing toward the end of composting period. The negative zeta potential values decreased from −8.855 to −20.4 mV due to organic degradation. The results show that phytotoxicity of the compost is decreasing as the GI increasing. Interestingly, the compost with GI of 127.26% was successfully produced.

KEYWORDS

- **composting**
- **palm oil mill effluent**
- **paper**
- **leaves**
- **phytotoxicity**
- **germination index**

REFERENCES

1. Abu Qdais, H.; Al-Widyan, M. Evaluating Composting and Co-Composting Kinetics of Various Agro-Industrial Wastes. *Int. J. Recyc. Org. Waste Agric.* **2016,** *5*, 273–280.
2. Ahmad, A.; Buang, A.; Bhat, A. H. Renewable and Sustainable Bioenergy Production from Microalgal Co-Cultivation with Palm Oil Mill Effluent (Pome): A Review. *Renew. Sustain. Energy Rev.* **2016,** *65*, 214–234.
3. Ahmed, Y.; Yaakob, Z.; Akhtar, P.; Sopian, K. Production of Biogas And Performance Evaluation of Existing Treatment Processes in Palm Oil Mill Effluent (Pome). *Renew. Sustain. Energy Rev.* **2015,** *42*, 1260–1278.
4. Alfano, G.; Belli, C.,; Lustrato, G.; Ranalli, G. Pile Composting of Two-Phase Centrifuged Olive Husk Residues: Technical Solutions And Quality of Cured Compost. *Bioresour. Technol.* **2008,** *99*, 4694–4701.
5. Arrigoni, J. P.; Paladino, G.; Laos, F. Feasibility And Performance Evaluation Of Different Low-Tech Composter Prototypes. *Int. J. Environ. Protect.* **2015,** *5*, 1.
6. Baharuddin, A. S.; Wakisaka, M.; Shirai, Y.; Abd-Aziz, S.; Abdul Rahman, N. A.; Hassan, M. A. Co-Composting of Empty Fruit Bunches and Partially Treated Palm Oil Mill Effluents in Pilot Scale. *Int. J. Agric. Res.* **2009,** *4*, 69–78.
7. Bazrafshan, E.; Zarei, A.; Kord Mostafapour, F.; Poormollae, N.; Mahmoodi, S.; Zazouli, M. A. Maturity and Stability Evaluation of Composted Municipal Solid Wastes. *Health Scope* **2016,** *5*.
8. Das M, U. H.; Singh, R.; Beri, S.; Mohan, K. S.; Gupta, V. C.; Adholeyaa. Co-Composting of Physic Nut (Jatropha Curcas) Deoiled Cake With Rice Straw And Different Animal Dung. *Bioresourtechnol* **2011,** *102*, 6541–6546.
9. Francou, C.; Linères, M.; Derenne, S.; Villio-Poitrenaud, M. L.; Houot, S. Influence of Green Waste, Biowaste And Paper–Cardboard Initial Ratios On Organic Matter Transformations During Composting. *Bioresour. Technol.* **2008,** *99*, 8926–8934.
10. Gómez-Brandón, M.; Lazcano, C.; Domínguez, J. The Evaluation of Stability and Maturity During The Composting of Cattle Manure. *Chemosphere* **2008,** *70*, 436–444.
11. Guerra-Rodríguez, E.; Diaz-Raviña, M.; Vázquez, M. Co-Composting Of Chestnut Burr and Leaf Litter with Solid Poultry Manure. *Bioresour. Technol.* **2001,** *78*, 107–109.

12. Himanen, M.; Hänninen, K. Composting of Bio-Waste, Aerobic and Anaerobic Sludges–Effect of Feedstock on The Process And Quality Of Compost. *Bioresour. Technol.* **2011,** *102*, 2842–2852.
13. Ishak, N. F.; Ahmad, A. L.; Ismail, S. Feasibility Of Anaerobic Co-Composting Empty Fruit Bunch With Activated Sludge from Palm Oil Mill Wastes for Soil Conditioner. *J. Phys. Sci.* **2014,** *25*, 77.
14. Jr, F. C. M.; Reddy, C. A.; Fomey, L. J. Yard Waste Composting: Studies Using Different Mixes Of Leaves And Grassin A Laboratory Scale System. *Compost Sci. Utiliz.* ***1993,*** 85–96.
15. Karnchanawong, S.; Nissaikla, S. Effects Of Microbial Inoculation on Composting of Household Organic Waste Using Passive Aeration Bin. *Int. J. Recyc. Organic Waste Agric.* **2014,** *3*, 113–119.
16. Krishnan, Y.; Bong, C. P. C.; Azman, N. F.; Zakaria, Z.; Othman, N. A.; Abdullah, N.; Ho, C. S.; Lee, C. T.; Hansen, S. B.; Hara, H. Co-Composting of Palm Empty Fruit Bunch and Palm Oil Mill Effluent: Microbial Diversity and Potential Mitigation of Greenhouse Gas Emission. *J. Cleaner Prod.* **2016**.
17. Kulcu, R.; Yaldiz, O. Determination of Aeration Rate and Kinetics of Composting Some Agricultural Wastes. *Bioresour. Technol.* **2004,** *93*, 49–57.
18. Kulikowska, D. Kinetics of Organic Matter Removal and Humification Progress During Sewage Sludge Composting. *Waste Manag.* **2016,** *49*, 196–203.
19. Li, J.; Lu, J.; Li, X.; Ren, T.; Cong, R.; Zhou, L. Dynamics of Potassium Release and Adsorption on Rice Straw Residue. *Plos One* **2014,** *9*, E90440.
20. Lin, C. A Negative-Pressure Aeration System for Composting Food Wastes. *Bioresour. Technol.* **2008,** *99*, 7651–7656.
21. Liu, K.; Price, G. Evaluation of Three Composting Systems for the Management of Spent Coffee Grounds. *Bioresour. Technol.* **2011,** *102*, 7966–7974.
22. López-López, N.; López-Fabal, A. Scientia Horticulturae Compost Based Ecological Growing Media According Eu Eco-Label Requirements. *Sci. Horticult.* **2016,** 212.
23. Manu, M.; Kumar, R.; Garg, A. Performance Assessment of Improved Composting System for Food Waste with Varying Aeration and Use of Microbial Inoculum. *Bioresour. Technol.* **2017**.
24. Moh, Y. C.; Manaf, L. A. Overview of Household Solid Waste Recycling Policy Status and Challenges in Malaysia. *Resourc. Conserv. Recyc.* **2014,** *82*, 50–61.
25. Mpob. *Overview of the Malaysian Oil Palm Industry*, 2016.
26. Onursal, E.; Ekinci, K. Co-Composting of Rose Oil Processing Waste with Caged Layer Manure and Straw or Sawdust: Effects of Carbon Source and C/N Ratio on Decomposition. *Waste Manag. Res.* **2015,** *33*, 332–338.
27. Oviedo-Ocaña, E.; Torres-Lozada, P.; Marmolejo-Rebellon, L.; Hoyos, L.; Gonzales, S.; Barrena, R.; Komilis, D.; Sanchez, A. Stability and Maturity of Biowaste Composts Derived by Small Municipalities: Correlation Among Physical, Chemical and Biological Indices. *Waste Manag.* **2015,** *44*, 63–71.
28. Paredes C, B. M.; Cegarra, J.; Roig, A.; Novarro, Af. *Nitrogen Transformation During the Composting of Different Organic Wastes;* Kluwer Academic Publishers: Dordrecht, 1996.

29. Petric I, H. A.; Avdić Ea. Evolution of Process Parameters and Determination of Kinetics for Co-Composting of Organic Fraction of Municipal Solid Waste with Poultry Manure. *Bioresour. Technol.* **2012,** *117*, 107–116.
30. Petric, I.; Mustafić, N. Dynamic Modeling the Composting Process of the Mixture of Poultry Manure and Wheat Straw. *J. Environ. Manag.* **2015,** *161*, 392–401.
31. Poincelot, R.; Day, P. Rates of Cellulose Decomposition During the Composting of Leaves Combined with Several Municipal and Industrial Wastes and Other Additives. *Compost Sci.* **1973**.
32. Raj, D.; Antil, R. S. Evaluation of Maturity and Stability Parameters of Composts Prepared from Agro-Industrial Wastes. *Bioresour. Technol.* **2011,** *102*, 2868–2873.
33. Rawoteea, S. A.; Mudhoo, A.; Kumar, S. Co-Composting of Vegetable Wastes and Carton: Effect of Carton Composition and Parameter Variations. *Bioresour. Technol.* **2017,** *227*, 171–178.
34. Saludes, R. B.; Iwabuchi, K.; Miyatake, F.; Abe, Y.; Honda, Y. Characterization of Dairy Cattle Manure/Wallboard Paper Compost Mixture. *Bioresour. Technol.* **2008,** *99*, 7285–7290.
35. Seng, Y. S. M. L. Palm Oil Mill Effluent (Pome) from Malaysia Palm Oil Mills: Waste or Resource. *Int. J. Sci. Environ.* **2013,** *2*, 1138–1155.
36. Storino, F.; Arizmendiarrieta, J. S.; Irigoyen, I.; Muro, J.; Aparicio-Tejo, P. M. Meat Waste as Feedstock for Home Composting: Effects on the Process and Quality of Compost. *Waste Manag.* **2016,** *56*, 53–62.
37. Sülük, K.; Tosun, İ.; Ekinci, K. Co-Composting of Two-Phase Olive-Mill Pomace and Poultry Manure with Tomato Harvest Stalks. *Environ. Technol.* **2017,** *38*, 923–932.
38. Sun, Z. -Y.; Zhang, J.; Zhong, X. -Z.; Tan, L.; Tang, Y. -Q.; Kida, K. Production of Nitrate-Rich Compost from the Solid Fraction of Dairy Manure by a Lab-Scale Composting System. *Waste Manag.* **2016,** *51*, 55–64.
39. Tabassum, S.; Zhang, Y.; Zhang, Z. An Integrated Method for Palm Oil Mill Effluent (Pome) Treatment for Achieving Zero Liquid Discharge–A Pilot Study. *J. Clean. Prod.* **2015,** *95*, 148–155.
40. Tiquia, S. M. Reduction of Compost Phytotoxicity During the Process of Decomposition. *Chemosphere* **2010,** *79*, 506–512.
41. Tiquia, S. M.; Tam, N. F. Y.; Hodgkiss, I. J. Effects of Composting on Phytotoxicity of Spent Pig-Manure Sawdust Litter. *Environ. Pollut.* **1996,** *93*, 249–256.
42. Unmar, G.; Mohee, R. Assessing the Effect of Biodegradable and Degradable Plastics on the Composting of Green Wastes and Compost Quality. *Bioresour. Technol.* **2008,** *99*, 6738–6744.
43. Villegas, M.; Huiliñir, C. Biodrying of Sewage Sludge: Kinetics of Volatile Solids Degradation Under Different Initial Moisture Contents and Air-Flow Rates. *Bioresour. Technol.* **2014,** *174*, 33–41.
44. Wong, J.; Mak, K.; Chan, N.; Lam, A.; Fang, M.; Zhou, L.; Wu, Q.; Liao, X. Co-Composting of Soybean Residues and Leaves in Hong Kong. *Bioresour. Technol.* **2001,** *76*, 99–106.
45. Yan, L.; Liu, Y.; Wen, Y.; Ren, Y.; Hao, G.; Zhang, Y. Role and Significance of Extracellular Polymeric Substances from Granular Sludge for Simultaneous Removal of Organic Matter and Ammonia Nitrogen. *Bioresour. Technol.* **2015,** *179*, 460–466.

46. Yang, L.; Zhang, S.; Chen, Z.; Wen, Q.; Wang, Y. Maturity and Security Assessment of Pilot-Scale Aerobic Co-Composting of Penicillin Fermentation Dregs (Pfds) with Sewage Sludge. *Bioresour. Technol.* **2016,** *204*, 185–191.
47. Yaser, A. Z.; Rahman, R. A.; Kali, M. S. Co-Composting of Palm Oil Mill Sludge-Sawdust **2007,** *10*, 4473–4478.
48. Zahrim, A. Palm Oil Mill Biogas Producing Process Effluent Treatment: A Short Review. *J. Appl. Sci.* **2014,** *14*, 3149–3155.
49. Zahrim, A.; Asis, T. Production of Non Shredded Empty Fruit Bunch Semi-Compost. *J. Inst. Eng. Malaysia* **2010,** *71*.
50. Zahrim, A. Y.; Leong, P. S.; Ayisah, S. R.; Janaun, J.; Chong, K. P.; Cooke, F. M.; Haywood, S. K. Composting Paper and Grass Clippings with Anaerobically Treated Palm Oil Mill Effluent. *Int. J. Recyc. Organic Waste Agric.* **2016,** *5*, 221–230.
51. Zahrim, A. Y.; Tahang, A. Production of Non Shredded Empty Fruit Bunch Semi-Compost. *Inst. Eng. Malaysia* **2010,** *71*, 11–17.
52. Zhang, L.; Zeng, G.; Dong, H.; Chen, Y.; Zhang, J.; Yan, M.; Zhu, Y.; Yuan, Y.; Xie, Y.; Huang, Z. The Impact of Silver Nanoparticles on the Co-Composting of Sewage Sludge and Agricultural Waste: Evolutions of Organic Matter and Nitrogen. *Bioresour. Technol.* **2017,** *230*, 132–139.
53. Zhou, C.; Liu, Z.; Huang, Z. -L.; Dong, M.; Yu, X. -L.; Ning, P. A New Strategy for Co-Composting Dairy Manure with Rice Straw: Addition of Different Inocula at Three Stages of Composting. *Waste Manag.* **2015,** *40*, 38–43.
54. Singh, R. P.; Hakimi Ibrahim, M.; Norizan, E.; Iliyana, M. S. Composting of Waste from Palm Oil Mill: a Sustainable Wastemanagement Practice. *Rev. Environ. Sci. Biotechnol.* **2010,** *9*, 331–344.
55. Wu, T. Y.; Mohammad, A. W.; Jahim, J. M.; Anuar, N. A Holistic Approach to Managing Palm Oil Mill Effluent (POME): Biotechnological Advances in the Sustainable Reuse of POME **2009,** *Biotechnol. Adv.* *27* (1), 40–52.
56. Polprasert, C. *Organic Waste Recycling*; Wiley Publishing: 1989.

CHAPTER 15

ALGAE: ROLE IN ENVIRONMENT SAFETY

RAJEEV SINGH[1], HEMA JOSHI[2], and ANAMIKA SINGH[3,*]

[1]*Department of Environmental Studies, Satyawati College, University of Delhi, Delhi, India*

[2]*Department of Botany, Hindu Girls College, Sonepat, Haryana, India*

[3]*Department of Botany, Maitreyi College, University of Delhi, Delhi, India*

[*]*Corresponding author. E-mail: arjumika@gmail.com*

ABSTRACT

Algae are photosynthetic thallophytic plants which are not differentiated into roots, stems, and leaves. They have chlorophyll as the main component and are photosynthetic. Along with chlorophyll a, they have chlorophyll b, c, d, and e. Reserve food material also varies and the presence of different types of chlorophyll and reserve food is the main criteria of algae classifications. Although at organizational level, algae are very simple, but economically they are very important. In this chapter, we are discussing some important uses of algae.

15.1 INTRODUCTION

The study of algae is known as Phycology and this word is derived from the Greek word "phykos," meaning seaweed or algae and logos which means science, so it is the science of algae. Phycology or algology is the study of the algae (singular, alga) and it deals with the morphology, taxonomy, phylogeny, biology, and ecology of algae. Algae are thallophytes, that is, they are not differentiated into roots, stems, and leaves, and have chlorophyll as a photosynthetic pigment. They lack sterile covering of cells around the reproductive cells. So, algae are photosynthetic thallophytes that show very low level of tissue differentiation. Algae can be broadly divided into macroalgae and microalgae. Seaweeds are larger algae's and are macroalgae. About 70 species are economically important and are used for food, food additives, animal feed, fertilizers, and biochemical. Microalgae can only be visualized under the microscope. They are important as breaking down sewage and generate methane and fuels for energy, improving soil fertility. Microalgae produce 10–20 times more biomass per unit area than any terrestrial crop. Algae act as an important component of natural economy in water and ocean as they provide them a photosynthetic system. They are found almost in all habitable environments like hot water spring, damp soil, ice, old wall, garbage, pits, etc. They acquire the place at the bottom of the food chain in all aquatic ecosystems. They are autotrophic and produce complex organic compounds from simple inorganic molecules using energy from light. Algae can be unicellular or multicellular and may be up to 150 feet long. Algae have wide economic importance. They are an important source of biochemical components like vitamins, minerals, proteins, fatty acid, etc. They affect man in various ways and it is not very easy to locate assembled information on their roles. Even a few algal cells produce phycotoxins and are produced by freshwater and marine algae are a serious hazard to water supplies, reservoirs, recreational beaches, etc. These toxins can be found in drinking water or may be accumulated through the food chain. World Health Organization recommended many countries to monitor algal toxins in freshwater and seafood. However, Algal cultures show several advantages over conventional plants and these are high output per hectare, thus minimizing land usage. Low water usage per unit of biomass, whole plant utilization, high protein, lipid, and vitamins output per hectare, carbon utilization and so on.[1–4]

In this chapter, we will discuss the economic importance of algae and a few of them are as follows:

- in medicine;
- as food and nutrition;
- in wastewater treatment;
- as biofertilizer;

- as source of energy;
- high value-added products from algae.

BOX 1: PHYCOTOXIN

A few algae play "nuisance role" where they tend to block water supply canals filtration unit due to eutrophication of reservoirs or canals or other water unit. Sometimes these algae are labeled as "phycotoxin" as they secrete some toxins which cause poisoning.

15.2 ROLE OF ALGAE IN MEDICINE

Algae have the potential to produce and benefit humans through the development of pharmaceutical and novel medical compounds. Algae are advantageous from the traditional medicines as they can synthesize bioactive compounds more quickly, easily, and cheaply. In spite of that, they have very less side effects and are pesticides free. These are unique qualities of algae which overcomes the consumers' limitations like intolerances and allergies.

Most prevalent diseases like malnutrition, nutritional anemia, xeropthelmia (due to vitamin A deficiency), and endemic goiter (iodine deficiency) can be treated easily by using tablespoon of algae.

15.2.1 PHYLUM CHLOROPHYTA

Species of Enteromorpha used to treat hemorrhoids, parasitic disease, goiter, asthma, coughing, and bronchitis. *Ulva* used to treat goiter, reduce fever, ease pain, induce urination *Codium*, and *Acetabularia* used to treat urinary diseases, treat edema.

15.2.2 PHYLUM PHAEOPHYTA

Laminaria used to treat goiter, urinary diseases and it is an ecbolic, contains iodine and potassium. *Sargassum* can be used to treat edema, cervical lymphadenitis, diminishes inflammation, reduces fever, and induces urination.

15.2.3 PHYLUM RHODOPHYTA

Chondrus, Porphyra, and *Eucheuma* are used to treat goiter, bronchitis, tonsillitis, and cough, *Gelidium* is a Laxative and can treat tracheitis, gastric diseases, and hemorrhoids. Agar is used to treat goiter, tonsillitis, bronchitis, asthma, cough, gastric diseases, and hemorrhoids. *Gracilaria* is used to treat goiter, edema, urinary diseases, can prevent ulcer, and also as agar extract. *Hypnea* is used to treat bronchitis, gastric diseases, and hemorrhoids. *Chondria* and *Grateloupia* are used as Ascaricide. *Centroceras* is used to cure gastro-intestinal intolerance. *Gloeopeltis* is used to treat goiter, tonsillitis, and bronchitis; prevents high blood pressure and scurvy.

15.3 ALGAE AS FOOD AND NUTRITION

Today, the whole word is suffering from obesity, high and low blood pressure, heart diseases, diabetes, vitamins, mineral deficiencies, etc. Most of the diseases are because of high caloric intake and junk food. There are many artificial products which contain added sugars and preservatives that cause severe health hazards. Algae, as a food, are a revolutionary step toward all these problems. It is an effective natural product enriched with many vitamins, minerals, and other important nutritional components that help persons resolving their diet problems. In coastal areas, seaweeds are cultivated on a large scale as they are used in human and animal nutrition. *Porphyra* sp., *Chondrus crispus, Himanthalia*, and *Undaria pinnatifida* are the few examples having low calories and high rate of vitamins, minerals, and dietic fibers.

Microalga biomass is available in the form of powder, tablets, capsules, liquids, and they are used in different food products like confectionaries, sweets etc. *Spirulina* and *Chlorella* are the two important genera used in human nutrition. *Spirulina* is considered as super food and dried *spirulina* is like eating cooked vegetables. Some seaweeds are also eaten as salad. Algal bioactive compounds are used to develop new drugs and health foods. Edible algae are a rich source of dietary fiber, minerals, proteins, etc. Marine algae are considered as rich source of anti-oxidants. Whereas, some brown algae are a rich source of antioxidants like phylopheophytin in *Eisenia bicyclis* (arame) and fucoxantinein in *Hijikia fusiformis* (hijiki). A few marine algae are stored after boiling and steaming and are soaked for 40 times volumes of water before being consumed. Agar is having gel properties and are extracted from various species of genus *Gracilaria* and used as food products like frozen foods, bakery icings, dessert gels, candies, and fruit juices. Chlorella algal strain has sugar and a property to be converted into fats and to be used as food taste enhancers. This is an innovative and revolutionary advancement that can resolve the problem caused by the current fat and sugar substitute that have problems including indigestibility and poor flavor. Algae-derived fats are healthy and have a property that it is palatable for baked goods and ice creams.

BOX 2: COMMERCIAL ALGAL PRODUCT AND USES

Spirulina (arthrospira): Human and animal nutrition, cosmetics (phycobiliproteins, powders, extracts, tablets, beverages, chips, pasta, and liquids extract).
Chlorella sp.: Human nutrition, aquaculture, cosmetics (tablets, powders, nectar, and noodles).
Dunaliella salina: Human nutrition, cosmetics (beta-carotene and powders).
Aphanizomenon flos-aquae: Human nutrition (capsules, crystals, and powder). *Haematococcus pluvialis*: Aquaculture and astaxnthin *Crypthecodinium cohnii* and *Shizochytrium* sp. DHA oil.

15.4 ROLE OF ALGAE IN WASTEWATER TREATMENT

Wastewater is a general term used to represent the water with poor quality that contains high proportion of pollutants and microbes. If wastewater is discharged into the water bodies, it can be hazardous to environment and health problems to human beings and animals too who will consume that water. Wastewater treatment is the major step toward reduction of pollutant and other substances that leads to further contamination of wastewater.

There are three steps in wastewater treatment:

- Primary treatment,
- Secondary treatment, and
- Tertiary treatment.

Primary treatment involves removal of solids, oil, and greasy waste from wastewater. Secondary treatment is also known as biological treatment and it involves exploitation of microorganisms for the elimination of chemicals present in wastewater. Tertiary treatment is the last step that involves elimination of microbes from wastewater before discharging into the river.

Domestic wastewater treatment and remediation is an expensive process and it needs significant time and planning for its success. Effluents produced from the secondary treatment plant contain more amount of nutrients like nitrogen and phosphorus and if these effluents are discharged into water bodies then they cause eutrophication and also influence the environment severely. Removal of these effluents is the most important task and it is an expensive process and causes sludge production. In developing countries, wastewater treatment is done by different methods. Waste stabilization of ponds, or lagoons, provide an ideal solution for wastewater treatment in developing countries and rural areas. In these ponds, there is a symbiotic relationship between bacterial consortiums and photoautotrophic microalgae.

There are several factors affecting the nutrient removal efficiency of immobilized microalgae:

- number of microalgae immobilized beads inoculated to treat wastewater,
- gel permeability for the nutrients present in the wastewater,
- cell density present in each bead,
- environmental factors,
- chemical components present in wastewater, and
- type of gel matrix used for immobilization.

Wastewater treatment by microalgae is an eco-friendly and inexpensive way of nutrient removal and biomass production. The microalgae grown in wastewater can be used as energy source, fertilizer, and fine chemicals production and as feed to animals. *Chlorella* sp. *Chlorella sorokiniana*, and *Chlorella vulgaris* are capable of removing nitrogen and phosphorous from effluents. *Azospirillum brasilense* is helpful because of its capacity of nutrient removal.

BOX 3: PHYCOREMEDIATION

The use of microalgae or macro algae to remove pollutants and nutrients from the wastewater is called phycoremediation.

15.5 ALGAE AS A BIOFERTILIZER

A biofertilizer is a substance that contains living microorganisms and is applied to seed, plant surfaces or soil, colonizes and promotes growth by increasing the supply or availability of primary nutrients to the host plant. Biofertilizers add nutrients through the natural process of nitrogen fixation, solubilizing phosphorus, and stimulating plant growth through the synthesis of growth-promoting substances. Biofertilizers belong to algal, fungal, or bacterial groups and are capable of improving chemical and biological characteristics of soil. Chemical fertilizers and pesticides are having side effects.

Azobacter, Azospirillium, blue–green algae, *Azolla, micorrhizae*, and *sinorhizobium* are used to benefit the crop production. *Azobactor,* blue green algae, *Rhizobium,* and *Azospirillium* are capable for the nitrogen fixation. *Rhizobium bacteria* are used to increase the nitrogen fixation capacity in leguminous plants. Blue green algae like *Nostoc, Tolypothrix,* and *Aulosira* helps in atmospheric nitrogen fixation and hence enrich the soil fertility. In the millets, sorghum, sugarcane, maize, and wheat field, *Azospirillium* is used. Cyanobacteria or Blue green algae are a group of microorganism, which plays an important role in the maintenance and buildup of soil fertility as they can fix atmospheric nitrogen. They help to improve the soil's aeration and water holding capacity. They also add biomass after their decomposition. *Azolla* is an aquatic fern, found in symbiotic association with Blue green algae. They are found in small and shallow water bodies as well as in rice fields. Biofertilizers are environment friendly and they do not damage the environment like chemical fertilizer does. They are cheap and can be used by farmers easily. They help in the improvement of soil fertility with time. Microorganisms help in the conversion of complex organic material into simple one. So, plant can easily take up the nutrients. They increase the phosphorus content of soil. They help in increasing the crop yield.

15.6 ALGAE AS A SOURCE OF ENERGY

Algae have the potential and it act as an important biological fuel. Biofuel is actually fuel derived from a renewable biological source, just opposite to fossil based fuel like oil, coal, and natural gases. Biofuels have advantages over fossil fuels, as they are renewable, they reduce dependence on nonrenewable fossil fuels and as they are produced domestically.

Biofuels burn cleanly, reduces greenhouse gas emission, particulates and finally reduces addition of pollutants in environment. There are a number of advantages in using algae as a biomass source such as corn, or oil palm. Algal cells are having a few advantages as it is having very rapid growth approximately thirty times faster than any terrestrial plant and produces more biomass in limited area. It is utilizing nutrients efficiently than any other plant groups. They are having faster growth rate so yield high within a short time (higher oil production).

15.6.1 TYPES OF ALGAL BIOFUELS

1. Bio oil,
2. Hydrogen gas,
3. Bio mass, etc.

1. Bio oil is mainly produced by thermochemical conversion and it is a process by which biomass can be converted into different types of fuels like char, oil, and gas in the absence of oxygen and at high temperature. The bio oil thus formed is composed of all the organic compounds in the algae like lipids, proteins, fibers, and carbohydrates and that is why it gives a higher yield as compared with a normal algae cells. Production of bio oil can be subdivided into pyrolysis and thermo chemical liquefication. Microalgae harvested from lakes are used to produce bio oil by fast pyrolysis and it also causes reduction of algae blooms. Oil obtained by pyrolysis has better chemical properties than any other oil sources like lignocelluloses.

2. Hydrogen gas is the most efficient and the cleanest burning fuel. Hydrogen is produced from different source of energy like fossil fuel, natural gas, and coal. Industrial hydrogen production includes coal gasification and steam reforming and it requires very high energy inputs. Algae can be grown to produce biological hydrogen or bio hydrogen. Algae *Chlamydomonas reinharditi* (a single cell green algae) is used for the production of hydrogen. Many experimental evidences have shown that algae culture medium, having poor sulfur content, will switch from the production of oxygen to the production of hydrogen. Hydrogensase is responsible for this process and it loses this function in the presence of oxygen. *Chlamydomonas moeursii* is also a good strain for the production of hydrogen.

3. Biomass: algal biomass obtained from microalgae mainly and it is dried alga and marketed in the form of powders, tablets in human food market. They mainly belong to the genus of *Spirulina, Chlorella, Dunaliella, Nostoc,* and *Aphanizomenon*.

15.7 HIGH VALUE-ADDED PRODUCTS FROM ALGAE

High value products are any product that when multiplied by its fraction in the algal biomass has a value of over US$10/kg (the present commercial production costs for microalgae) example small molecules and polymers.

15.7.1 SMALL MOLECULES

Algae can be used to generate some important hydrocarbons, ethanol, biogas, or hydrogen. Substantial progress has been made in strain identification, lipid extraction, and fuel refining.The L fraction (a lignin related fraction) was suggested as a component for making adhesives, plastics, pharmaceuticals or pesticides. The microalgae *Dunaliella salina* can contain up to 40% of its dry weight as glycerol.

15.7.2 POLYMERS

Colloids such as alginates, carrageenan, or agars, were used since long. These polymers are either located in the cell walls or within the cell serving as a storage material. Marine algae mostly contain sulfated polysaccharides in their cell walls. Other polymers like cyanophycin, multi-L-arginyl-poly-L-aspartic acid are also produced by cyanobacteria.

a) **Hydrocolloids:** They are group of phycocolloid polymers. They include the alginates, carrageenans and agar. They together constitute the major industrial products derived from algae. Macro-algae are the raw materials for the hydrocolloids production however certain land plants can also produce polymers with similar properties. They are used in food and industrial products to thicken, emulsify, and stabilize. Hydrocolloids dissolve in warm water get liquefy and after cooling it forms gel, which is used for various applications.

b) **Ulvan:** It is a group of polymers that can be extracted from the cell walls of green seaweeds of family ulvales. Ulvan are composed of repeating sequences of rhamnose, glucusonic acid, iduronic acid, xylose, and sulfate.

c) **Colorants**: Carotenoids are produced by microalgae approximately 40 carotene and xanthophylls were isolated and characterized. Beta-carotene is found in almost all algal spp as well as in other plants. Other carotenoid is canthaxanthin, zeaxanthin, and lycopene. Phycobilins or phycobiliproteins are used as fluorescent markers in cell and molecular biology. They are water soluble accessory pigments. Phycobilins are also used as colorants for food and cosmetic products. Blue phycobilin obtained from *Arthrospira* is widely used as cosmetics color.

15.8 ALGAE AS FODDER FOR FISH AND ANIMALS

Algae are the most available primary producers in the aquatic environment all through the seasons for fish that take them, an increase in the algal flora directly food content to the fish. The addition of super phosphate fertilizers in adequate concentration in a fish pond to increase the algal flora, increase the nutritional quantity to the algae. So addition of super phosphate fertilizers is mostly used in experimental fish ponds. Many different algae play a vital role in aquaculture. On adding microalgae to larval fish culture tanks, a number of benefits can be seen like bumping against the wall of tanks.

Some examples algae used as fish feed are as follows:

a) **Azolla:** It is a floating aquatic macrophyte (plants that floats on water surface usually with submerged roots) has very high protein content (from 19% to 30%). It is also rich in essential amino acid (e.g., lysine) which is required for animal nutrition. It is also used as livestock, poultry, and fish farmers.

b) **Duckweeds:** they are small (1–5 cm) free floating aquatic plants distributed widely. They serve as nutrient pumps, also reduce the eutrophication effects and provide oxygen from their photosynthesis activity. Duckweeds accumulate large amounts of minerals in their tissues. They are also rich source of nitrogen, phosphorus, potassium, and calcium. They are also used as a source of food for fish in fresh form or in combination with other feed ingredients. They are also feed as a dried meal ingredient in pelleted diets.

c) **Water hyacinths:** *Eichhornia crassipes* is commonly known as water hyacinth and it is a floating aquatic macrophytes consisting of long, pendant roots, rhizomes, stolons, leaves, inflorescence, and fruit clusters. It is widely used as fodder for domestic animals. It is also a good food for fishes, used to produce biogas, removes heavy metals and phenols from polluted waters to protect fish from dying. It has very high moisture content. Its components are hemi-cellulose (22–44.4%), cellulose (17.8–31%), lignin (7–26.36%), and magnesium (0.17%).

A few species like *Rhodymenia palmata* are used as food for sheep in Narvey. *Laminaria saccharina, Pelvitia, Ascophyllum*, etc. species are used as food for cattle. *Palmaria palmate* (a red alga) is rich in protein and harvested for food. It was appreciated by sheep and goats in Gotland (Sweden) and by cows in Brittany (France).

15.9 CONCLUSION

It is clear that algae are an up-coming solution for a vast array issues. Furthermore, algae are not only a solution to health problem but it also promotes regular healthy functioning of body. They help to increase immune system, maintains healthy functioning of cardiovascular system, respiratory system, nervous system, etc. Their use as food make it easier to have a healthy diet enriched with vitamins, minerals, and other important nutrients. Algae provide a natural way to treat wastewater at a low price and are an effective method than conventional methods. Bio fertilizers have various benefits. Accessing nutrients for current intake as well as residual, different bio fertilizer also provides growth-promoting factors to plants and some have been successfully facilitating composting and effective recycling of solid waste. Beside all these algae provide various benefits; it is a good source of energy, a natural colorants, cosmetics, etc.

The cultivation of algae with so many benefits needs to be promoted and improved using modern techniques.

KEYWORDS

- **algal cell**
- **algae as fodder**
- **as medicine**
- **sewage treatment**
- **pollution control**

REFERENCES

1. Lee, R. E. *Phycology;* Cambridge University Press, 2008.
2. Nabors, Murray W. *Introduction to Botany;* Pearson Education, Inc.: San Francisco, CA, ISBN 987-0-8053-4416-5, 2004.
3. Lewis, Charlton T.; Chales Short. *Alga;* Clarendon Press: Oxford, ISBN 978-0-19-864201-5, 1879.
4. Mumford, T. F.; Miura, A. Porphyra as Food: Cultivation and Economic. In *Algae and Human Affairs;* Lembi, C. A.; Waaland, J. R., Eds., Cambridge University Press. ISBN 978-0-521-32115-0, 1988; pp 87–117.

CHAPTER 16

FUNCTIONALITY FEATURES OF CANDELILLA WAX IN EDIBLE NANOCOATINGS

OLGA B. ALVAREZ-PEREZ[1], MIGUEL ÁNGEL DE LEÓN-ZAPATA[1], ROMEO ROJAS MOLINA[1], JANETH VENTURA-SOBREVILLA[1], MIGUEL A. AGUILAR-GONZÁLEZ[2], and CRISTÓBAL NOÉ AGUILAR[1,*]

[1]*Food Research Department, Group of Bioprocesses and Bioproducts, School of Chemistry, Universidad Autonoma de Coahuila, 25280 Saltillo, Coahuila, México*

[2]*Center for Research and Advanced Studies of the National Polytechnic Institute, CINVESTAV-IPN, Unit Saltillo, Coahuila, México.*

Corresponding author. E-mail: cristobal.aguilar@uadec.edu.mx

ABSTRACT

The use of lipids in the development of edible coatings is a viable alternative to conserve natural and fresh products, allowing the extension of the shelf life quality without causing any harm to the environment, compared to films prepared from synthetic plastics. By this reason, the most important lipid in this application is the wax, because it possesses the necessary properties to perform this function. Since its nature, the fruits tend to have a layer of wax that confers protection against pathogens, environmental, pests, etc. In this review, we report the importance of candelilla wax as a natural resource from the Mexican semi-arid region, as well as their uses in the development of edible coatings, because it is a biodegradable and edible material approved by the FDA. Its importance as an essential component of an edible coating made with this type of biomaterial is based in its action as a plasticizer. Also, we describe some reports of the development of nutraceutical edible coatings formulated with the addition of natural phenolic antioxidants.

16.1 INTRODUCTION

The concern for the mass production of waste from packaging materials has meant a decisive shift in the vision toward the use of biodegradable materials, especially those coming from agricultural surpluses,[43] due to the growing demand by consumers of food made from natural products, has led to the innovation of new conservation technologies at agro-industrial level, which help prolong the shelf life of fresh fruit. However, the use of natural coatings cannot, nor intends to, replace the usage of traditional packaging materials, but it is necessary to take into account its functional characteristics and the possible advantages of behavior in certain applications.[2]

The technical challenges involved in producing food and preserving them with stable quality, indicate that the use of this type of coatings will be greater than what it currently is; however, despite the fact that the technical information available for the elaboration of edible covers is wide, it is not universal for all products, which implies a challenge for the development of specific coatings and films for each food.[35] In the particular case of fruits and vegetables for fresh consumption, the edible coatings provide an additional protective cover whose technological impact is equivalent to that of a modified atmosphere, which consists of a thin and continuous layer, made of materials that can be ingested, and provides a barrier to moisture, oxygen and solutes, this can completely cover the food or can be placed in the components of the product[32] and must ensure the stability of the food and prolong its useful life.

According to the conditions of storage of fruits and vegetables should be considered some factors whether mechanical or chemical involved in the design of films,[32] therefore represent a storage alternative for products that can be consumed in fresh.[35] The edible coatings that are being tested in post-harvest are mixed formulations of lipid compounds and hydrocolloids, whose main priority is the preservation and protection of products of plant origin, against microbial contamination generated during manipulation.[28] The use of candelilla wax has been reported as a basic component in the elaboration of edible coatings evaluating different factors involved in the conservation of fresh fruits. The candelilla wax is extracted from the wild plant *Euphorbia antisyphilitica* Zucc., it represents one of the main biopolymers that present biodegradable and edible properties, its structure is amorphous and its hardness is of an intermediate degree between that of the wax of carnauba and that of Bee.[33]

Candelilla wax is considered a GRAS substance by the FDA, so it has multiple applications in the food industry,[18] that is why it is used for the elaboration of edible roofs, providing one of the best barriers permeable to moisture and gases that are product of the metabolism of the fruit. Candelilla is a perennial plant that develops in semi-desert climates, almost devoid of leaves. The candelilla is one of the plants that grow in the wild with greater number of applications of use. It is reproduced by both aerial and underground stem shoots and seed. The collection of the candelilla plant for the production of natural wax has been one of the most important economic activities in five states of the Mexican Republic. It is currently being used in more than 20 different industries around the world, mainly in the United States, the European Union, and Japan. Its distinctive properties confer on it the category of raw material essential for the manufacture of cosmetics, inks, paints, adhesives, coatings, brighteners and polishes, electrical insulators, integrated circuits, chewing gums, fruit coatings for export purposes, thinners and hardeners of other waxes, candles against insects, among others (National Forestry Commission 2009).

At the pharmaceutical level, the candelilla is recognized in several therapeutic properties. Currently in Mexico, there are research projects aimed at technological improvement in the process of extraction and purification of timber and non-timber forest products with high commercial value, in particular, the Department of Food Research of the Faculty of Chemical Sciences of the Autonomous University of Coahuila have reported the elaboration of edible roofs with the addition of natural antioxidants produced by fungal fermentation in solid medium, they have presented positive results for the conservation of fresh fruits,[38] providing a barrier against pathogenic microorganisms that may possibly damage the fruit[39] and give the consumer a health benefit, thanks to the presence of antioxidants, which have anti-tumor, anticarcinogenic, and anti-cancer properties.[38]

16.2 NATURAL WAXES USED IN THE FORMULATION OF EDIBLE NANOCOATINGS

The use of waxes to cover the fruits by immersion, is one of the oldest methods, practiced for the first time in China,[26] since the early 12th century,[30] retarding perspiration in lemons and oranges, as they are more effective in blocking the migration of moisture, specifically during seasonal changes and continues to be used in other types of fruits,[26] being the candelilla one of the most resistant.[9]

Recently, De Leon-Zapata et al. (2018) demonstrated that emulsions of candelilla wax can be nano-structures used to prolong the shelf life of apples at industrial level. Natural waxes applied to fresh perishable products to reduce respiration are: beeswax, carnauba wax, candelilla wax, and rice bran wax,[9] also paraffin waxes are some of the waxes prepared and used in the elaboration of edible coatings, which are also used as microencapsulation agents, specifically for substances that provide fruit smells and flavors.[43] Edible waxes are significantly more resistant to moisture transport than most other lipid or non-lipid films,[9] in addition to preventing the softening caused by enzymatic hydrolysis of plant cells and membrane components during the cutting process,[44] however, it is important that wax covers in fresh or perishable fruits is not completely waterproof, which causes anaerobes favoring the physiological disorders that shorten the half-life.[45] Waxing is a conservation technique widely used by marketers, supermarkets and exporters in the world, whose method generates a barrier of protection between the product and the environment to prevent the fruit from breathing less or deteriorate faster, this wear is characterized by the loss of moisture or dehydration of horticultural products and is a deterioration factor so we must try to maintain an optimum quality of the product.[45]

16.3 CANDELILLA WAX (*Euphorbia antisyphilitica Zucc.*)

The candelilla wax is extracted from the wild plant *Euphorbia antisyphilitica* Zucc., which is formed by esters of long-chain fatty acids that create a protective surface in the plant.[38] It is insoluble in water, but highly soluble in acetone, chloroform, benzene, and other organic solvents.[33]

It presents a wide variety of applications, being currently used in more than 20 different industries around the world (Table 16.1).

Its distinctive properties confer on it the category of essential material for the manufacture of cosmetics, inks, adhesives, coatings, emulsions, polishes, and pharmaceutical products. In the cosmetics industry, for being a good plasticizer,[11] it is also used in the manufacture of chewing gum, in the smelting, molding industry, in manufacturing various products in the electronic and electrical industries. There are many other applications where it is currently used, including cardboard coatings, crayon manufacturing, paints, wax candles, lubricants, paper coatings, anticorrosives, waterproofing, and Fireworks.[11] Candelilla's wax is recognized by the Food and Drug Administration of the United States of America (FDA), as a natural safe-GRAS substance, generally recognized as safe-for application in the food industry, therefore it is widely used in various sectors of the branch.[18] Because candelilla wax is an edible wax, it is being used for the elaboration of natural coatings that can retard the ripening and ageing of fruits and vegetables, maintaining a controlled atmosphere on the exterior surface, which allows the protection of the product in the face of environmental, transport and storage conditions.[33]

TABLE 16.1 Commercial Importance of Mexican Candelilla Wax.

Importing country	Imported candelilla wax (2016, USD)	Increase in purchase in the last 5 years (%)
Germany	11,000.00	23.3
China	9900.00	11.0
Italy	5600.00	23.6
France	5200.00	10.0
Spain	3800.00	27.5

Source: International Trade Center.

16.4 EFFECT OF CANDELILLA WAX AS PLASTICIZER ON THE FUNCTIONALITY OF AN EDIBLE COATING

Plasticization is a very important factor in the formulation, since they affect the mechanical properties[25] and physical of the coating (elasticity, flexibility, permeability, and wettability),[35] because they alter the mobility of the chain, the diffusion coefficients of gas or water and the structure of the films,[25] reducing the intermolecular forces between the polymer chains and increasing the free volume,[36] consequently, there is more space for water molecules to migrate, as well as hydrophilic plasticizers such as glycerol, are compatible with the polymeric material that forms the film and increase the absorption capacity of polar molecules such as water.[17] The increase in permeability with the plasticizer content may be related to the hydrophilicity of the plasticizer molecule,[5] because the permeability to the water vapor increases as the plasticizer content increases, however up to 30% glycerol content, that increase is relatively mild, later observed a more pronounced increase.[8]

Cellulose-based coatings, are very efficient barriers to the permeability of oxygen and their property of barrier to the water vapor,[29] these can be improved with the addition of lipids as plasticizers,[29] since they generally increase the permeability of the same.[36] The application of a lipid layer on the surface of fruits replaces the natural waxes of the cuticle, which may have been partially removed during washing.[9] The edible wax covers of candelilla have different functional properties, because when mixed with oils and polymers of high molecular weight as natural gums, they have an effect on the fruit to be coated, avoiding weight loss,[8] the use of candelilla wax on combined edible roofs has been amply evidenced by Bosquez-Molina et al., who demonstrated that the covers with this material and rubber mesquite create a modified atmosphere inside the fruit, to retard the process of maturation and senescence in a way similar to that of a controlled atmosphere that is much more expensive,[8] also avoids an increase in the production of ethylene and the hauling of additives that retard the discoloration and microbial growth (Gahouth 1991),[2] allow to control the respiration of the product, providing better permeability and texture, since it modifies the mechanical properties; Fulfilling the function of Plasticizer,[3] weakening the intermolecular forces between the polymer chains, increasing the flexibility of the coating.[8] The use of a hydrophilic or hydrophobic plasticizer will produce a coating with similar characteristics (Kester et al. 1986). Lipids such as candelilla wax are good plasticizers, so this natural wax product has shown its advantages over most synthetic waxes used in this industry,[11] they are compounds of low volatility and function as plasticizers, which are added to the coating,[28] considering themselves two forces, one among the forming molecules of the film, called cohesion and another in the coating and substrate, called adhesion,[24] in order to reduce the fragility, increasing the flexibility, hardness, and resistance to cutting, as they decrease the intramolecular forces of the polymer chains, thus producing a decrease in the strength of cohesion, tension and in the vitreous transition temperature.

Candelilla wax covers, among others, can be used as a support when adding preservatives or other additives on the surface of foodstuffs, mainly fresh fruits and vegetables to prolong periods of post-harvest storage, which consist of an emulsion made of waxes and oils in water, which are sprinkled in the fruits to improve their appearance, brightness, color, softness, control its maturity, and retard the loss of water.[15] The emulsion originated in the elaboration of an edible cover based on candelilla wax must present an adequate homogenization of the system and in this way guarantee the uniformity in the size and distribution of the particles of the dispersed phase,[8] as it will be reflected in the final barrier properties such as water vapor permeability and gases.[8] It is important to know the volumetric fraction of the dispersed phase of the wax emulsion of candelilla, as it influences much on the appearance.

16.5 FUNCTIONAL EDIBLE COATINGS BASED ON CANDELILLA WAX SUPPLEMENTED WITH PHENOLIC ANTIOXIDANTS

Additives are used to impart mechanical, nutritional, and organoleptic properties to edible roofs,[42] these may be of the plasticizer type (polyhydric alcohols, oils, and fatty acids), surfactant and emulsifier type (fats, oils, and polyethylene glycol) chemical preservatives (benzóico acid, sodium benzoate, sorbic acid, potassium sorbate, and propionic acid),[2] as well as antimicrobial agents, antioxidants, dyes, flavorings, and calcium as a firming agent of cell membranes, among others,[42] can be applied to control and modify surface conditions, reducing some of the degrading reactions.[13,19] The maintenance of microbial stability can be obtained using edible coatings with antimicrobial action and combined with refrigeration and controlled atmosphere. For fruits, waxes are usually used with addition of sorbic acid and sorbates as antifungal[13,19]

The influence of a given additive will depend on its concentration, chemical structure, degree of dispersion in the film and degree of interaction with the polymer.[28] Some chemicals and natural products are used as antioxidants or as microbicides; ascorbic acid, citric, and lemon juice are used as antioxidants and salts of 5–acetyl–8–hydroxyquinoline or strong inorganic acids such as H_2SO_4 or H_3PO_4 are used as microbicides with very good results. The use of these substances does not prevent or retard the maturation so, other methods should be used together for this purpose, such as the application of gamma radiation that modifies the ripening time of the fruit according to its dose; however, some defects have been detected, such as the darkening of the pulp becoming coffee or by contrast discoloration. Good results have been reported when radiation is less than 7 J/kg.[20]

Zhang et al.[46] reported that the use of ascorbic acid, isoascorbic acid, and acetyl cysteine reduced darkening in litchi.[10] They used combinations of 4–hexylresorcinol, isoascorbic acid, $CaCl_2$, and acetyl cysteine to reduce changes in apple slices. Luo et al.[31] controlled darkening in apple slices using 4-hexylresorcinol mixed with ascorbic acid. Baldwin et al.[2] observed better protection with the addition of ascorbic acid in edible roofs. Ruiz-Cruz et al.[37] showed the positive effect of different antioxidants (independent and mixed) in inhibiting the darkening of cut fresh pineapple, because when antioxidants are used in combination with other technologies: treatments with heat, modified and controlled atmospheres, edible covers, gamma radiation, and electromagnetic pulses the darkening in fruits is inhibited.

Saucedo-Pompa et al.[38] reported for the first time the use of gallic acid, ellagic acid, and aloevera in the formulation of an edible cover based on candelilla wax, as anti-darkening additives of fruits, showing excellent results, even the controls with the cover without antioxidants were better compared with the fruits without edible cover.[38] The addition of ellagic acid and aloevera as natural antioxidants to the edible candelilla wax cover showed positive results when applied to fresh fruits, this due to the protective barrier that represents the cover of wax of candelilla as physical barrier of the coating, allowing greater control of gases, greater permeability and therefore the better control in the respiration of the fruit and in turn the addition of antioxidants, they intervene inhibiting microbial growth, as well as the possibility of providing a benefit to the health of the consumer, due to the anti-cancer and anti-tumor properties that present this type of antioxidants.[38] Saucedo-Pompa et al.[39] reported that when applying the edible cover based on candelilla wax with the addition of ellagic acid and aloevera to freshly cut avocados decreases the aqueous activity of the fruit, these results indicate that for freshly cut fruits reduction in weight loss is very important and the use of edible covers carrying natural antioxidants is an excellent tool to control weight loss[1,21,23] The apple slices with the edible covers with ellagic acid and aloevera kept to a greater extent the initial firmness, these covers had a protective effect on the firmness of freshly cut fruits. The texture of freshly cut fruits was improved with the application of edible candelilla wax covers, these results are similar to those reported by Ghaouth et al.,[22] who applied edible covers in tomato. Saucedo-Pompa et al.[39] reported antifungal properties of ellagic acid and aloe vera as part of the formulation of a candelilla wax-based edible cover which was applied in avocados, its functionality allowed to increase the resistance to the invasion of common phytopathogenic fungi and prolong its shelf life, improving the physical and chemical quality of the product. According to the results obtained by Saucedo-Pompa et al.,[39] the concentration of antioxidant influences the speed of water loss, as it is directly proportional to the increase in the concentration of antioxidant.

At the beginning of the new millennium, a new era in the area of food and nutrition sciences has become increasingly intense: the area of food–medicine interaction increasingly recognized as the "functional foods" that accepts the role of food components, as essential nutrients for the maintenance of life and health and as non-nutritional compounds but that contribute to prevent or retard the chronic diseases of the ripe age. Initially regarded as a passing curiosity, the idea of food formulation based on health benefits that its non-nutritional components could provide to the consumer, it has become an area of great interest today for large food companies,[6] so the addition of natural antioxidant additives from the group of phenols such as ellagic acid and aloevera to an edible cover, they represent a viable alternative for the development of nutraceutical roofs and enter in the field of functional foods.[37] In addition to offering a novel and comfortable presentation for the consumption of antioxidants by consumers and the VES improve the quality of shelf life in fruits[37] and avoid losses, by the attack of microorganisms, this due to the antifungal and antibacterial activity of the antioxidant additives.[38]

16.6 PHENOLIC COMPOUNDS AS NATURAL ANTIOXIDANTS AND THEIR FUNCTIONAL ACTIVITY IN THE EDIBLE COATING

Phenols are also antioxidants and as such trap free radicals, preventing them from joining and damaging the molecules of deoxyribonucleic acid (DNA), a critical step in initiating carcinogenic processes, they also prevent lipid peroxidation, which, being free radicals can cause structural damage to normal cells, interfering with the transport of molecules through these membranes affecting cell growth and proliferation.[41] These phytonutrients, include a large group of compounds that have been subject to extensive research as preventive agents of diseases, which protect plants against oxidative damage and carry the same function in the human organism, whose main characteristic is its ability to block the action of specific enzymes that cause inflammation, but also modify the metabolic steps of prostaglandins and thus protect the agglomeration of platelets, according to data obtained from experimental studies, it seems that there are some possible mechanisms for the action of phenols, inhibiting the activation of carcinogens and therefore block the initiation of the process of carcinogenesis.[27] Phenolic compounds come from barks, stems, leaves, flowers, organic acids present in fruits, and phytoalexins produced in plants,[7] such as caffeic, chlorogenic, p-coumaric, hydroxicinnamic, cinnamic, ferulic, and quinic acids that are present in plants which are used as spices, these acids present antimicrobial activity so they can retard the rot of fruits and vegetables, in fact it has been shown that tannins and tannic acid also present antimicrobial activity.[7] Antimicrobial additives can inactivate essential enzymes, reacting with the cell membrane or altering the function of the genetic material, it has been observed that the pH and temperature affect the antimicrobial activity of these compounds.[14]

The antioxidant compounds prevent the negative effects of free radicals on tissues and fats, reducing the risk of cardiac disturbances by avoiding the oxidation and cytotoxicity of LDL in vitro, decreasing the aterogenicidty.[4,40] Vitamins C, E, and β-carotene that prevent the oxidation of the LDL fraction of cholesterol, reduce the risk of coronary alterations, as well as possessing anti-cancer properties, whose protection measure consists of increasing the ingestion of fruits and vegetables, as well as foods containing antioxidant nutrients to protect from oxidation to LDL mentioned and thus avoid its oxidative modification and atherogenic formation.[40]

Biomaterials such as candelilla wax have had a significant role in the coating of food, as it is a natural and biodegradable material, so it does not harm the environment besides being a natural safe-GRAS, generally recognized as safe-recognized by the FDA, for its application in the food industry and can be used as an alternative in the elaboration of natural edible covers, reducing the use of synthetic polymers as derivatives of petroleum, which, when discarded, present a slow degradation in the environment compared with biomaterials, representing serious pollution problems. The edible wax covers of candelilla, allow controlling the respiration of the product, providing better permeability and texture, since it modifies the mechanical properties; fulfilling the function of plasticizer. In addition, additives such as nutrients, dyes, antioxidants, antimicrobials can be added, which, as additives, make the edible cover a functional food.

At present, the realization of edible roofs is of lesser proportion compared with the elaboration of synthetic roofs, but the main advantage of edible covers made with biomaterial such as candelilla wax, are easy to produce and quickly biodegradable, so it is a natural technique of conservation of fresh fruits.

It is important to consider the elaboration and commercialization of edible candelilla wax covers with the addition of antioxidants such as ellagic acid and aloe vera for the conservation of fresh fruit products in post-harvest, since it will allow to extend its shelf life avoiding microbial contamination, due to the antioxidant potential of ellagic acid and aloe vera, in addition to providing a benefit to the health of the consumer thanks to the anti-tumor characteristics, anticarcinogenic, and anti-cancer agents that present these antioxidants of natural origin.

ACKNOWLEDGMENTS

This work is part of the project "design of a high-performance process in the extraction of high quality candelilla wax and formulation of products of final use from the prepared wax," financed by the project CONAFOR-CONACYT-S0002-2008-C01-91633.

KEYWORDS

- **coatings**
- **wax**
- **candelilla**
- **biodegradable**
- **edible**

REFERENCES

1. Báez, R.; Bringas, E.; González, G.; Mendoza, T.; Ojeda, J.; Mercado, J. Comportamiento postcosecha del mango 'Tommy Atkins' tratado con agua caliente y ceras. *Proc. Interamer. Soc. Trop. Hort.* **2001,** *44*, 39–43.

2. Baldwin, E. A.; Nisperos-Carriedo, M. O.; Baker, R. A. Use of Edible Coatings to Preserve Quality of Lightly (and Slightly) Processed Products. *Crit. Rev. Food Sci. Nut.* **1995,** *35*, 509–524.
3. Baldwin, E. A.; Nisperos-Carriedo, M. O.; Hagenmaier, R. D.; Baker, R. A. Using Lipids in Coatings for Food Products. *Food Technol.* **1997,** *51* (6), 56–61, 64.
4. Bello, J. Principales ámbitos clínicos de aplicación de los alimentos funcionales o nutracéuticos. *Alimentación Equipos y Tecnología* **1997,** *16*, 43–48.
5. Bertuzzi, M. A.; Armada, M.; Gottifredi, J. C.; Aparicio, A. R.; Jiménez, P. *Estudio de la permeabilidad al vapor de agua de films comestibles para recubrir alimentos*; Congreso Nacional de Ciencia y Tecnología: Buenos Aires, Argentina, 2002, pp 220.
6. Best, D. All Natural and Nutraceutical. *Prep. Foods* **1997,** *166*, 32–38.
7. Beuchat, L. R. Control of Food Borne Pathogens and Spoilage Microorganisms by Naturally Occurring Antimicrobials, Chapter 11. In *Microbial Food Contamination*; Wilson, C. L., Droby, S., Ed.; CRS Press: London, UK; 2001, 149–169.
8. Bósquez-Molina, E.; Vernon, E. J. *Efecto del Glicerol, Sorbitol y Calcio en la Permeabilidad al Vapor de Agua de Películas a Base de Goma de Mezquite. XXV Encuentro Nacional AMIDIQ. Resúmenes (ALI-21)*; 4-7 Mayo: Puerto Vallarta, México, 2004.
9. Bósquez-Molina, E.; Vernon-Carter, E. J. Efecto de Plastificantes y Calcio en la Permeabilidad al Vapor de Agua de Películas a Base de Goma de Mezquite y Cera de Candelilla. *Revista Mexicana de Ingeniería Química* **2005,** *4* (002), 157–162.
10. Buta, G. J.; Moline, H. E.; Spaulding, D. W.; Wang, C. Extending Storage Life of Fresh-cut Apples Using Natural Products and their Derivates. *J. Agric. Food Chem.* **1999,** *47*, 1–6.
11. Cenamex (Ceras naturales mexicanas SA de CV), 2009 http://www.cenamex.com.mx
12. Centro Internacional de Comercio, 2009. www.intracen.org
13. Cuq, B.; Gontard, N.; Guilbert, S.; Edible Films and Coatings as Active Layers. In: *Active Food Packaging*; Rooney, M. L., Ed.; Blackie Academic & Professional: London, 111–135.
14. Davison, P. M. Chemical Preservatives and Antimicrobial Compounds. In: *Food Microbiology and Fronteirs*; Doley, M. P., Beuchat, L. R., Montville, T. J., Eds;ASM Press: Washington DC, 1997, 520–566.
15. Debeaufort, F.; Quezada-Gallo, J. A.; Voilley, G. Edible Films and Coatings: Tomorrow´s Packagings: A Review. *Crit. Rev. Food Sci.* **1998,** *38* (4), 299–313.
16. De León-Zapata Miguel, A.; Ventura-Sobrevilla Janeth, M.; Salinas-Jasso Thalia, A.; Flores-Gallegos Adriana, C.; Rodríguez-Herrera, Raul; Pastrana-Castro, Lorenzo; Rua-Rodríguez, María Luisa; Aguilar Cristóbal, N. Changes of the Shelf Life of Candelilla Wax/Tarbush Bioactive Based-nanocoated Apples at Industrial Level Conditions. *Scientia Horticulturae* **2018,** *231*, 43–48.
17. Fennema, O.; Donhowe, I. G.; Kester, J. J. Lipid Type and Location of the Relative Humidity Gradient Influence on the Barrier Properties of Lipids to Water Vapor. *J. Food Eng.* **1994,** *22* (1–4), 225–239.
18. FDA (U.S. Food and Drug Administration), 2009. http://www.fda.gov/
19. Fernandez, M. Review: Active Packaging of Foods. *Food Sci. Technol. Int.* **2000,** *6*, 97–108.

20. Fira. Situación y Perspectivas Económicas de la Producción de Aguacate en México; Banco de México, S. A; División de Planeación, 1997, 62–68.
21. Ghaouth, E. L.; Arul, J.; Ponnampalam, R. Use of Chitosan Coating to Reduce Water Loss and Maintain Quality of Cucumber and Bell Pepper Fruits. *J. Food Proc. Preserv.* **1991,** *15*, 359–368.
22. Ghaouth, E. L.; Ponnampalam, R.; Castaigne, F.; Arul, J. Chitosan Coating to Extend the Storage Life of Tomatoes. *Hort. Sci.* **1992,** *27*, 1016–1018.
23. Gonzales-Aguilar, G. A.; Monroy-Garcinia, I. N.; Goycoolea-Valencia, F.; Diaz-Cinco, M. E.; Ayala-Zavala, J. F. Cubiertas comestibles de quitosano. Una alternativa para prevenir el deterioro microbiano y conservar la calidad de papaya fresca cortada. Proceedings of the Simposium "Nuevas tecnologías de conservación y envasado de frutas y hortalizas. Vegetales frescos cortados" La Habana, Cuba, 2005; pp 121–133.
24. Guilbert, S.; Biquet, B. Technology and Application of Edible Protective Films. In *Food Packaging and Preservation.* Mathlouthi, M., Ed.; Elsevier: Londres, 1986.
25. Guilbert, S.; Gontard, N.; Morel, M. H.; Chalier, P.; Micard, V.; Redl, A. Formation and Poperties of Wheat Gluten Films and Coatings. In *Protein-based Films and Coatings*; Gennadios, A., Ed.; CRC Press: Boca Raton, FL, pp 69–122.
26. Hagenmaier, R. A Comparison of Ethane, Ethylene and CO_2 Peel Permeance for Fruit with Different Coatings. *Postharv. Biol. Technol.* **2005,** *37* (1), 56–64.
27. Hertog, M. G. Dietary Antioxidant Flavonoids and Risk of Coronary Heart Disease: The Zutphen Elderly Study. *Lancet* **1993,** *342*, 1007–1011.
28. Kester, J.; Fennema, O. Edible Films and Coatings: A Review. *Food Technol.* **1986,** *40*, 47–59.
29. Koelsch, C. Edible Water Vapor Barriers: Properties and Promise. *Trends Food Sci. Technol.* **1994,** *5*, 76–81.
30. Krochta, J.; Baldwin, E.; Nisperos-Carriedo, M. *Edible Coatings and Films to Improve Food Quality*; Technomic Publishing Company: New York, 1994; pp 1344.
31. Luo; Barbosa-Canovas, G. V. Inhibition of apple-slice browning by 4-hexylresorcinol. In *Enzymatic browning and its prevention*;America Chemical Society; Washington DC, USA, 240–250.
32. Miranda, M. Comportamiento de películas de Quitosán compuesto en un modelo de almacenamiento de aguacate. *Revista de la Sociedad Química de México* **2003,** *47* (4), 331–336.
33. Multiceras, 2009 http://www.multiceras.com.mx/pro-candelilla.htm
34. Park, H. J.; Chinnan, M. S. Gas and water vapor barrier properties of edible films from protein and cellulosic materials. *J. Food Eng.* **1993,** *25*, 497–507.
35. Park, H. J. Development of Advanced Edible Coatings for Fruits. *Trends Food Sci. Technol.* **1999,** *10*, 254–260.
36. Pascat, B. Study of some factors affecting permeability. In *Food Packaging and Preservation. Theory and Practice;* Mathlouthi, M., Ed.; Elsevier Applied Science Pub.: London, 1986; 7–24.
37. Ruiz-cruz, S.; Gonzáles-Aguilar, G. A. Efecto de agentes antioxidantes en atmósferas modificadas en la calidad de rodajas de piña fresca. CIAD. Tesis de maestría, Hermosillo, Sonora, México, 2002.

38. Saucedo-Pompa, S.; Jasso-Cantu, D.; Ventura-Sobrevilla, J.; Sáenz-Galindo, A.; Aguilar-Gonzales, C. N. Effect of Candelilla Wax With Natural Antioxidants On The Shelf Life Quality of Cut Fresh Fruits. *J. Food Qual.* **2007,** *30,* 823–836.
39. Saucedo-Pompa, S.; Rojas-Molina, R.; Aguilera-Carbo, A.; Saenz-Galindo, A.; De La Garza, H.; Jasso-Cantú, D.; Aguilar-Gonzales, C. N. Edible Film Based on Candelilla Wax to Improve the Shelf Life and Quality of Avocado. Food Res. Int. **2009,** *42,* 511–515.
40. Seelert, K. Antioxidants in the Prevention of Atherosclerosis and Coronary Heart Disease. *Internist Prax* **1992,** *32,* 191–199.
41. So, F. V.; Guthrie, N.; Chambers, A. F.; Moussa, M.; Carroll, K. K. Inhibition of Human Breast Cancer Cell Proliferation and Delay of Mammary Tumorigenesis by Flavonoids and Citrus Juices. *Nutr. Cancer* **1996,** *26,* 167–181.
42. Soliva-Fortuny, R. C.; Martín-Belloso, O. Evaluation of Zein Films as Modified Atmosphere Packaging for Fresh Broccoli. *J. Food Sci.* **2001,** *66* (8), 1108–1111.
43. Tharanathan, R. Biodegradable Films and Composite Coatings: Past, Present and Future. *Crit. Rev. Food Sci. Technol.* **2003,** *14,* 71–78.
44. Varoquax, P.; Lecendre, I.; Varoquax, M.; Souty, M. Changes in Firmness of Kiwifruit After Slicing (Perte de fermeté du fiwiaprés découpe). *Science des Aliments* **1990,** *10,* 127–139.
45. Villada, H.; Acosta, H. A.; Velasco, R. Biopolímeros naturales usados en empaques biodegradables. *Temas agrarios* **2007,** *12* (2), 5–13.
46. Zhang, D.; Quantick, P. Effects of Chitosan Coatings on Enzymatic Browning and Decay During Postharvest Storage of Litchi (*Litchi chinensis* Sonn.) Fruit. *Postharvest. Biol. Technol.* **1997,** *12,* 195–202.

CHAPTER 17

PHYSICAL–CHEMICAL PROPERTIES OF THE TABLETS AS THE MAIN FACTORS OF LOCAL INTERACTION AND MECHANISM OF ACTION OF DRUGS BY INGESTION

A. L. URAKOV[1,2,3,*], N. A. URAKOVA[2,3], and V. B. DEMENTIEV[1]

[1]*Udmurt Federal Research Center, Ural Division, Russian Academy of Sciences, Department of Modeling and Synthesis of Technological Processes, Izhevsk, Russia*

[2]*Institute of Thermology, Department of Drug Research and Development, Izhevsk, Russia*

3Izhevsk State Medical Academy, Department of General and Clinical Pharmacology, Izhevsk, Russia

[*]*Corresponding author. E-mail: urakoval@live.ru*

ABSTRACT

The definition and theoretical fundamentals of a new scientific trend—physical–chemical pharmacology are considered on the example of the mechanical, physical, chemical, and physical–chemical properties of modern pharmaceutical tablets. The theoretical, laboratorical, and experimental methods for the processes prognosis action of medications as well as for the local physical–chemical systems reactivity estimation are discussed. It is shown that modern tablets are significantly different from natural food clumps, which people swallow after chewing of bread. The fact is that today tablets are prepared by pressing, and therefore represent artificial stones. Proposed to change, the technology of production of tablets to tablet is fully consistent with artificial food lumps. The basic mechanical, physical, chemical, and physical–chemical parameters of natural food clumps are determined. The harmful effects influence of sizes, shapes, hardness values, solubility, osmotic, and acid activity tablets on the healthy and diseased teeth, different dental constructions, and on mucous membranes of the mouth and stomach is shown. It is shown that the danger of modern tablets for people due to their "wrong" mechanical, physical, chemical, and physical–chemical properties. It is shown that for the acquisition of high security, pharmaceutical tablets must be in the form of olive fruit, being soft and porous like chewed bread, and not be sinking like stones, but to be floating, like the foam.

17.1 INTRODUCTION

Currently, we are witnessing the formation of a new direction in the pharmacology associated with explanation of the mechanism of direct action of drugs on the basis of physical–chemical processes occurring during local interactions, "the pill" and "solutions" with various organs and tissues rather than "pure" chemicals.[1–7] It is shown that the first place in the mechanism of local action of finished drugs are the mechanisms of nonspecific local irritant action.[8–16] All this is easily explained by the fact that God did not create people for an introduction stones in the cavity of his mouth and stomach, and not for the introduction by injection of solutions of salts, acids, and alkalis in the blood, in skeletal muscle, in subcutaneous fat and in the skin.[3]

Let us consider modern tablets with the natural position of the advisability of introducing into the human body. It is very strange, but modern tablets are manufactured by compressing powders, as they were 100 years ago. Therefore, the tablets are artificial stones. Typically, these artificially made stones look like white or gray circular shaped disks. Pharmaceutical tablets are very similar to the chalk, clay, and/or a compressed lump of salt. Therefore, all modern tablets are harder than high quality food, so they sink in gastric juice. However, man is not a bird, whose stomach is designed to stones in order to grind solid food in the stomach in the absence of teeth in the oral cavity.

From the Bible, we know that God did not create people to swallow pills in the form of artificial stones. That is why the human stomach is not adapted to stones of any size and shape, including silica sand, gravel, and river pebbles and pressed pharmaceutical tablets. Moreover, people can remain healthy by ingesting a few pieces of sand grains, pebbles and crumbs of gravel the size of pharmaceutical tablets. However, a person cannot remain healthy, ingesting a comparable number of tablets of most modern medicines. On the one hand, modern tablets are as hard as rocks, so when you chew tablets a man can break the jaw, denture, tooth, crown, implant, clips, seals, and damage the gums, tongue, and palate. On the other hand, modern tablets are as aggressive as chemical reagents, salts, acids, and alkalis.[3]

Earlier, we proved that pills of most modern medicines can have overly strong and acidic hyperosmotic activity, so they can dehydrate and/or cauterize the cells of the mucous membranes and/or dissolved mineral salts of the tooth enamel. All this contributes to the development of drug iatrogenic in the form of gingivitis, stomatitis, gastritis, stomach ulcers, and caries.[10,15]

Due to the fact that we cannot stop the production of pharmaceutical tablets and cannot prevent a "pill epidemic" in society to us remains nothing how to look for ways of "survival" of people under the influence of pills–aggressors. It is therefore important to understand the real, not fantastic mechanisms of mechanical, physical, chemical, and physical–chemical aggressive actions pharmaceutical tablets on hard and soft tissues of the oral cavity and stomach.

The results of our research showed that many of these unfortunate consequences of tablets local action are easily detected with a thermal imager according to local hyperthermia.[8,14,17,18,19]

17.2 MATERIALS AND METHODS

In the laboratory were studied specific gravity, shape, color, weight, diameter, height, volume, acidity, and osmolarity of more than 50 tablets of various tablets by different pharmaceutical companies worldwide. Conventional methods and equipment for the pharmaceutical tablets for quality control were used for this purpose. In addition to the study, the generally accepted indicators of quality tablets were measured in hardness of tablets at the specific load by Rockwell hardness test method in Brunnel scale (in HB units).

In parallel experiments was studied the dynamics of food, food clumps, and pills movement inside the gastric cavity, and analyzed viscosity (hardness), temperature, osmolarity, and acidity of gastric contents. These studies were carried out in model conditions, in which the role of the stomach was used as a transparent, colorless plastic container of 1000 mL. To mimic food products, container was introduced in 180 g of oatmeal, and 150 mL of milk and/or water. After eating and tablets, we added 150 mL of natural gastric juice, pH 0.8–1.2. All substances were introduced into the container at a temperature of +37°C. In a parallel series of experiments in the stomach, cavity was input pieces of chalk, clay, gravel, river pebbles, and pharmaceutical tablets of similar size. In addition, the movement dynamics of food and tablets inside the body was studied, and analyzed viscosity (hardness), temperature, osmolarity, and acidity of gastric contents.

17.3 RESULTS

It was found that the modern standard quality control of the tablets does not include their specific gravity, size, shape, hardness value, the local acidic, and osmotic activity. So every manufacturer has the right to issue tablets with any hardness, with osmotic and acid activity. In this regard, almost all modern tablets have high physical–chemical aggressiveness and therefore have a marked local irritant and cauterizing effect. All this contributes to the development of drug iatrogenic in the form of gingivitis, stomatitis, gastritis, stomach ulcers, and dental caries.

The results of experiments using transparent plastic container showed that modern pharmaceutical tablets, pebbles, sand, chalk, clay, food, and water very quickly fall down to the bottom of an artificial gastric reservoir. Moreover, natural and artificial stones (modern tablets) sink in water and in gastric juice. It is established that specific gravity of all modern tablets greater than 1 g/cm^3 and, therefore, all the pills drown in gastric juice, water, and milk. In addition, we saw that these items do not move inside the cavity of the stationary container, regardless of the volume of injected liquid in it.

However, all the items can move inside the cavity, if the container is moved from place to place or inverted. In particular, the results of our research showed that if the container is vertical and motionless, all the modern pills very quickly fall down bottom and arrive in one place. Then pills lie motionless, like a river rock in a glass of water, despite the fact that the volume of fluid may increase. It was found that the increased water volume in the cavity changes the position of the tablets only in case if they are floating on the liquid surface (Fig. 17.1).

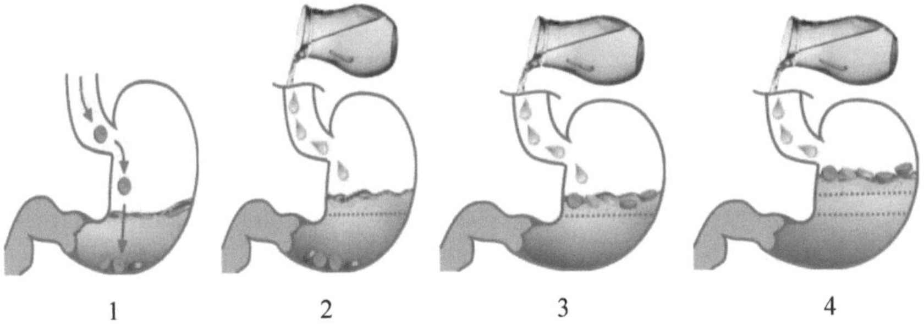

FIGURE 17.1 (See color insert.) Influence of gravity and water on placement in the stomach cavity stones and modern tablets (1 and 2) and floating tablets, which is a dry foam (RU Patent 2254121) (3 and 4).

In this regard, all modern tablets are sinking like pebbles or pieces of chalk and clay. To eliminate this drawback in Russia, "Floating tablet" has been invented.[20] In order for the tablet to not sink in water and float on the liquid surface, it was proposed to make it porous and filled with air like dry foam.

Previously, we showed that as people swallow the pills while standing or sitting, all the pills fall on the same part of the stomach, namely gastric mucosa in region of pylorus. However, unlike the pebbles and crushed stones, modern tablets have aggressive impact on gastric mucosa. They corrode the stomach wall, and may lead to ulcers.[3]

The pills that are the most well-known "pests" of our health cause stomach ulcers, today brazenly called nonsteroidal anti-inflammatory drugs (NSAIDs). Aspirin remains the sales leader of this group of drugs almost 100 years!

However, the results of our research show that this group of drugs is the real "inflammatory" and cauterizing pharmaceuticals. The results of studies on healthy adult volunteers have shown that the location of the tablets of acetylsalicylic acid, metamizole sodium, or ketorola on lip, gingiva, or tongue cause a local irritant and/or cauterizing effect after 2–5 min of interaction.[3]

However, it was first predicted that traditional medicines are poisons. The fact that even Paracelsus showed "everything is a poison, and nothing is devoid of toxicity. Only one dose makes the poison unnoticed." In addition, the truth is that drugs of this group are traditionally used in experimental pharmacology for the simulation of ulcers of the stomach and intestines of animals!

Thus, it is not surprising that our results refute the safety of modern pharmaceutical tablets. Our studies showed that all modern tablets have a pronounced mechanical, physical, chemical, and physical–chemical aggressiveness in a local continuous interaction with the tissues of our body. The reasons for this are the following factors: the "wrong" shape, high hardness and strength, poor solubility, high hyperosmotic activity, and acid activity.

We explored the shape and size of the modern pharmaceutical tablets and the subjects are swallowing people after a thorough chewing of fresh bread in his mouth. This subject is known in science is called "food bolus." It turned out that natural food bolus has the shape of a fruit of an olive tree with a maximal diameter of 1 cm and a maximum length of 2.5 cm.[8,18] This food olive has a dark color, almost zero stiffness, medium elasticity, is porous, has a specific gravity less than 1 g/cm^3 and is devoid of osmosis and osmotic aggressiveness in relation to the contents of the mouth and stomach.

Previously, we have shown that the local irritant effect of tablets is due to their dehydrating effect on the cells of the mucous membrane at the site of contact. Such damaging effects of the tablets stem from the fact that they have a high concentration of ingredients that occur in solution at the site of contact. The fact is that high concentrations of ingredients make liquid very salty and hyperosmotic. Therefore, the longer the local interaction of pills, the more inevitable they cause necrosis of those tissues with which they come in contact.

Parallel to this, we investigated a range of shapes and sizes modern tablets. The results showed that the current tablets do not have standard forms and size. Moreover, it was found that the sizes and shapes of modern pharmaceutical tablets do not have optimal values for the normal functioning of the digestive system. After determining the mechanical and physical–chemical properties of natural food crumbs and comparing them with similar properties of pharmaceutical tablets, it becomes obvious that tablets in the form of artificial stone washers are not needed patients. These tablets are beneficial for those who produce them and who trade them. Moreover, today there is no standard, not only in the form of tablets, but also on their sizes. Therefore, different manufacturers are releasing tablets in different forms and with different sizes.

In particular, our results showed that modern tablets differ in diameter and height by upto three times, and by upto 10 times in volume.

Besides, we found that there is no standard for chewing resistance and disintegration of tablets in gastric juice. So they are manufactured with varying chewing resistance and disintegration characteristics.

In our study of the specific deforming pressure of 19 pharmaceutical tablets, it was found that the value of the specific deforming pressure is in the range from 0.03 ± 0.0001 N/mm^2 [tablet xefocam (lornoxicam 4 mg, Nycomed) to 160 ± 0.3 N/mm^2 (tablet of ketorol, Dr. Reddy's, India)].

In the group of relatively soft tablets includes the following tablets: xefocam, aceclofenac, fenigidin, dexamethasone, acetylsalicylic acid, no-shpalgin, eleflox, mirlox, analgin, nurofen, and ofloxacin. The "soft" pills (with values of hardness at least 2 n/mm^2) proved to be tablets of acetylsalicylic acid, analgin, aertal, nurofen, and no-shpalgin.

The group of relatively solid pills includes diclofenac sodium, nimesulide, pektusin, sulfasalazin, fromilid, prednisolone, ampisid, and ketorol. The "hard" pills (with values of hardness of more than 70 n/mm^2) proved to be tablets of diclofenac sodium, nimesulide, pektusin, sulfasalazine, fromilid, prednisolone, ampyside, and ketorol.

In other words, our results indicate that the current tablets differ in strength when chewed more than 5000 times (Fig. 17.2).

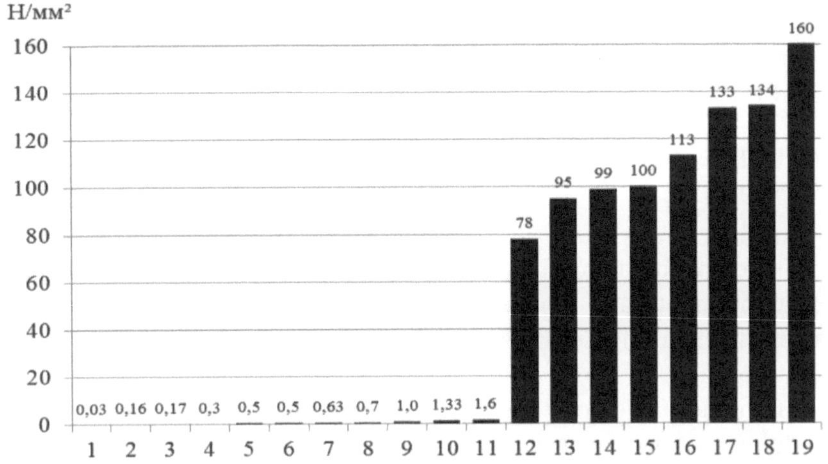

FIGURE 17.2 Hardness 19 tablets of various drugs. Designations of tablets: (1) xefocam, (2) aceclofenac, (3) fenigidin, (4) dexamethasone, (5) acetylsalicylic acid, (6) no-shpalgin, (7) eleflox, (8) mirlox, (9) analgin, (10) ofloxacin, (11) nurofen, (12) diclofenac sodium, (13) nimesulide, (14) pektusin, (15) sulfasalazin, (16) fromilid, (17) prednisolone, (18) ampisid, (19) ketorol.

On the other hand, all modern tablets are fundamentally different from natural food lumps (Fig. 17.3).

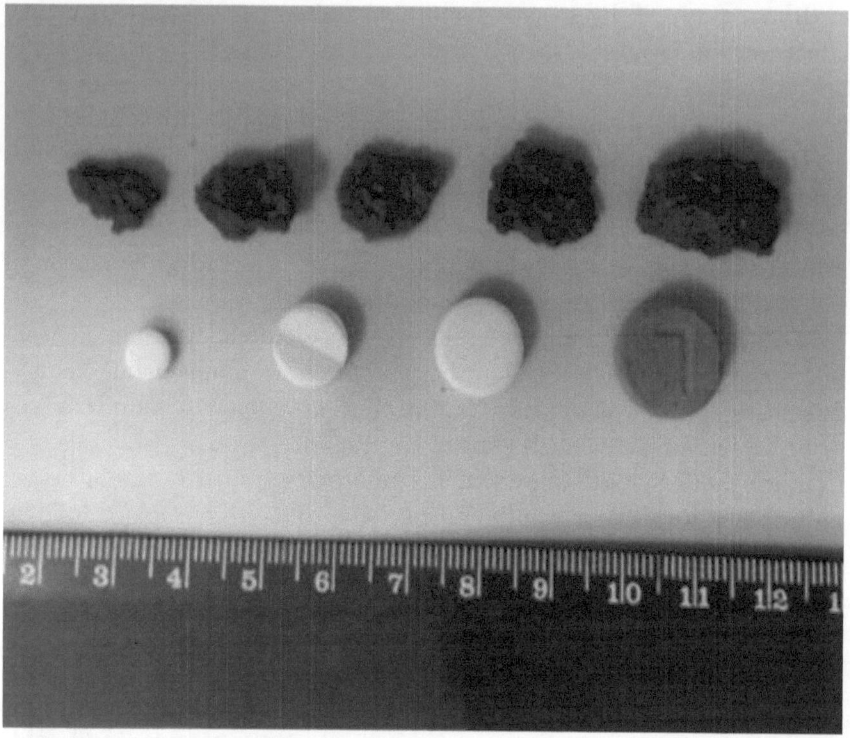

FIGURE 17.3 (See color insert.) Range value of diameter and volume of tablets and food bolus.

Therefore, modern tablets have a wide range of aggressive actions inside the mouth and stomach. No coincidence that modern tablets can damage teeth, fillings, crowns, dentures, implants, braces, and other dental structures.

In another series of experiments, which were carried out in model conditions at a temperature of +37°C, a mixture of cereals, gastric juice, and water had the following physical–chemical characteristics:

- the viscosity is in the range of 200–500 centipoise,
- acid activity in the range pH 4.5–8.0,
- osmotic activity in the range 240–340 mOsmol/L of water.

Then, after the introduction of the mixture in 20 small river pebbles in the form of washers with a diameter of about 6–20 mm, height 2–6 mm and volume of 0.1–1.0 cm,[3] these physical–chemical properties of the content has remained virtually unchanged for hours.

Other results we received after the introduction of a similar mix of cereals, gastric juice, and water for 20 tablets of different quality of medicines of similar shapes and sizes. Thirty minutes after introducing them to this mix, the contents had the following physical–chemical characteristics:

- viscosity- 100–300 centipoise;
- the value of the acidity- pH of 6.0–7.1;
- the value of osmotic activity- 240–340 mOsmol/L of water.

Thirty minutes after administration of 20 tablets to empty plastic containers (model the introduction of "on an empty stomach"), the tablets were broken in half. The contents of the container were slurry with the remnants of solid tablets and had the following physical–chemical properties:

- the viscosity value 0–10 centipoise;
- the value of the acidity- pH 2.0–3.3;
- the value of osmotic activity- 340–600 mOsmol/L of water.

The results are presented in Table 17.1.

TABLE 17.1 The Influence of River Pebbles and Tablets of NSAIDs on Some Physical–Chemical Properties of Stomach Contents.

Physical–chemical properties	The mixture (oatmeal, gastric juice, and water)	The mixture (oatmeal, gastric juice, and water) + 20 small river pebbles	The mixture (oatmeal, gastric juice, and water) + 20 NSAIDS tablets	Gastric juice + + 20 NSAIDS tablets
Viscosity (centipoise)	200–500	200–500	100–300	0–10
Acidity (pH)	4.5–8.0	4.5–8.0	6.0–7.0	2.0–3.3
Osmotic pressure (mOsmol/L water)	240–340	240–340	240–340	340–600

Thus, pharmaceutical companies produce drugs to be ingested in the form of tablets, physical, and physical–chemical properties which dramatically differ from analogous properties of natural food lumps, in which we make the bread in my mouth before swallowing. It is shown that now considered "normal" form, color, size, volume, specific gravity, hardness, osmotic, and acid activity tablets violate the physical and physical–chemical properties of the contents of the oral cavity and stomach. Moreover, the "wrong" physical and physical–chemical parameters of modern tablets reduce the safety of drugs for the digestive system because it make all tablets without exception nonspecific physical and physical–chemical aggressiveness at intake.

17.4 CONCLUSIONS

The results show that people are not birds. Therefore, pharmaceutical stones are hurting and not helping the human digestive system. Therefore, the human digestive system happily accept pills only if their physical and physicochemical properties will not differ corresponding properties of natural food lumps produced in the oral cavity of the quality of food.

KEYWORDS

- **human**
- **ingestion**
- **pills**
- **drugs**
- **physical–chemical properties**

REFERENCES

1. Urakov, A. L.; Strelkov, N. S.; Lipanov, A. M.; Dementiev, V. B.; Urakova, N. A., et al. Physical–chemical and Hydrodynamical Ways of Increasing Safety of Intestine. *Chem. Phys. Mesoscopics* **2007,** *9* (3), 231–238.
2. Urakov, A. L. The Change of Physical–chemical Factors of the Local Interaction with the Human Body as the Basis for the Creation of Materials with New Properties. *Epitőanya J. Silicate Based Composite Mater.* **2015,** *67* (1) 2–6.

3. Urakov, A.; Urakova, N.; Reshetnikov, A.; Kasatkin, A.; Kopylov, M.; Baimurzin, D. About What is Happening in the Stomach After Swallowing Human River Pebbles, Gravel, Chalk, Clay and Tablets Drugs. *Epitőanyag J. Silicate Based Composite Mater.* **2016,** *68* (4), 110–113.
4. Kasatkin, A.; Urakov, A. *Why the Drug Solutions May Cause Inflammation at the Injection Site,* Proceedings of 6th World Congress on Medicinal Chemistry and Drug Design, Milan, Italy, June 7–8, 2017; p 78.
5. Urakov, A.; Urakova, N.; Reshetnikov, A.; Kopylov, M.; Chernova, L. Solvents of Pus-medicines with Physical–chemical Aggressive Action. IOP Conf. Series: *J. Phys. Conf. Series* **2017,** *790*, 012033.
6. Urakov, A.; Urakova, N.; Kasatkin, A.; Reshetnikov, A. Infrared Thermography Skin at the Injection Site as a Way of Timely Detection Injection Disease. *Thermol. Int.* **2015,** *25*, (1), 30.
7. Urakov, A. L. Development of New Materials and Structures Based on Managed Physical–chemical Factors of Local Interaction. IOP Conf. Ser.: Mater. Sci. Eng., 2016, 123, 012008.
8. Urakov, A., et al. Artificial Food Bolus and Method for Instant Assessment of Dento-facial Health with Using Artificial Food Bolus. Patent RF N 2533840, 2014.
9. Urakov, A. L.; Strelkov, N. S.; Urakova, N. A., et al. Ultrasound as a Method of Study Travel Medical Imaging Solid Forms in the Stomach. *Exp. Clin. Gastroenterol.* **2008,** *2*, 27–29.
10. Urakov, A.; Urakova, N.; Reshetnikov, A.; Kopylov, M.; Kasatkin, A.; Baymurzin, D.; Gabdrafikov, R. The Facts that the Physical–chemical Properties of Modern Tablets Distinguish them from Natural Food Lumps. IOP Conf. Series: *Mater. Sci. Eng.* **2017,** *175*, 012012.
11. Urakov, A.; Urakova, N. Rheology and Physical–chemical Characteristics of the Solutions of the Medicines. *Conf. Series: J. Phys.* **2015,** *602*, 012043.
12. Urakov, A. L.; Urakova, N. A. Temperature of the Site of Injection in Subjects with Suspected "Injection's Disease". *Thermol. Int.* **2014,** *2*, 63–64.
13. Urakov, A. L.; Urakova, N. A. Thermography of the Skin as a Method of Increasing Local Injection Safety. *Thermol. Int.* **2013,** *23* (2), 70–72.
14. Urakov, A. L. Thermology is the Basis of Medicine Since Ancient Times. *Thermol. Int.* **2017,** *27, 2*, 78–79.
15. Kasatkin, A. A.; Urakov, A. L.; Lukoyanov, I. A. Nonsteroidal Anti-inflammatory Drugs Causing Local Inflammation of Tissue at the Site of Injection. *J. Pharmacol. Pharmacother.* **2016,** *7* (1), 26–28.
16. Urakov, A. L.; Urakova, N. A.; Chernova, L. V.; Fischer, E. L.; Nasyrov, M. R. Infrared Thermography Forearm Skin in Places Intradermal Injections of Blood or Solutions of Drugs Before and After the Appearance of the Bruise. *Thermol. Int.* **2015,** *25* (2), 66–67.
17. Urakov, A. L.; Kasatkin, A. A.; Urakova, N. A.; Urakova, T. V. Cold Sodium Chloride Solution 0.9% and Infrared Thermography Can be an Alternative to Radiopaque Contrast Agents in Phlebography. *J. Pharmacol. Pharmacother.* **2016,** *7* (3), 138–139.
18. Reshetnikov, A.; Urakov, A.; Kasatkin, A.; Soiher, M. G.; Kopylov, M. Artificial Food Lump from Porous Neoprene and the Method of Its Use for the Evaluation

of Adaptation Patients to the Dental Constructions. IOP *Conf. Ser.: Mater. Sci. Eng.* **2016,** *123*, 012007.
19. Urakov, A. L.; Ammer, K.; Urakova, N. A.; Chernova, L. V.; Fisher, E. L. Infrared Thermography Can Discriminate the Cause of Skin Discolourations. *Thermol. Int.* **2015,** *25* (4), 209–215.
20. Urakova, N. A.; Muravtseva, O. V.; Urakov, A. L.; Ovchinnikova, E. N.; Shcherbakova, N. V.; Tulenkov, A. M.; Pertseva, N. A. Floating Tablet. Patent RF N 2254121, 2005.

CHAPTER 18

RECENT ADVANCES IN BIOORGANISM-MEDIATED GREEN SYNTHESIS OF SILVER NANOPARTICLES: A WAY AHEAD FOR NANOMEDICINE

DEBARSHI KAR MAHAPATRA[1,*] and SANJAY KUMAR BHARTI[2]

[1]*Department of Pharmaceutical Chemistry, Dadasaheb Balpande College of Pharmacy, Nagpur 440037, Maharashtra, India*

[2]*Division of Pharmaceutical Chemistry, Institute of Pharmaceutical Sciences, Guru Ghasidas Vishwavidyalaya (A Central University), Bilaspur 495006, Chhattisgarh, India*

[*]*Corresponding author. E-mail: mahapatradebarshi@gmail.com*

ABSTRACT

Nanotechnology is an emerging branch of science utilizing technological advances of nanomaterials of size 1–100 nm. The nanotechnology in biomedical research and clinical practices emerged as nanomedicine that makes a major impact on human health. Nowadays, nanomaterials are increasingly used in therapeutics, diagnostics, theranostics, and targeted drug delivery due to their unique and specific function at the cellular, atomic, and molecular levels. Silver and silver containing compounds have been used as therapeutic agents since ages. With the pace of time, silver nanoparticles (AgNPs) have gained utmost position due to the wide range of pharmacological activities such as anticancer, anti-inflammatory, antiplatelet, antimicrobial, antiparasitic, and antiviral, etc. AgNPs have traditionally been synthesized using wet chemical techniques, where the chemicals used are often toxic, produces by-products, expensive, and flammable. In this book chapter, ecofriendly, simple, and size/shape-controlled biosynthesis of silver nanoparticles using seaweeds, algae, bacteria, and fungi has been described along with the plausible mechanism of AgNPs formation.

18.1 INTRODUCTION

Nanotechnology has revolutionized all fields across the globe and opened up several frontiers in nanobiotechnology,[1] nanopharmacotherapeutics,[2] material and applied sciences.[3] Recently, silver nanoparticles (AgNPs) gained key importance in biomedical sciences,[4] drug delivery,[5] catalysis,[6] optics,[7] and photoelectrochemistry.[8] These silver nanomaterials have advantages due to their extremely small size and a very high surface-to-volume ratio which contributes chemical and physical attributes in their mechanical, steric and biological properties, catalytic activity, thermal and electrical conductivity, optical absorption, and melting point when compared with the bulk of the same material.[9-10] This enables designing and production of the same materials with novel characteristics by simply controlling of shape and size at the nanometer scale, which ultimately influences their applications ranging from tumor therapy imaging,[11] biosensing,[12] electronics,[13] bactericidal activity,[14] and many more. The nanotechnology in biomedical research and clinical practice emerged as nanomedicine that could potentially make a major impact on human health. AgNPs are increasingly used in therapy, diagnostics, imaging, and targeted drug delivery due to their specific function at the cellular, atomic, and molecular levels. Nanomedicine has potential to integrate diagnostics/imaging with therapeutics and facilitates the development of theranostics for personalized medicine.[15]

In recent years, synthesis of AgNPs is a dynamic field in academics and applied research. Generally, the processes employed for the synthesis of nanoparticles are broadly classified into two types: the "top-down" process and "bottom-up" process. In "top-down" approach, the bulk material breaks down into particles at the nanoscale with various lithographic procedures like grinding, milling, etc., whereas in "bottom-up" approach, the atoms bring together to form a nucleus which grows eventually into a particle of nanosize.[16] A variety of chemical and physical processes have been reported for the synthesis of metallic nanoparticles. However, various methods exist, but many problems are associated with these processes like utilizing toxic solvents, production of harmful by-products, expensive, etc.[17] The chemicals hydrazine hydrate, hydrogen peroxide, sodium borohydride, polyvinylpyrrolidone, dimethylformamide, and ethylene glycol are generally used for the reduction of metal salts, but it gets absorbed on the surface of nanoparticles formed, thereby producing toxicity. Thus, there is a need to explore ecofriendly green synthetic protocols for the synthesis of AgNPs.

Nature has provided abundant biological resources to synthesize AgNPs ecofriendly.[18] With the advancement in technologies, a new way for research and development in the field of biology toward nanomedicine has been established.[19] The use of biological resources in the synthesis of nanomaterials is rapidly developing due to their growing success, ease of formation of nanoparticles, economic, and ecofriendly in nature.[20] Over the past few years, purified extracts of bacteria, fungi, algae, seaweeds, and viruses have received adequate attention for the development of energy proficient, nontoxic, economic AgNPs.[21]

Typically, a bioorganism-mediated metallic salt reduction process occurs when the aqueous extract of bioorganism reacts with an aqueous solution of the metal salt (here, $AgNO_3$). The complete reaction occurs at room temperature within a few minutes. Due to the presence of a wide variety of chemicals, the bioreduction process is relatively complex. Biological synthesis of nanoparticles using plant extracts is relatively scalable and less expensive compared with microbial processes.[22] The nature of plant extract, its concentration, the concentration of metal salt, pH, temperature, and contact time affect the rate of production, quantity, and characteristics of the nanoparticles. The source of the plant extract also influences the characteristics of nanoparticles because of varying concentrations and combinations of organic reducing agents.[23] For the green synthesis of nanoparticles, the following points must be considered for (1) selecting of appropriate bioresources for the reduction of metal salts and (2) providing optimal reaction conditions for nanoparticle formation.[24]

This book chapter focuses primarily on bioorganism-mediated synthesis of AgNPs and their probable mechanism(s) of the formation.

18.2 BIOORGANISM-MEDIATED SYNTHESIS OF AgNPs

18.2.1 BACTERIUM-MEDIATED SYNTHESIS OF AgNPs

Synthesis of AgNPs employing bacteria as reducing principle is an exciting approach. A number of prokaryotic bacteria have been utilized to synthesize AgNPs intracellularly or extracellularly by enzymatic processes.[25] In the former, the formation of AgNPs is due to their chemical detoxification as well as due to energy-dependent ion efflux from the cell by membrane proteins that function either as ATPase or as the chemiosmotic cation or proton antitransporters. In case of intracellular production, the accumulated particles are of the particular dimension and with less polydispersity. In order to release the intracellularly synthesized nanoparticles, additional processing steps such as ultrasound treatment or reaction with suitable detergents are required. In extracellular synthesis, the cell wall reductive enzymes or soluble secreted enzymes are involved in the reductive process of metal ions, and then it is obvious to find the metal nanoparticles extracellularly. The microbial-mediated synthesis of metal nanoparticles depends upon the localization of the reducing components of the cell.[26]

Slawson et al. (1992) discovered few silver-resistant bacteria strain *Pseudomonas stutzeri* AG259 from silver mines which accumulated AgNPs of size 35–46 nm within the periplasmic space. When this bacterium was placed in a concentrated aqueous solution, particles of larger size (200 nm) were formed. In order to survive in environments containing high levels of metals, some microbes (esp. bacteria) have adapted mechanisms to cope with them.[27]

These mechanisms involve altering the chemical nature of the toxic metal so that it no longer causes toxicity, resulting in the formation of nanoparticles of the metal concerned. Thus, nanoparticle formation is the "by-product" of a resistance mechanism against a specific metal, and this can be used as an alternative way of producing nanoparticles.[28]

The rapid formation of AgNPs from Ag^+ by culture supernatants of *Klebsiella pneumonia*, *Escherichia coli*, and *Enterobacter cloacae* is of considerable interest.[29] The reduction of aqueous Ag^+ ions with the cell filtrate of *Rhodobacter sphaeroides* has been reported.[30] The size of AgNPs was controlled by the specific activity of nitrate reductase in the cell filtrate. Juibari et al. (2011) isolated thermophilus bacteria from geothermal hot springs and identified as *Ureibacillus thermosphaericus*, which showed high potential for AgNP biosynthesis with the extracellular mechanism. Biosynthesis reactions were conducted using the culture supernatant at different temperatures (60–80°C) and silver ion concentrations (0.001–0.1 M). The experimentation showed that pure spherical nanoparticles in the range of 10–100 nm was produced, and the maximum nanoparticle production was achieved using 0.01 M $AgNO_3$ at 80°C.[31] Kalimuthu et al. (2008) synthesized silver nanocrystals of size 50 nm from bacterium

Bacillus licheniformis, which reduced silver ions into AgNPs as indicated by the change in color from whitish–yellow to brown.[32] Ali et al. (2011) reported the extracellular biosynthesis of AgNPs of size 100–200 nm using marine cyanobacterium, *Oscillatoria willei* NTDM01, which reduces silver ions and stabilizes the AgNPs by a secreted protein.[33] The biosynthesis of AgNPs has been successfully conducted using *Plectonema boryanum* UTEX 485, a filamentous *cyanobacterium*, reacted with aqueous $AgNO_3$ solutions (560 mg/L Ag) at 25–100°C for upto 28 days have been reported.[34] Saifuddin et al. (2009) described a novel combinatorial synthesis approach for the synthesis of metallic nanostructures of silver (Ag) of size 5–60 nm, by using a combination of culture supernatant of *Bacillus subtilis* and microwave irradiation in water in the absence of a surfactant or soft template. Similarly, few bacterial species like *Morganella, Bacillus*, etc., have been utilized explicitly for the synthesis of AgNPs[35] (Table 18.1).

TABLE 18.1 Bacterium-Mediated Production of AgNPs.

S. no.	Bacteria	Size (in nm)	Application
1.	*Acetobacter xylinum*	NA	NA
2.	*Aeromonas* sp. SH10	6.4	NA
3.	*Bacillus licheniformis*	50	NA
4.	*Bacillus* sp.	5–15	NA
5.	*Bacillus subtilis*	5–60	NA
6.	*Bacillus thuringiensis spore*	15	Bactericidal
7.	*Brevibacterium casei*	10–50	Anticoagulant
8.	*Corynebacterium* sp. SH09	10–15	NA
9.	*Escherichia coli*	50	NA
10.	*Klebsiella pneumoniae, Escherichia coli and Enterobacter cloacae*	50–100	NA
11.	*Klebsiella pneumoniae, Escherichia coli and Enterobacter cloacae*	28.2–122	NA
12.	*Lactobacillus* sp.	20–50	NA
13.	*Morganella* sp.	20–30	NA
14.	*Oscillatoria willei*	100–200	NA
15.	*Plectonema boryanum*	1–200	NA
16.	*Pseudomonas aeruginosa*	13	NA
17.	*Pseudomonas stutzeri* AG259	< 200	NA
18.	*Rhodobacter sphaeroides*	9.56	NA
19.	*Ureibacillus thermosphaericus*	10–100	NA

NA: not available.

The exact mechanism of AgNP formation is still not clear but a few kind of research strongly support the role of nitrate reductase as the key enzyme for nanoparticle formation in bacteria, where bioreduction of $AgNO_3$ could be associated with metabolic processes utilizing nitrate by reducing nitrate to nitrite and ammonium.[34]

18.2.2 FUNGUS-MEDIATED SYNTHESIS OF AgNPs

The use of bioorganisms for the production of various therapeutically active products is an attractive process developed in the 21st century. Nowadays, the production of metal nanoparticles from their corresponding metal salts by fungi is gaining importance. It is believed that the reduction of the metal ions occurs by an enzymatic process. It is creating an avenue for the fungi-mediated synthesis of nanomaterials over a range of chemical constituents. Some of the most popular species include *Fusarium*, *Aspergillus*, and *Penicillium*.

Ahmad et al. (2003) observed that aqueous silver ions, when exposed to the fungus *Fusarium oxysporum*, are reduced in solution, thereby leading to the formation of an extremely stable silver hydrosol. The AgNPs of size 5/15 nm are stabilized in solution by proteins secreted by the fungus.[36] Similarly, extracellular production of AgNPs of size 20–50 nm by several strains (O6 SD, 07 SD, 534, 9114, and 91248) of the fungus *Fusarium oxysporum* have been reported. The authors concluded that the reduction of the metal ions occurs by a nitrate-dependent reductase and a shuttle quinone extracellular process.[37]

Similar biosynthesis of highly stable spherical AgNPs of size 10–60 nm by fungus *Fusarium semitectum* from silver nitrate solution has been reported.[38] Likewise, an efficient, ecofriendly and simple process of AgNP biosynthesis (extracellularly) by *Fusarium solani* have been reported by Ingle et al. (2009).[39] The authors described that the AgNPs were found to be quite stable in solution due to the capping of AgNPs by proteins secreted by the fungus.

In a study, *Aspergillus clavatus*, an endophytic fungus was challenged with 1 mM $AgNO_3$ solution, resulted in the formation of extracellular, polydispersed spherical, or hexagonal particles of size 10–25 nm and having antimicrobial activity against *Candida albicans*, *Pseudomonas fluorescens*, and *Escherichia coli*.[40] Bhainsa and D'Souza (2006) investigated extracellular biosynthesis of AgNPs using *Aspergillus fumigates* where author reported the formation of well-dispersed silver nanoparticles of size 5–25 nm.[41] Jain et al. (2011) demonstrated an ecofriendly and low-cost protocol for the synthesis of AgNPs using the cell-free filtrate of *Aspergillus flavus* NJP08. The extracellular proteins present in cell extract were found to be responsible for the synthesis and stability of AgNPs.[42]

Kathiresan et al. (2009) reported in vitro biosynthesis of AgNPs from $AgNO_3$ as a substrate by *Penicillium fellutanum* isolated from coastal mangrove sediment. The authors explored a single prominent protein band with molecular weight of 70 kDa in the culture filtrate, which was secreted out of the fungal biomass and was believed to be involved in the reduction of the silver ions.[43] AgNPs of size 52–104 nm were biologically synthesized using filamantous fungi *Penicillium* sp. isolated from the soil samples. The cell filtrate of *Penicillium* sp., when challenged with 1 mM of silver nitrate, resulted in a change of the mixture from colorless to orange–brown indicated the synthesis of AgNPs in the reaction mixture.[44] Similar experiments for AgNPs biosynthesis from *Penicillium* sp. was demonstrated by Sadowski et al (2008).[45]

Generally, the synthesis of AgNPs takes place in two steps: firstly, reduction of bulk silver ions into AgNPs and, secondly, capping of the synthesized nanoparticles. The mechanism of fungal-mediated synthesis occurs in the following steps; former step involves trapping of Ag^+ ions at the surface of the fungal cells and subsequently a 32 kDa protein which may be a reductase secreted by the fungal isolate specifically reduce silver ions into AgNPs. The later step involves 35 kDa proteins which bind with nanoparticles and confer stability.[39] The protein–nanoparticle interactions can play an important role in providing stability to nanoparticles. The fungal proteins proffer key role in the reduction of the metal salts into their corresponding elemental form. The proteins present in *Volvariella volvacea* are such example of natural bioreductant. In *Fusarium oxysporum*, bioreduction takes place by α-NADPH-dependent sulfite reductase (35.6 kDa) and phytochelatin components,[46] whereas, in *Penicillium brevicompactum*, compactin serves as the bioreductant.[47]

A large fungus-mediated synthesis of nanoparticles has been reported. Fungi like *Cladosporium cladosporioides*, *Trichoderma Reesei*, *Chrysosporium tropicum*, *Cochliobolus lunatus*, etc., have been reported to contribute biofabrication of AgNPs. The biosynthesized metal nanoparticles have exhibited potential cytotoxic, immunomodulatory, larvicidal activities. The most common fungi used for the biosynthesis of AgNPs are listed in Table 18.2.

TABLE 18.2 Fungus-Mediated Synthesis of AgNPs.

S. no.	Fungi	Size (in nm)	Application
1.	*Alternaria alternata*	20–60	Fungicidal
2.	*Aspergillus clavatus*	10–25	Bactericidal
3.	*Aspergillus flavus*	17	NA
4.	*Aspergillus fumigatus*	5–20	NA
5.	*Aspergillus niger*	20	NA
6.	*Aspergillus terrus*	2.5	NA
7.	*Bipolaris tetramera*	54.78–73.49	Bactericidal and immunomodulatory
8.	*Chrysosporium tropicum*	20–50	Larvicidal
9.	*Cladosporium cladosporioides*	10–100	NA
10.	*Cochliobolus lunatus*	3–21	Larvicidal
11.	*Coriolus versicolor*	25–75	NA
12.	*Cryphonectria* sp.	30–70	Bactericidal
13.	*Fusarium acuminatum*	5–40	Bactericidal
14.	*Fusarium oxysporum*	5–15	NA
15.	*Fusarium oxysporum*	20–50	NA
16.	*Fusarium semitectum*	10–60	NA
17.	*Fusarium solani*	5–35	NA
18.	*Hormoconis resinae*	20–80 (triangle) 10–20 (spherical)	NA
19.	*Humicola* sp.	5–25	Cytotoxicity
20.	*Neurospora crassa*	11	NA
21.	*Penicillium brevicompactum*	23–105	NA
22.	*Penicillium fellutanum*	5–25	NA
23.	*Penicillium* sp.	52–104	NA
24.	*Penicillium* sp.		NA
25.	*Phaenerochaete chrysosporium*	50–200	NA
26.	*Phoma glomerata*	60–80	Bactericidal
27.	*Pleurotus djamor*	5–50	Cytotoxicity
28.	*Pleurotus sajor caju*	5–50	Bactericidal
29.	*Trichoderma asperellum*	13–18	NA
30.	*Trichoderma reesei*	5–50	NA
31.	*Trichoderma viride*	5–40	NA
32.	*Verticillium*	25	NA
33.	*Volvariella volvacea*	20–150	NA
34.	Yeast strain MKY3	2–5	NA

NA: not available.

18.2.3 ALGAE-MEDIATED SYNTHESIS OF AgNPs

The extensive literature on various bioorganisms has revealed that marine has abundant resources comprising of diverse biomolecules for successful biosynthesis of AgNPs. Rajesh et al. (2012) demonstrated a simple and ecofriendly biosynthesis of AgNPs of size 28–41 nm at room temperature using *Ulva fasciata* crude ethyl acetate extract as reducing and capping agent. The presence of 1-(hydroxymethyl)–2, 5, 5, 8a-tetramethyl decahydro-2-napthalenol as reducing agent and hexadecanoic acid was found to be a stabilizing agent. The AgNPs inhibited the growth of *Xanthomonas campestris* pv. *malvacearum*, with a MIC value of 40.00 ± 5.77 μg/mL.[48] Similarly, fabrication of AgNPs using marine macroalgae *Ulva reticulate* has been reported.[49]

Merin et al. (2010) employed four algal species: *Chaetoceros calcitrans, Chaetoceros salina, Isochrysis galbana, and Turbinaria conoides* for the synthesis of AgNPs. The AgNPs of size 53–72 nm demonstrated prospective antimicrobial activity against *E. coli*, *P. aeruginosa*, *P. vulgaris,* and *Klebsiella* sp. Similarly, bactericidal activity has been reported by AgNPs of size 96 nm biosynthesized from algae *Turbinaria conoides.*[50] Kannan et al. (2013) showed the formation of AgNPs (3–44 nm) by the reduction of the aqueous silver metal ions using macro alga *Chaetomorpha linum* extract. The authors reported that few water-soluble compounds such as amines, peptides, flavonoids, and terpenoids present in *C. linum* extract is responsible for bioreduction.[51] The extract of brown Australasian marine alga *Cystophora moniliformis* was utilized to synthesize AgNPs of size 50–100 nm. Spherical and smaller particles appeared at low temperatures, and agglomeration and increased size of the particles (<2 μm) were observed at higher temperatures[52] (Table 18.3).

18.2.4 SEAWEED-MEDIATED SYNTHESIS OF AgNPs

The literature highlights the use of extracts of seaweeds for the production of AgNPs. Aqueous extract of seaweed *Gelidiella* sp. has been employed for synthesizing 40–50 nm AgNPs which exhibited potent cytotoxic activity against Hep2 cell line[53]. Murugan et al. (2011) reported the seaweed-mediated synthesis of AgNPs by using the frond extract of *Caulerpa scalpelliformis*. The toxicity of the formed AgNPs was assessed against the filarial vector *Culex quinquefasciatus.*[54] The formation of

AgNPs by the reduction of aqueous silver metal ions during exposure to both fresh and dry seaweed extracts of *Codium capitatum* has been reported.[55] Antimicrobial activity of AgNPs of size 20 nm synthesized from an aqueous extract of red seaweed *Gelidiella acerosa* and *Sargassum tenerrimum* has also been reported.[56] A list of seaweeds used to synthesize AgNPs has been provided in Table 18.4.

TABLE 18.3 Algae-Mediated Production of AgNPs.

S. no.	Algae	Size (in nm)	Application
1.	*Chaetoceros calcitrans, Chaetoceros salina, Isochrysis galbana*, and *Taterillus gracilis*	53–72	Bactericidal
2.	*Chaetomorpha linum*	3–44	NA
3.	*Chlorococcum humicola*	2–16	Bactericidal
4.	*Cystophora moniliformis*	50–100	NA
5.	*Turbinaria conoides*	96	Bactericidal
6.	*Ulva fasciata*	28–41	Bactericidal
7.	*Ulva reticulate*	-	Bactericidal

NA: not available.

TABLE 18.4 Seaweeds-Mediated Synthesis of AgNPs.

S. no.	Seaweeds	Size (in nm)	Application
1.	*Caulerpa scalpelliformis*	20–35	Larvicidal
2.	*Codium capitatum*	3–44	NA
3.	*Gelidiella acerosa*	22	Fungicidal
4.	*Gelidiella* sp.	40–50	Cytotoxicity
5.	*Sargassum tenerrimum*	20	Bactericidal
6.	Water hyacinth	5.69 ± 5.89 nm (pH 4), 4.53 ± 1.36 nm (pH 8), and 2.68 ± 0.69 nm (pH 11)	NA

NA: not available.

18.3 CONCLUSION

Nature has provided abundant bioresources that can reduce silver ion into the AgNP. In the field of nanotechnology, a reliable and ecofriendly process for the synthesis of AgNPs is the foremost demand. In order to achieve this goal, the use of natural sources for the green synthesis of AgNPs becomes crucial. This approach has several advantages over conventional methods as an efficient, cost-effective, and environmentally safe. However, there are a few drawbacks associated with the biosynthesis of nanoparticles using green chemistry approach. It is relatively a slow process, slightly difficult to control the size/shape of nanoparticles using biological sources, the exact mechanism for nanoparticles formation is unknown and nonspecific conjugation of phytoconstituents/proteins during the synthesis of AgNPs. The microorganism (bacteria and fungi)-mediated synthesis is relatively easy, but there is a need to explore the biochemical and molecular mechanism of nanoparticle synthesis by these organisms. Algae and seaweeds are relatively newer bioresources and are not explored adequately and need further research. The wide variety of biological activities and imaging property of biosynthesized AgNPs may provide a basis for the development of future nanomedicine. Additionally, purification and proper characterization of these AgNPs are the essential steps to be taken into consideration before the nanomaterials to be used commercially in healthcare. In spite of all major challenges and issues, the biosynthesized AgNPs may be potential nanomedicine for the treatment of various diseases in near future.

KEYWORDS

- **bioorganism**
- **biosynthesis**
- **green synthesis**
- **mechanism**
- **nanoparticle**
- **silver**

REFERENCES

1. Sarikaya, M.; Tamerler, C.; Jen, A. K. Y.; Schulten, K.; Baneyx, F. Molecular Biomimetics: Nanotechnology Through Biology. *Nat. Mater.* **2003,** *2*, 577–585.
2. Kim, G. J.; Nie, S. Targeted Cancer Nanotherapy. *Mater Today* **2005,** *8* (8), 28–33.

3. Nalwa, H. S. *Handbook of Nanostructured Materials and Nanotechnology*. Academic Press, 1999.
4. Wang, A. Z.; Gu, F.; Zhang, L.; Chan, J. M.; Radovic-Moreno, A.; Shaikh, M. R.; Langer, R. S.; Farokhzad, O. C. Biofunctionalized Targeted Nanoparticles for Therapeutic Applications. *Expert Opin. Biol. Ther.* **2008,** *8* (8), 1063–1070.
5. Singh, R.; Lillard, Jr. J. W. Nanoparticle-based Targeted Drug Delivery. *Exp. Mol. Pathol.* **2009,** *86* (3), 215–223.
6. Nutt, M. O.; Heck, K. N.; Alvarez, P.; Wong, M. S. Improved Pd-on-Au Bimetallic Nanoparticle Catalysts for Aqueous-phase Trichloroethene Hydrodechlorination. *Appl. Catal. B Environ.* **2006,** *69*, 115–125.
7. Haynes, C. L.; McFarland, A. D.; Zhao, L.; Duyne, R. P. V.; Schatz, G. C. Nanoparticle Optics: The Importance of Radiative Dipole Coupling in Two-dimensional Nanoparticle Arrays. *J. Phys. Chem. B* **2003,** *107*, 7337–7342.
8. Sheeney-Haj-Ichia, L.; Pogorelova, S.; Gofer, Y.; Willner, I. Enhanced Photoelectrochemistry in CdS/Au Nanoparticle Bilayers. *Adv. Funct. Mater.* **2004,** *14* (5), 416–424.
9. Rao, C. N. R.; Müller, A.; Cheetham, A. K. *Nanomaterials Chemistry: Recent Developments and New Directions*. John Wiley & Sons, 2007.
10. Rao, C. N. R.; Müller, A.; Cheetham, A. K. *The Chemistry of Nanomaterials: Synthesis, Properties and Applications*; John Wiley & Sons, 2006.
11. Singh, R.; Nalwa, H. S. Medical Applications of Nanoparticles in Biological Imaging, Cell Labeling, Antimicrobial Agents, and Anticancer Nanodrugs. *J. Biomed. Nanotechnol.* **2011,** *7* (4), 489–503.
12. Luo, X.; Morrin, A.; Killard, A. J.; Smyth, M. R. Application of Nanoparticles in Electrochemical Sensors and Biosensors. *Electroanal* **2006,** *18* (4), 319–326.
13. Huang, D.; Liao, F.; Molesa, S.; Redinger, D.; Subramanian, V. Plastic-compatible Low Resistance Printable Gold Nanoparticle Conductors for Flexible Electronics. *J. Electrochem. Soc.* **2003,** *150* (7), G412–G417.
14. Mahapatra, D. K.; Bharti, S. K.; Asati, V. Nature Inspired Green Fabrication Technology for Silver Nanoparticles. *Curr. Nanomed.* **2017,** *7* (1), 5–24.
15. Patra, C. R.; Mukherjee, S.; Kotcherlakota, R. Biosynthesized Silver Nanoparticles: A Step Forward for Cancer Theranostics? *Nanomedicine* **2014,** *9* (10), 1445–1448.
16. Tsakalakos, T.; Ovid'ko, I. A.; Vasudevan, A. K. *Springer Science and Business Media*, 2003.
17. Anastas, P.; Eghbali, N. Green Chemistry: Principles and Practice. *Chem. Soc. Rev.* **2010,** 39, 301–312.
18. Pavani, K. V.; Kumar, N. S.; Gayathramma, K. Plants as Ecofriendly Nanofactories. *J. Biosci.* **2012,** *6* (1), 1–6.
19. Malik, P.; Shankar, R.; Malik, V.; Sharma, N.; Mukherjee, T. K. Green Chemistry Based Benign Routes for Nanoparticle Synthesis. *J. Nanopart.* **2014.**
20. Kumar, V.; Yadav, S. C. Plant Mediated Synthesis of Silver and Gold Nanoparticles and Their Applications. *J. Chem. Technol. Biotechnol.* **2008,** *84* (2), 154–157.
21. Telrandhe, R.; Mahapatra, D. K.; Kamble, M. A. Bombax Ceiba Thorn Extract Mediated Synthesis of Silver Nanoparticles: Evaluation of Anti-staphylococcus Aureus Activity. *Int. J. Pharmaceut. Drug Anal.* **2017,***5* (9), 376–379.

22. Mahapatra, D. K.; Tijare, L. K.; Gundimeda, V.; Mahajan, N. M. Rapid Biosynthesis of Silver Nanoparticles of Flower-like Morphology from the Root Extract of Saussurea lappa. *Res. Rev. J. Pharmacog.* **2018,** *5* (1), 20–24.
23. Dwivedi, A. D.; Gopal, K. Plant Mediated Biosynthesis of Silver And Gold Nanoparticle. *J. Biomed. Nanotech.* **2011,** *7* (1), 163–164.
24. Iravani, S. Green Synthesis of Metal Nanoparticles Using Plants. *Green Chem.* **2011,** 13, 2638–2650.
25. Klaus-Joerger. T.; Joerger, R.; Olsson, E.; Granqvist, C. Bacteria as Workers in the Living Factory: Metal-Accumulating Bacteria and Their Potential for Materials Science. *Trends Biotechnol.* **2001,** *19* (1), 15–20.
26. Narayanan, K. B.; Sakthivel, N. Biological Synthesis of Metal Nanoparticles by Microbes. *Adv. Colloid Interface Sci.* **2010,** *156*, 1–13.
27. Slawson, R. M.; Trevors, J. T.; Lee, H. Silver Accumulation and Resistance in Pseudomonas stutzeri. *Arch. Microbiol.* **1992,** *158*, 398–404.
28. Pantidos, N.; Horsfall, L. E. Biological Synthesis of Metallic Nanoparticles by Bacteria, Fungi and Plants. *J. Nanomed. Nanotechnol.* **2014,** *5* (5), 233.
29. Shahverdi, A. R.; Minaeian, S.; Shahverdi, H. R.; Jamalifar, H.; Nohi A. Rapid Synthesis of Silver Nanoparticles Using Culture Supernatants of Enterobacteria: A Novel Biological Approach. *Process Biochem.* **2007,** *42*, 919–923.
30. Bai, H.; Yang, B.; Chai, C.; Yang, G.; Jia, W.; Yi, Z. Green Synthesis Of Silver Nanoparticles Using Rhodobacter Sphaeroides. *World J. Microbiol. Biotechnol.* **2011,** *27*, 2723–2728.
31. Juibari, M. M.; Abbasalizadeh, Gh. S.; Jouzani, S.; Noruzi, M. Intensified Biosynthesis of Silver Nanoparticles Using a Native Extremophilic Ureibacillus Thermosphaericus Strain. *Mater. Lett.* **2011,** *65*, 1014–1017.
32. Kalimuthu, K.; Babu, R. S.; Venkataraman, D.; Bilal, M.; Gurunathan, S. Biosynthesis of Silver Nanocrystals by Bacillus Licheniformis. *Colloids Surf B Biointerfaces* **2008,** *65*, 150–153.
33. Ali, D. M.; Sasikala, M.; Gunasekaran, M.; Thajuddin, N. Biosynthesis and Characterization of Silver Nanoparticles Using Marine Cyanobacterium, Oscillatoria Willei Ntdm01. *Digest J. Nanomater. Biostr.* **2011,** *6* (2), 385–390.
34. Lengke, M. F.; Fleet, M. E.; Southam, G. Biosynthesis of Silver Nanoparticles by Filamentous Cyanobacteria from a Silver(I) Nitrate Complex. *Langmuir* **2007,** *23*, 2694–2699.
35. Saifuddin, N.; Wong, C. W.; Yasumira, A. A. N. Rapid Biosynthesis of Silver Nanoparticles Using Culture Supernatant of Bacteria with Microwave Irradiation. *Euro. J. Chem.* **2009,** *6*(1), 61–70.
36. Ahmad, A.; Mukherjee, P.; Senapati, S.; Mandal, D.; Khan, M. I.; Kumar, R.; Sastry, M. Extracellular Biosynthesis of Silver Nanoparticles Using the Fungus Fusarium oxysporum. *Colloids Surf B Biointerfaces* **2003,** *28*, 313–318.
37. Durán, N.; Marcato, P. D.; Alves, O. L.; de Souza, G. I. H.; Esposito, E. Mechanistic Aspects of Biosynthesis of Silver Nanoparticles by Several Fusarium Oxysporum Strains. *J. Nanobiotechnol.* **2005,** *3*, 8.
38. Basavaraja, S.; Balaji, S. D.; Lagashetty, A.; Rajasab, A. H.; Venkataraman, A. Extracellular Biosynthesis of Silver Nanoparticles Using the Fungus Fusarium Semitectum. *Mater. Res. Bull. 2008*, *43*, 1164–1170.

39. Ingle, A.; Rai, M.; Gade, A.; Bawaskar, M. Fusarium Solani: A Novel Biological Agent for the Extracellular Synthesis of Silver Nanoparticles. *J. Nanopart. Res.* **2009,** *11*, 2079–2085.
40. Verma, V. C.; Kharwar, R. N.; Gange, A. C. Biosynthesis of Antimicrobial Silver Nanoparticles by the Endophytic Fungus Aspergillus Clavatus. *Nanomedicine* **2010,** *5* (1), 33–40.
41. Bhainsa, K. C.; D'Souza, S. F. Extracellular Biosynthesis of Silver Nanoparticles Using the Fungus Aspergillus Fumigates. *Colloids Surf B Biointerfaces* **2006,** *47*, 160–164.
42. Jain, N.; Bhargava, A; Majumdar, S.; Tarafdar, J. C.; Panwar, J. Extracellular Biosynthesis and Characterization of Silver Nanoparticles Using Aspergillus Flavus NJP08: A Mechanism Perspective. *Nanoscale* **2011,** *3*, 635–641.
43. Kathiresan, K.; Manivannan, S.; Nabeel, M. A.; Dhivya, B. Studies on Silver Nanoparticles Synthesized by a Marine Fungus, Penicillium Fellutanum Isolated from Coastal Mangrove Sediment. *Colloids Surf B Biointerfaces* **2009,** *71*, :133–137.
44. Naveen, K. S. H.; Kumar, G.; Karthik, L.; Bhaskara Rao, K. V. Extracellular Biosynthesis of Silver Nanoparticles Using the Filamentous Fungus Penicillium sp. *Arch. Appl. Sci. Res.* **2010,** *2* (6), 161–167.
45. Sadowski, Z.; Maliszewska, I. H.; Grochowalska, B.; Polowczyk, I.; Koźlecki T. Synthesis of Silver Nanoparticles Using Microorganisms. *Mater. Sci.-Poland* **2008,** *26* (2), 419–424.
46. Kumar, S. A.; Abyaneh, M. K.; Gosavi, S. W.; Kulkarni, S. K.; Pasricha, R.; Ahmad, A.; Khan, M. I. Nitrate Reductase-mediated Synthesis of Silver Nanoparticles from AgNO3. *Biotech. Lett.* **2007,** *29* (3), 439–445.
47. Shaligram, N. S.; Bule, M.; Bhambure, R.; Singhal, R. S.; Singh, S. K.; Szakacs, G.; Pandey, A. Biosynthesis of Silver Nanoparticles Using Aqueous Extract from the Compactin Producing Fungal Strain. *Process Biochem.* **2009,** *44*, 939–943.
48. Rajesh, S.; Raja, D. P.; Rathi, J. M.; Sahayaraj, K. Biosynthesis of Silver Nanoparticles Using Ulva Fasciata (Delile) Ethyl Acetate Extract and Its Activity Against Xanthomonas campestris pv. Malvacearum. *J. Biopest.* **2012,** *5*, 119–128.
49. Devi, J. S.; Bhimba, B. V. Antimicrobial Potential of Silver Nanoparticles Synthesized Using Ulva Reticulata. *Asian J. Pharm. Clin. Res.* **2014,** *7* (2), 82–85.
50. Merin, D. D.; Prakash, S.; Bhimba, B. V. Antibacterial Screening of Silver Nanoparticles Synthesized by Marine Micro Algae. *Asian Pac. J. Trop. Med.* **2010,** *3* (10), 797–799.
51. Kannan, R. R. R.; Arumugam, R.; Ramya, D.; Manivannan, K.; Anantharaman, P. Green Synthesis of Silver Nanoparticles Using Marine Macroalga Chaetomorpha Linum. *Appl. Nanosci.* **2013,** *3*, 229–233.
52. Prasad, T. N. V. K. V.; Kambala, V. S. R.; Naidu, R. Phyconanotechnology: Synthesis of Silver Nanoparticles Using Brown Marine Algae Cystophora Moniliformis and Their Characterization. *J. Appl. Phycol.* **2013,** *25*, 177–182.
53. Vivek, M.; Kumar, P. S.; Steffi, S.; Sudha, S. Biogenic Silver Nanoparticles by Gelidiella acerosa Extract and their Antifungal Effects. *Avicenna J. Med. Biotech.* **2011,** *3* (3), 143–148.

54. Murugan, K.; Benelli, G.; Ayyappan, S.; Dinesh, D.; Panneerselvam, C.; Nicoletti, M.; Hwang, J. S.; Kumar, P. M.; Subramaniam, J.; Suresh, U. Toxicity of Seaweed-synthesized Silver Nanoparticles Against the Filariasis Vector Culex Quinquefasciatus and Its Impact on Predation Efficiency of the Cyclopoid Crustacean Mesocyclops Longisetus. *Parasitol. Res.* **2015,** *114* (6), 2243–2253.
55. Kannan, R. R. R.; Stirk, W. A.; Staden, J. V. Synthesis of Silver Nanoparticles Using the Seaweed Codium capitatum P.C. Silva (Chlorophyceae). *S. Afr. J. Bot.* **2013,** *86,* 1–4.
56. Kumar, P.; Selvi, S. S.; Prabha, A. L.; Kumar, K. P.; Ganeshkumar, R. S.; Govindaraju, M. Synthesis of Silver Nanoparticles from Sargassum Tenerrimum and Screening Phytochemicals for Its Antibacterial Activity. *Nano. Biomed. Eng.* **2012,** *4* (1), 12–16.

INDEX

C

D

E